EAI/Springer Innovations in Communication and Computing

Series Editor

Imrich Chlamtac, European Alliance for Innovation, Ghent, Belgium

The impact of information technologies is creating a new world yet not fully understood. The extent and speed of economic, life style and social changes already perceived in everyday life is hard to estimate without understanding the technological driving forces behind it. This series presents contributed volumes featuring the latest research and development in the various information engineering technologies that play a key role in this process. The range of topics, focusing primarily on communications and computing engineering include, but are not limited to, wireless networks; mobile communication; design and learning; gaming; interaction; e-health and pervasive healthcare; energy management; smart grids; internet of things; cognitive radio networks; computation; cloud computing; ubiquitous connectivity, and in mode general smart living, smart cities, Internet of Things and more. The series publishes a combination of expanded papers selected from hosted and sponsored European Alliance for Innovation (EAI) conferences that present cutting edge, global research as well as provide new perspectives on traditional related engineering fields. This content, complemented with open calls for contribution of book titles and individual chapters, together maintain Springer's and EAI's high standards of academic excellence. The audience for the books consists of researchers, industry professionals, advanced level students as well as practitioners in related fields of activity include information and communication specialists, security experts, economists, urban planners, doctors, and in general representatives in all those walks of life affected ad contributing to the information revolution.

Indexing: This series is indexed in Scopus, Ei Compendex, and zbMATH.

About EAI - EAI is a grassroots member organization initiated through cooperation between businesses, public, private and government organizations to address the global challenges of Europe's future competitiveness and link the European Research community with its counterparts around the globe. EAI reaches out to hundreds of thousands of individual subscribers on all continents and collaborates with an institutional member base including Fortune 500 companies, government organizations, and educational institutions, provide a free research and innovation platform. Through its open free membership model EAI promotes a new research and innovation culture based on collaboration, connectivity and recognition of excellence by community.

Muhammet Nuri Seyman
Editor

2nd International Congress of Electrical and Computer Engineering

Springer

EAI
RESEARCH MEETS INNOVATION

Editor
Muhammet Nuri Seyman
Department of Electrical and Electronics
Engineering
Bandırma Onyedi Eylül University
Bandirma, Türkiye

ISSN 2522-8595 ISSN 2522-8609 (electronic)
EAI/Springer Innovations in Communication and Computing
ISBN 978-3-031-52759-3 ISBN 978-3-031-52760-9 (eBook)
https://doi.org/10.1007/978-3-031-52760-9

This Springer imprint is published by the registered company Springer Nature Switzerland AG
The registered company address is: Gewerbestrasse 11, 6330 Cham, Switzerland

Paper in this product is recyclable.

Preface

We are pleased to present the proceedings of the 2nd International Congress of Electrical and Computer Engineering (ICECENG'23). This conference was organized online from November 22 to 25, 2023, in Bandirma, Turkey. Presenting the proceedings of ICECENG'23 to the authors and delegates is a significant honor for us, and we trust that you will find them to be valuable, captivating, and motivating.

This conference brought together researchers, developers, and students from diverse global communities, including computing, artificial intelligence, security, signal processing, and telecommunications. Additionally, ICECENG'23 serves as a platform for discussing the issues, challenges, opportunities, and findings in electrical and computer engineering research. Our International Conference efforts to address these needs, exploring the processes, actions, challenges, and outcomes of learning and teaching.

We extend our sincere thanks to the individuals, institutions, and companies whose contributions played a pivotal role in the success of this inaugural edition of ICECENG'23. We are also grateful to all the reviewers for their constructive comments on the papers. Our acknowledgment extends to Springer for publishing the proceedings of ICECENG'23 and the Microsoft Conference Management (CMT) system for providing a valuable platform for conference administration.

With the conclusion of this second edition, we hope that ICECENG'23 serves as a catalyst for future collaborations, projects, and subsequent editions of the conference.

Bandirma, Türkiye

Muhammet Nuri Seyman

Contents

Part I
Artificial Intelligence

Model Agnostic Knowledge Transfer Methods for Sentence Embedding Models

Kadir Gunel and Mehmet Fatih Amasyali

1 Introduction

The most accepted method of knowledge transfer from one space to another is knowledge distillation. This process focuses on transferring information from a more complex architecture (teacher) to its simpler architecture (student). The difference between the two models lies in the number of parameters, such as the number of network layers and the size of weight matrices. The student model's objective is to learn the same (or similar) representations as its teacher model for the same inputs [1]. This operation can be time-consuming in larger models and energy-deficient due to the training requirement of the student model. Additionally, the newly trained student model might still be slow during the prediction phase.

Unlike the knowledge distillation methods in the literature, this chapter considers use cases where applying distillation methods is practically impossible. Such cases might include situations where model parameters are partially shared with the end user or when models require to be reimplemented using the same programming language and deep learning framework.

With these assumptions, this chapter focuses on knowledge transfer between two entirely different model architectures: *FastText* and *sBert*. These two models share nothing but the selected language (English) for training. Their architectures, training data, and output vector dimensions are all different. Additionally, the model parameters of FastText word embeddings are not shared with the end user, and both models are written in different programming languages.

From a model perspective, FastText uses its trained word embeddings by taking the average of all the words in the sentence (*bag-of-words*) for calculating its

K. Gunel (✉) · M. F. Amasyali
Yildiz Technical University, Istanbul, Turkey
e-mail: kadir.gunel@std.yildiz.edu.tr; amasyali@yildiz.edu.tr

M. N. Seyman (ed.), *2nd International Congress of Electrical and Computer Engineering*, EAI/Springer Innovations in Communication and Computing,
https://doi.org/10.1007/978-3-031-52760-9_1

sentence vectors. This allows FastText to generate its sentence embeddings very rapidly compared to sBert. Conversely, it has a significant loss in NLP tasks. On the other hand, sBert needs its transformer architecture in order to produce its sentence embeddings. Even though this impedes its generation speed, it has better performance on different NLP tasks. This indicates that sBert has a superior sentence representation power compared to FastText.

For transferring knowledge from sBert toward FastText, we construct and evaluate our proposed models based on their sentence-level and word-level representations. The sentence-level model aims to minimize the distance between FastText sentence representations and sBert sentence representations by minimizing the distance from FastText toward sBert. This method uses a projection matrix followed by a feed-forward network. The projection matrix captures the linear relations, whereas the feed-forward network aims to capture non-linear relationships. The word-level model operates only on a subset of FastText's word embedding space and uses the sBert sentence vectors as auxiliary information. This method builds a language model that uses recurrent neural network (*RNN*). Its aim is to control the correctness of word predictions by making the generated sentence vectors as close as possible to sBert. Both proposed models rely on the representation distillation method, where the student model seeks to imitate the representations of its teacher model [2]. The information transfer operation is done by using the WMT's EN-ES dataset. For validating the proposed methods, various NLP tasks are considered by using the tasks in SentEval datasets. These tasks include natural language inference (NLI), semantic task similarities (STS), classification, and probing. A total of 26 datasets are used for evaluating the proposed models.

Since KD methods also consider the alignment of feature spaces, we find it appropriate to evaluate the alignment relation between FastText and sBert sentence embeddings. Due to that, we used an unsupervised algorithm for checking the sentence alignment. The proposed methods in this chapter solely use the output layer of the sBert model. The rest of the chapter is structured as follows:[1]

- Propose two methods for knowledge transfer between two sentence embedding models that are completely different from each other (FastText and sBert): *distance minimization* and *perplexity minimization* methods (Sect. 3).
- Provide brief information about how sentence representations of both models are calculated, datasets used for knowledge transfer, and data for evaluating newly formed sentence representations (Sect. 4).
- Present the results of the proposed models using two different groups of datasets. The first dataset (WMT) is used for knowledge transfer, while the second dataset (SentEval) demonstrates how knowledge transfer can positively impact the representation power of the weaker model (Sect. 5).
- Discuss the relationship between knowledge transfer of different model architectures and alignment (Sect. 6).

[1]The source code for the experiments conducted in this study is available under MIT license and can be accessible at https://github.com/kadir-gunel/iceceng23

2 Related Work

There is a rich body of literature on the formation of sentence vector representations. These methods range from averaging pretrained word embeddings [3], subtracting the principal components [4], and mimicking the idea of the skip-gram method [5] which was later extended by Logeswaran and Lee [6] for quick training. Lastly, Reimers and Gurevych [7] proposed sentence embeddings which rely on training a siamese network built on top of Bert word embeddings [8].

Knowledge distillation is another well-studied area in NLP models for transferring knowledge from a more complex architecture to its simpler version. All these approaches use the same architecture between the teacher and the student model. The most common way of applying knowledge distillation is to use either logits (pre-SoftMax) outputs of the teacher and force the student model to produce similar outputs or apply the same technique layer-wise. Mao and Nakagawa [9] combine both methods for obtaining language-agnostic sentence embeddings. Sahlgren [10] uses ensemble method to boost sentence embeddings of the student model. However, none of the mentioned methods try to transfer knowledge between different architectures. Recently, Liu et al. [11] showed a model for transferring knowledge from a transformer model to a convolutional model for image classification tasks. Their proposed model is based on the alignment of student and teacher feature spaces in two projected feature spaces.

In contrast to previous works done in both sentence embeddings and knowledge transfer, we directly use the outputs of both models (generated by different architectures) without updating the original architecture of the student model.

3 Methodologies

We consider two *sentence embedding spaces* models for this problem: *FastText* and *sBert*. Both produce their own representation of the same sentence. Our aim is to *reshape* the less informative space (*FastText*) by injecting information from the more informative space (*sBert*). In order to do so, two methods are proposed:

1. *Distance Minimization.*
2. *Perplexity Minimization.*

The first method attempts to solve the problem by rotating the representational space of the weaker model toward the robust model. This method involves two steps. The first step is to find an orthogonal rotation matrix (W) which is a linear projection method, and in the second step, a one-layer feed-forward neural network is applied to exploit the non-linear relationships between the two spaces. This second step can also be viewed as a fine-tuning stage.

The second method aims to solve the same problem by directly injecting representational power on the word level.

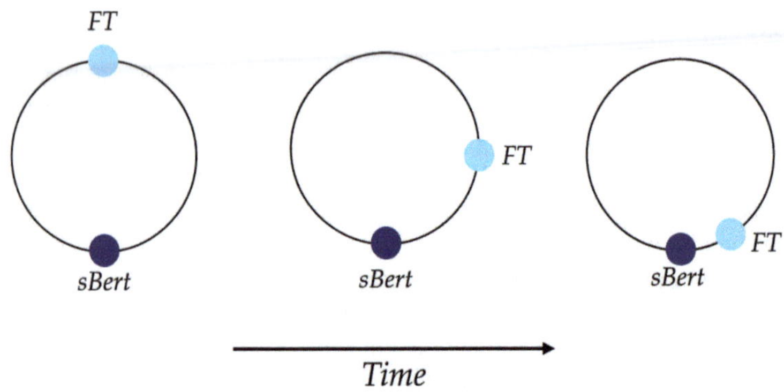

Fig. 1 The three figures depict the distance minimization procedure. Two sentence vectors represented by a light blue point (FastText sentence) and a dark blue point vector (sBert sentence). Both representations belong to the same sentence. The objective is to minimize the distance between the two data points by moving FastText sentence representation toward sBert

Both mentioned methods rely on the representation distillation method where both approaches aim to align their representations with the teacher model (*sBert*).

3.1 Distance Minimization

Consider Fig. 1 where two sentence representations of the same sentence are shown as two data points. Since both models are trained on different architectures, their sentence vectors should point in different directions. Yet, these points are projected on different spaces both should be correlated due to the semantics of the given sample.

One possible approach to increase this correlation might be to project both spaces on a common space and minimize the distance between their representations.

For minimizing the distance between two vector representations, two methods are used in sequence:

1. Orthogonal Mapping.
2. Feed-Forward Neural Network (*FFNN*).

The process of Orthogonal Mapping involves the rotation of sentence embeddings from one space toward the other using a linear regression method. Inspired by the work of Artetxe et al. [12] they align two word embedding spaces in different languages by using the well-known Procrustes method. We adopted the same rotation method mainly for two reasons:

1. The output spaces of FastText(300) and sBert(768) do not have equal dimensions. By using the orthonormal matrix property, the dimensional difference can be alleviated without sacrificing the representation power of sBert. This property can be seen as supervised dimension reduction.

2. The aim is to make the representations of the weaker model as close as possible to its teacher. We think that an alignment technique could help.

After applying orthogonal projection, a feed-forward network utilizes the obtained rotated sentence embeddings as its input and their associated sentences as its output. It attempts to leverage non-linear relationships between them by minimizing the Euclidean distance between two representation spaces. This second step can be described as a fine-tuning process.

3.2 Perplexity Minimization

Consider the noisy channel model where a source wants to send a message (information) to a target that the target does not have. The target tries to decode the given message according to its internal knowledge. At the end of the message decoding process on the target side, the target wants to have the same information as the source. Knowledge distillation can also be viewed as a noisy channel problem where the student tries to have the same information as the teacher model. From the NLP perspective, this can be interpreted as *perplexity* (*PP*). *PP* is a measurement that quantifies how well a probabilistic model predicts a sample. In our case, sBert is a better probabilistic model compared to FastText which makes its representational power more robust. To realize information transfer, we train a neural language model (*NLM*) that takes the sBert representations as the input, decodes these representations by updating the currently available FastText word embeddings, and finally checks the generated representation with the original sBert representation by using a distance metric (Fig. 2). Using sBert embeddings in initial states and FastText

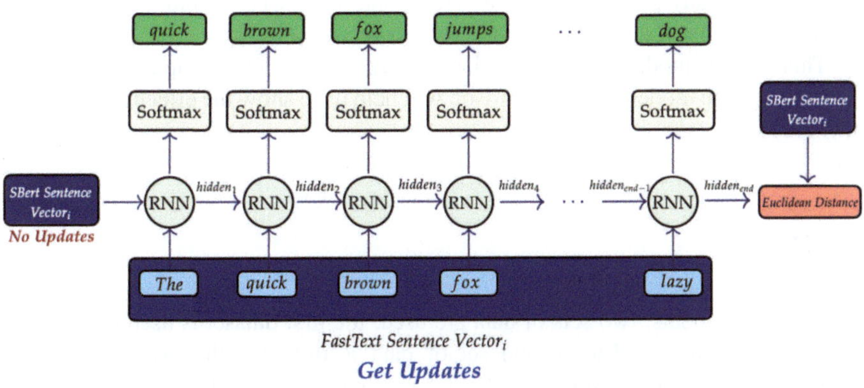

Fig. 2 Given a sentence to the model, the initial states are initialized with sBert sentence embeddings. The model consumes all the words by minimizing the cross-entropy loss. At the final stage, when it produces the last hidden state, it minimizes the Euclidean distance between the last hidden state and the corresponding sBert sentence embedding

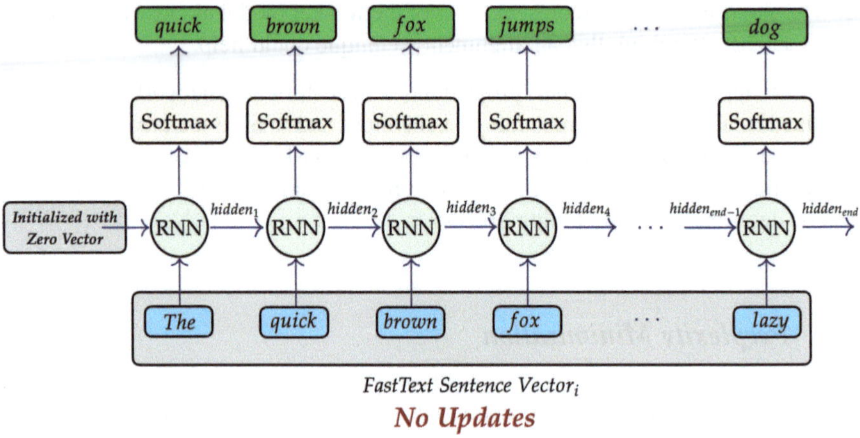

Fig. 3 To compare our model, a standard language model with an RNN is built. This model neither updates its embedding space nor uses any additional information from sBert

word embeddings in the final state of this model can gain more information about the sBert representations:

$$PP(NLM) = exp(crossentropy(NLM)) \tag{1}$$

A neural language model uses only cross entropy loss to achieve the decoding process which is directly related to perplexity (Eq. 1). Since sBert representations will be used at the end, there is a need for an additional loss function. By adding the Euclidean distance, the final representations of the language model can be compared with the original sBert sentence representation.

To compare the effect of the proposed method, another language model is built without using any sBert information. Additionally, FastText word embeddings do not receive updates during the training of this model (Fig. 3).

This second model should have larger perplexity values compared to the proposed language model. This comparison will help us to show the effectiveness of our methodology.

4 Data

For the experiments, two sets of data are used: the first dataset is used only for the *information transfer*. The second set of data is used for the evaluation of the transferred sentence embeddings. As our task requires operating in one language only, English datasets are used for all the experiments.

4.1 Preprocessing

For all datasets, all sentences are tokenized and lowercased by using Moses tokenizer before their embeddings are calculated.

4.2 Sentence Embedding Models

FastText calculates a sentence representation by using the so-called *bag of words* approach. It first converts each word in the sentence to its vector form by using subword tokenization. Then taking the average of these word vectors yields a sentence representation. *sBert*, on the other hand, produces its sentence embeddings by using the *Bert* model which also uses the subword tokenization method for handling vocabulary words. It then applies average pooling operation for generating sentence embeddings [7].

4.3 Data for Transferring Knowledge

To transfer *knowledge* from *sBert* to *FastText* sentence embeddings, we decided to use WMT's English–Spanish news corpus. This corpus contains nearly 2 million sentences about news around the world. For all experiments, only *English* sentences that are longer than ten words are used. To make proper comparison, both proposed methods use the same training and testing data: 285 thousand sentences for training, 15 thousand sentences for testing; a total of 300 thousand sentences.

4.4 Data for Evaluating Transfer

SentEval[2] [13] dataset is used for evaluating the produced sentence representations. This data toolkit focuses on four main tasks: classification, sentence natural language inference (NLI), textual similarity (STS), and probing. Total of 26 datasets. The classification datasets are used for evaluating models on tasks such as sentiment analysis, question types, and product reviews NLI tasks determine the models' ability to assess the logical relationship between two sentences. On the other hand, STS is used for measuring the similarity degree between a pair of sentences based on their underlying meanings. The goal is to evaluate the cosine similarity score between a sentence pair with human score through correlation. The probing is the

[2] https://github.com/facebookresearch/Sentval

task of evaluating the model's capability of capturing linguistic knowledge such as the ability to correctly predict sentence length, word order.

5 Experiments and Results

The first set of experiments is about knowledge transfer via *Distance Minimization* and *Perplexity Minimization*. The second set of experiments is only about the evaluation of generated sentence representations on SentEval datasets. The output of the first set of experiments is used as input for the second set of experiments.

5.1 Distance Minimization

This experiment aims to rotate sentence embeddings from *FastText* toward *sBert* in hope of getting similar sentence representations like sBert. In order to do so, a rotation matrix W is obtained by using orthogonal mapping. After many failed attempts to decrease the distance between two spaces, we decided to use the orthogonality features of the rotation matrix. The first feature that we used is the transpose of the rotation matrix. Instead of rotating FastText representations toward sBert, we rotated sBert toward FastText and called them $\text{sBert}_{\text{reduced}}$. Hence, both embedding spaces have equal dimensions (300). Yet, our results obtained from this did not improve the distance minimization method. Then we decided to exploit the second feature of orthogonality. Since the summation of two or more orthogonal matrices yields a new orthogonal matrix. We decided to divide the training data into many parts and calculate separate rotation matrices (W_i) for each group. Then we sum up all these W_i matrices to obtain a single(*global*) rotation matrix W. This operation yields much better results compared to finding the single rotation matrix W.

For capturing better non-linear relationships between two sentence embedding spaces, one layered feed-forward network is trained after obtaining the rotated sentence embeddings. The network is trained using the Adam optimizer with a learning rate of 3e−4 for 100 epochs.

The obtained *Euclidean* and *Cosine* distances on the test set from WMT can be seen in Table 1. Since the orthogonal mapping operation is described as a *linear*

Table 1 The first column shows the minimized distances by applying the orthonormal rotation matrix. The second column shows the obtained results of sentence vectors that are first passed from rotation operation and then fine-tuned by a neural network

	Rotation	Fine-tuning
Euclidean distance	0.632	0.402
Cosine distance	0.139	0.088

mapping, the summation of many different W orthogonal matrices yields a *non-linear mapping*.

It can be concluded that fine tuning the obtained sentence embeddings can reduce both distance metrics by a large margin on average.

5.2 Perplexity Minimization

To show the effect of the information transfer by the perplexity minimization method, two separate *language models* are trained on a corpus composed of a 300 thousand sentences from WMT English–Spanish dataset. The first trained model does not update the word embeddings, whereas the second model updates its word embedding space according to *sBert* sentence representations. The vocabulary size used for training is composed of 44,000 words.

Both language models are composed of a single layer of *RNN* with a batch size of 128. We trained these models using the Adam optimizer with a learning rate of $5e{-}4$ for a total of 5 epochs and took the best-resulted models according to the test set. After training, only words that are used for training language models will be used for the sentence evaluation tasks.

Table 2 shows the result of each model. Updating *FastText* word embeddings under the supervision of *sBert* sentence representations shows improved *perplexity* results compared to the vanilla version.

To show the effect of the proposed methods visually, a sample sentence from the WMT test set is generated for all proposed models. Then by using the UMAP algorithm, all high-dimensional sentence vectors are reduced to two-dimensional data (Fig. 4).

5.3 Sentence Evaluation Tasks

This subsection shows the results of proposed approaches on the SentEval library. In order to make a fair comparison, it is decided to use the same parameters that are described on their repository page. Hence, for tasks that require classification models batch size is set to 64, Adam is used as optimizer, epoch is set to 4, and each experiment is done with the ten-fold cross-validation method.

Table 2 First row shows the results of the language model where embedding space is not updated. The second row shows the updated version of embedding space under the supervision of *sBert* sentence representations

Model	PP	XE
LM (*w/o updates*)	48.32	3.878
LM (*w/updates*)	28.53	3.351

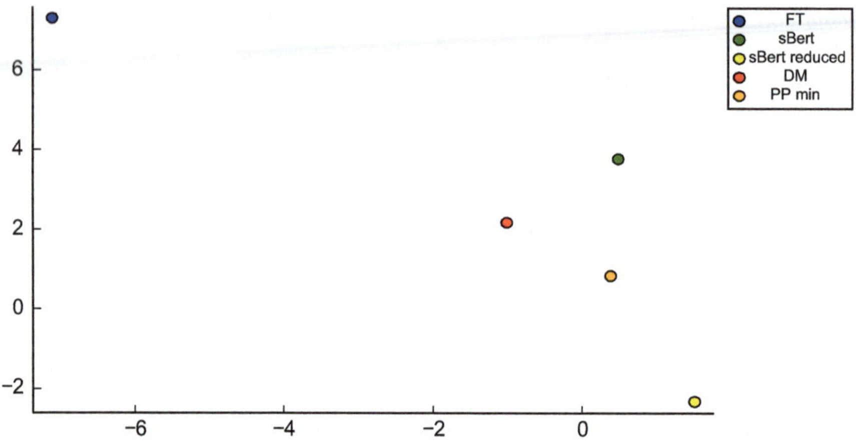

Fig. 4 Different sentence embedding models representing the same sentence on a two-dimensional space: "We are going to work seriously to ensure that the Europeans come to see being European as an advantage and that European citizens have certain fundamental rights that are connected to the union's institutions"

For evaluating *sBert* sentence embeddings, we implemented our script by using the Sentence-Transformers Python package.[3] For *FastText* sentence embeddings, their bag of words script is executed inside the toolbox.

The *PP* minimization experiment updates 44,000 words from the *FastText* model, we decided to show the effect of only 44,000 words as $PP\ Min_{subset}$ and remove the old representations from the original *FastText* embedding space and replace them with these updated 44,000 words (named as *PP Min* in results tables). This second approach can be seen as *hybrid* embeddings. Hence, we show three results from two different models.

From Tables 3, 4 and 5, it can be concluded that the *Distance Minimization* approach fails to improve the results in general. On the other hand, by using the *Perplexity Minimization* technique, *FastText* embeddings get an improvement over the original ones. For classification tasks, three out of six results are enhanced. For natural language inference tasks, two out of three results are improved. For semantic textual similarity tasks, five out of seven are improved. Finally, for probing tasks, five out of ten are enhanced.

[3]https://www.sbert.net/index.html

Table 3 Results of classification tasks including natural language inference tasks (right most three results)

Model	MR	CR	SUBJ	MPQA	SST2	TREC	MRPC	SICK-E	SNLI
sBert	85.05	86.86	93.97	89.32	88.76	94.0	75.07	81.86	78.1
sBert (*reduced*)	85.05	86.86	93.97	89.32	88.74	94.0	75.07	81.86	78.1
FT (*Avg.*)	**76.98**	76.45	90.4	**87.72**	**82.21**	77.8	72.52	75.62	**67.69**
FT (*DM*)	71.63	74.60	87.78	84.35	78.36	67.8	72.23	77.99	62.70
FT (*PP min*)	75.48	**77.35**	**90.63**	86.91	79.57	**81.8**	**73.45**	**78.55**	66.02
FT (*PP min subset*)	73.93	**77.11**	90.35	86.52	77.98	70.8	72.52	**78.22**	62.86

Results better than FT Embeds (*Avg.*) are shown in bold

Table 4 Results for semantic textual similarity

Model	STS12	STS13	STS14	STS15	STS16	STSB	SICK-R
sBert	0.73/0.73	0.78/0.77	0.80/0.77	0.82/0.83	0.82/0.83	0.82/0.83	0.87/0.82
sBert (*reduced*)	0.72/0.70	0.78/0.78	0.81/0.78	0.82/0.83	0.82/0.83	0.83/0.84	0.88/0.83
FT (*Avg.*)	0.48/0.49	0.44/0.45	0.54/0.54	0.58/0.59	0.49/0.54	**0.69/0.67**	**0.80/0.72**
FT (*DM*)	0.45/0.47	0.38/0.39	0.45/0.47	0.50/0.52	0.43/0.49	0.59/0.57	0.76/0.70
FT (*PP min*)	**0.56/0.56**	**0.56/0.56**	**0.65/0.62**	**0.70/0.70**	**0.64/0.66**	0.68/0.66	**0.80/0.72**
FT (*PP min subset*)	**0.53/0.54**	**0.54/0.54**	**0.62/0.63**	**0.67/0.68**	**0.58/0.60**	0.64/0.63	0.76/0.68

Reported numbers are *Pearson* (left) and *Spearman* (right) correlation coefficients. Each task has several subtasks, these are the averaged results of all. Results better than FT Embeds (*Avg.*) are shown in bold

6 Further Analysis: Importance of Alignment

Alignment is the process of finding a mapping function in which a representation, such as a word vector, from a *source space* finds the most related neighbor in a *target space*. This can be done by using either supervised [14] or unsupervised learning [15].

Since our approach focuses only on the outputs of the models in use and not on the internal structures. We find it appropriate to discuss the relationship in the alignment of vector representations.

By using the vecmap tool[4,5] of Artetxe et al. [16], it is noted that representations of the same samples from both spaces (neither $FastText_{original}$ and *sBert* nor $FastText_{DM}$ and *sBert*) do not align with each other. On the other hand, original sentence representations of sBert on the WMT dataset can be aligned perfectly with their reduced form.

[4] https://github.com/artetxem/vecmap

[5] Our implementation can be found in our shared repository.

Table 5 Results for probing tasks

Model	SLen	WC	Depth	Const	BShift	Tense	SNum	ONum	SOMO	CInv
sBert	49.93	61.59	27.91	57.09	75.71	87.99	98.98	78.07	61.98	61.05
sBert (*reduced*)	49.93	61.59	27.91	57.09	75.71	87.99	79.98	78.07	61.98	61.05
FT (*Avg.*)	56.76	36.32	**32.99**	**61.15**	**50.61**	**85.64**	78.06	78.15	49.28	**52.78**
FT (*DM*)	49.9	28.03	30.19	53.30	49.66	83.04	75.59	75.74	49.71	52.03
FT (*PP min*)	**57.64**	**85.27**	32.76	60.46	50.54	84.83	**79.4**	**79.55**	49.24	51.89
FT (*PP min subset*)	**56.79**	**72.18**	32.46	39.24	49.9	76.9	73.14	77.97	**50.18**	51.7

Results better than FT Embeds (*Avg.*) are shown in bold

One possible explanation would be FastText and sBert come from different distributions. And a linear projection matrix cannot capture the non-linear relationships between them. Instead, it just prioritizes the features of the higher dimensional space(sBert), but still, the distributional difference is not eliminated. In fact, it distorts the input space (FastText). The SentEval task is a proof of this phenomenon. The reduced sBert vectors can get near-exact results just like their original sBert vectors, whereas the $FastText_{DM}$ gets the lowest scores in every task. On the other hand, the perplexity minimization method uses a more complicated structure in terms of architecture and objective functions, yet the alignment between its updated FastText embeddings and sBert causes a slight increase compared to the DM method.

The outcome of the alignment process shows that the distributional differences cause misalignment which impedes information transfer. This phenomenon is more observable in the DM method. On the other hand, even though the PP minimization method performs better, there is still a performance gap compared to sBert vectors.

7 Conclusion

This chapter proposes model-agnostic solutions for situations where applying knowledge distillation techniques is impractical. Our methods consider use cases where model parameters are not shared with the end user and/or models are implemented using different deep learning frameworks, which impedes parameter sharing.

The experiments showcased rely directly on the outputs of the embedding spaces without interfering with the architectures of both embedding models. Based on the obtained results, directly attempting to minimize the Euclidean distance has no positive effect. Conversely, attempting to minimize the *perplexity* with the help of the Euclidean distance assists in transferring knowledge.

There are two major findings in this chapter. The first one is the importance of the distributional difference between operated spaces affects the alignment. If the distributions are incompatible with each other, then transferring knowledge gets harder. Our results show that eliminating this conflict using a linear approach does not help, instead using more complicated architectures seems to be a better option. The performance gap still exists in most of the cases.

The second finding is about the usage of dimensionality reduction operation by utilizing orthogonal mapping. Even though its linear nature impedes rotation between different distributions, it helps to reduce the dimensions of the higher dimensional space to its lower space. Our obtained results show that this reduction operation has no effect on information loss.

The misalignment issue between the two spaces appears to hinder the full potential of the applied methods. Therefore, for future research, we will investigate ways to *calibrate* the representation spaces. Potential advances in calibration can provide a solution for knowledge transfer between different spaces, which, consequently, entails distillation between different architectural models.

References

1. Hinton, G., Vinyals, O., Dean, J.: Distilling the Knowledge in a Neural Network (2015)
2. Sanh, V., Debut, L., Chaumond, J., Wolf, T.: Distilbert, a Distilled Version of Bert: Smaller, Faster, Cheaper and Lighter (2020)
3. Bojanowski, P., Grave, E., Joulin, A., Mikolov, T.: Enriching word vectors with subword information. CoRR abs/1607.04606. (2016) http://arxiv.org/abs/1607.04606
4. Arora, S., Liang, Y., Ma, T.: A simple but tough-to-beat baseline for sentence embeddings, International Conference on Learning Representations (2017)
5. Kiros, R., Zhu, Y., Salakhutdinov, R., Zemel, R.S., Torralba, A., Urtasun, R., Fidler, S.: Skip-thought vectors. CoRR abs/1506.06726. (2015) http://arxiv.org/abs/1506.06726
6. Logeswaran, L., Lee, H.: An Efficient Framework for Learning Sentence Representations (2018), https://arxiv.org/abs/1803.02893
7. Reimers, N., Gurevych, I.: Sentence-bert: Sentence embeddings using siamese bert networks. CoRR abs/1908.10084. (2019) http://arxiv.org/abs/1908.10084
8. Devlin, J., Chang, M.W., Lee, K., Toutanova, K.: Bert: Pre-Training of Deep Bidirectional Transformers for Language Understanding (2019)
9. Mao, Z., Nakagawa, T.: Lealla: Learning lightweight Language-Agnostic Sentence Embeddings with Knowledge Distillation (2023)
10. Sahlgren, F.C.M.: Sentence Embeddings by Ensemble Distillation (2021).
11. Liu, Y., Cao, J., Li, B., Hu, W., Ding, J., Li, L.: Cross-Architecture Knowledge Distillation (2022). https://doi.org/10.48550/ARXIV.2207.05273. https://arxiv.org/abs/2207.05273
12. Artetxe, M., Labaka, G., Agirre, E.: Generalizing and improving bilingual word embedding mappings with a multi-step framework of linear transformations. In: AAAI, pp. 5012–5019 (2018)
13. Conneau, A., Kiela, D.: Senteval: An evaluation toolkit for universal sentence representations (2018), https://arxiv.org/abs/1803.05449
14. Joulin, A., Bojanowski, P., Mikolov, T., Jegou, H., Grave, E.: Loss in translation: Learning bilingual word mapping with a retrieval criterion. In: Proceedings of the 2018 Conference on Empirical Methods in Natural Language Processing (2018)
15. Artetxe, M., Labaka, G., Agirre, E.: A Robust Self-Learning Method for Fully Unsupervised Cross-Lingual Mappings of Word Embeddings (2018), arXiv preprint arXiv:1805.06297
16. Artetxe, M., Labaka, G., Agirre, E.: Learning bilingual word embeddings with (almost) no bilingual data. In: Proceedings of the 55th Annual Meeting of the Association for Computational Linguistics, vol. 1, pp. 451–462 (2017)

Prediction of Heart Attack Risk with Data Mining by Using Blood Tests and Physical Data

Osman Ali Waberi and Şükrü Kitiş

1 Introduction

Millions of people die every year from cardiovascular diseases, which have been the number one killer globally for many years. According to World Health Organization (WHO) data, the causes of death of the majority of people who die are sudden onset diseases such as heart attack or stroke [1, 2].

It was reported that around 17.9 million people die each year due to cardiovascular disease by the World Health Organization (WHO) [3, 4]. It was also reported that of the 17.3 million deaths caused by heart disease in 2008, an estimated 7.3 million were due to coronary heart disease. A prediction by WHO states that around 23.6 million people would die due to heart disease by 2030 [4, 5].

The definitive diagnosis of heart disease requires a doctor's examination supported by X-rays, electrocardiograms, echocardiograms, physical exercise, blood tests, and angiography. However, especially the increase in clinical data makes the diagnosis of heart disease more and more difficult [6].

Data mining is the analysis of data in a data warehouse in order to create meaningful rules, perform data clustering, and reveal whether it belongs to a certain class or not, with a specific goal [7]. Millions of data in databases are meaningless on their own. These data become valuable when information extraction is performed, and patterns in the database are uncovered. For example, a patient's clinical laboratory results are just data. If early diagnosis of any disease can be made from these data, then these data become meaningful and become information. For this purpose, the process of knowledge discovery from data is called data mining [7].

O. A. Waberi · Ş. Kitiş (✉)
Kütahya Dumlupinar University, Kütahya, Turkey
e-mail: sukru.kitis@dpu.edu.tr

M. N. Seyman (ed.), *2nd International Congress of Electrical and Computer Engineering*, EAI/Springer Innovations in Communication and Computing,
https://doi.org/10.1007/978-3-031-52760-9_2

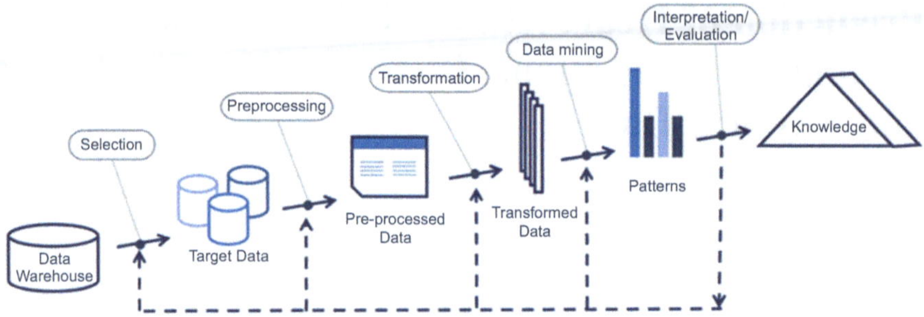

Fig. 1 Data mining process according to Fayyad [7, 16, 17]

Data mining is the process of searching and analyzing data to discover meaningful rules and patterns [8]. Data mining is also known as knowledge extraction, data archaeology, knowledge discovery or knowledge discovery from databases (KDD), data/pattern analysis, and data mining [9]. KDD is the process of finding hidden patterns for better decision-making. In fact, data mining is only one step in the KDD process, but in the literature, data mining and KDD are considered the same [10]. Data mining includes techniques such as association analysis, predictive modeling, anomaly detection, and clustering [8, 9].

Data mining referring to the relationships and rules that enable predictions to be made from large amounts of data using computer programs [11] is also considered as a data analysis process [12] that brings into the open information that raw data alone cannot provide [13]. Fayyad [14] defined it as the emergence of valid, reliable, potentially useful, previously unknown, and understandable patterns from databases. Using data mining techniques will result in falling costs, rising revenues, increasing productivity, emerging new opportunities, making new discoveries, and uncovering frauds [15].

Fayyad et al. outlined the steps of data mining as shown in Fig. 1 [7, 16, 17]. Fayyad et al. explained the selection, preprocessing, transformation, data mining, and interpretation in Fig. 1.

Han et al. also outlined the steps of the data mining process as shown in Fig. 2 [7, 18, 19]. Han et al. explained cleaning and integration, selection and transformation, data mining, evaluation, and presentation in Fig. 2.

The data preprocessing steps mentioned in Figs. 1 and 2 can be listed as follows [9]:

(a) Data Cleaning

In this step, missing data are filled, deviating data arw detected, inconsistencies in the data are eliminated, and deviated data are completely discarded [7].

(b) Data Merge

In this step, data from databases and various information sources are combined, and residual data are removed. Schema merge errors occur when data from different databases are merged into a single database. For example, in one database, entries may be written as "consumer-ID," while in another, they may

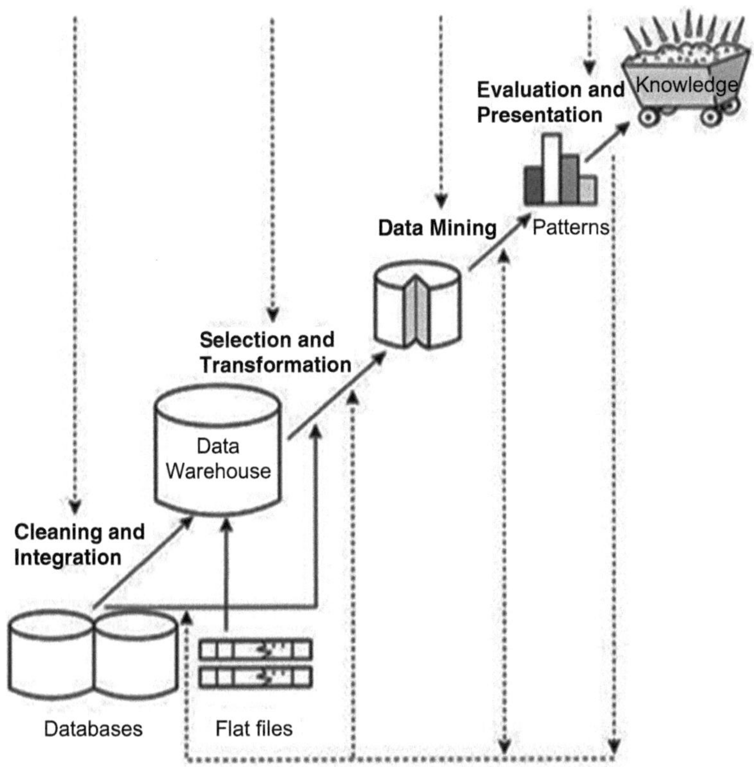

Fig. 2 Data mining according to Han [7, 18, 19]

be written as "consumer-number." To avoid such errors, metadata is used. Databases and data warehouses usually have metadata [19].

(c) Data Conversion

In this step, normalization and generalization are performed to make the data more understandable [7].

(d) Data Reduction

Since there are many features that affect the solution of the problem in this step, the problem of finding the ones that affect the result the most (feature selection) arises. This problem is solved by feature selection for data reduction, feature extraction, dimension reduction, and data integration [7].

There are many methods and algorithms for data mining. Many of these methods are statistical in nature. Data mining models can be basically grouped as follows [7, 20, 21]:

(a) Classification.
(b) Clustering.
(c) Association rules (Fig. 3).

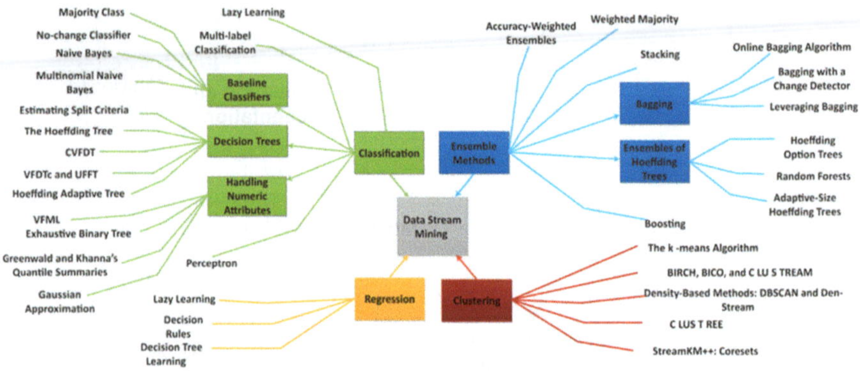

Fig. 3 Data mining models and algorithms [7, 20, 21]

2 Materials and Methods

2.1 Data Set

The cardiovascular disease dataset consists of 70,000 records, 11 features, and a target [22]. There are 3 types:

Subjective: information given by the patient.
Examination: medical examination.
Objective: factual information (Tables 1 and 2).

Objective property was defined in the first 4 variables:

1st Age, type int, expressed as day
2nd Height, also of type is int, expressed in cm
3rd Weight, type int, expressed in kg
4th Gender type expressed as categorical 1: boy, 2: girl.

5 to 8 Test features were defined:

5th Systolic blood pressure, type int
6th Diastolic blood pressure, type int
7th Cholesterol, type categorically expressed as 1: normal, 2: above normal, 3: high
8th Glucose, type categorically expressed as 1: normal, 2: above normal, 3: high.

Subjective characteristics 9 through 11 were defined:

9th Cigarette, type binary 0:no, 1:yes
10th Alcohol intake, type binary 0:no, 1:yes
11th Physical activity, type binary 0:no, 1:yes.

And finally, the goal variable was specified:

Presence of cardiovascular disease, type binary 0:no, 1:yes.

Table 1 Dataset [23]

Features name	Variable	Value type
Age	Age	Day
Height	Height	Cm
Weight	Weight	Kg
Gender	Gender	Categorical
S. Blood Press.	Ap_hi	Int.
D. Blood Press.	Ap_lo	Int.
Cholesterol	Cholesterol	1: normal, 2: above normal, 3: well above
Glucose	Glu	1: normal, 2: above normal, 3: well above
Smoking	Smoke	Binary
Alcohol	Alco	Binary
Physical activity	Active	Binary
Presence of CVD's	Cardio	Binary

Table 2 Dataset attributes description [24]

Serial number	Variable description
1	Age-int (days); min: 10798, max: 23713, mean:19468.866, StdDev:2467.252
2	Height-int (cm); min:55, max:250, mean:164.359, StdDev:8.21
3	Weight-float (kg); min:10, max: 200, mean:74.206, StdDev:14.396
4	Gender-categorical code (f = female, m = male)
5	Ap_hi-int; min:-150, max:16020, mean:128.817, StdDev:154.011
6	Ap_lo-int; min:-70, max:11000, mean:96.63, StdDev:188.473
7	Cholesterol; (1 = Normal, 2 = above normal, 3 = well above normal)
8	Gluc; (1 = Normal, 2 = above normal, 3 = well above normal)
9	Smoke-binary; (1 = smoker, 0 = non-smoker)
10	Alco-binary; (1 = yes, 0 = no)
11	Active-binary;(active = 1, inactive = 0)
	Target-binary; (1 = presence, 0 = absence of cardiovascular disease)

2.2 *Processes and Methods*

Naive Bayes Algorithms This algorithm is a categorization/classification algorithm. This classification determines the class/category of data presented to the system by a series of calculations defined according to probabilistic principles. The data presented must have a class/category. In this classification, how it is classified is more important than what is classified. In other words, rather than the data type and what it is, it is important how a proportional relationship is established between these data [25]. In the Naive Bayes classifier, it is assumed that one attribute value does not contain information about another attribute value, that attributes are independent of each other, and that all attributes are important. The advantages of the

Naive Bayes classifier are that it produces high accuracy, precise and fast results; the algorithm is simple and understandable, works well in most cases, and is easy to perform. The disadvantage of the Naive Bayes classifier is that variables are dependent on each other in real life [7].

J48 Decision Tree Algorithm The J48 algorithm aims to optimize the decision tree and find the entropy value of the variable by making use of Shannon's Information Theory. This algorithm first calculates the entropy value for each target variable/class and the information value for each prediction variable/class. It then calculates information gain for each prediction variable/class. The purpose of these calculations is to determine the class that collects the most information [25].

Function Simple Logistic Algorithms Logistic model trees were expressed as very precise and compact classifiers [26]. The major disadvantage of them iscomputational complexity of building logistic regression models on trees. To avoid overfitting these models, we address this issue by using the AIC criterion [27] rather than cross-validation. In addition, a weight correction heuristic is used, which provides significant acceleration. Also, a weight correction heuristic is used, which provides significant acceleration. Comparing the training time and accuracy of the new induction process with the original process on various datasets and showing that the training time is often reduced, while classification accuracy is only slightly reduced are performed by our study [28]. Cross-validation leading to automatic attribute selection is carried out for an optimal number of LogitBoost iterations to perform [29].

3 Research Results and Discussion

3.1 Literature

Taşçı et al. eliminated values in a dataset of 85,000 cases according to some optimization methods and used a total of 303 cases with 13 features for classification. According to the results, they found that different methodologies were successful with different criteria, and when the criteria were averaged, k-NN was the most accurate and gave the best result (accuracy rate 61%) [25].

Erkuş made performance comparisons using data mining methods to contribute to the early diagnosis of cardiovascular diseases. Of the 604 patients in the dataset, 297 were diagnosed with CVD, and 307 were not diagnosed with CVD. He created 3 separate datasets using statistical grouping techniques and made separate evaluations. Among the classifiers he used, the Hidden Naive Bayes (HNB) algorithm gave the best performance with 84.8% success rate [30].

Çilhoroz and Çilhoroz, 2021 examined the factors affecting mortality due to cardiovascular diseases with data from the OECD database. They used the Least Squares (LS) method in the study. As a result of the study, they found that smoking and alcohol use have positive effects on CVD-related deaths [31].

Kim et al. applied Logistic Regression, K-Nearest Neighbor, Decision Trees, Random Forest, Extra Trees, XGBoosting, Gradient Boosting, AdaBoost, Support Vector Machines, and Multilayer Perceptrons methods to find the machine learning algorithm that builds the best prediction model for NHIS health screening dataset. According to the results, they concluded that the best prediction model was established with XGBoosting, Gradient Boosting, and Random Forest algorithms [32].

In a study performed [33], six different machine learning algorithms were run on 12 different datasets. The accuracy rate obtained on the heart disease dataset was 59.72%.

A cardiovascular disease prediction model using a machine learning approach was offered by Geetha et al. [4, 34]. Using the KNN algorithm to classify heart disease was performed. In total, 87% accuracy was acquired by this model.

3.2 Data Cleaning

Getting rid of ; with Java program. Since the Weka program does not accept the; the editing of the; was performed with the JAVA program (Fig. 4).

Weka program works with arff type files. Files are saved as csv type with the excel program and changed as arff type with weka program. Weka program files don't work with; punctuation. This punctuation was performed with a Java program (Fig. 4).

Getting rid of outliers. Outliers graphics (Fig. 5) were obtained from the "visualize all" section when the weka program run.

The outliers shown in Fig. 5 were removed with the IQR method as shown in Fig. 6 and the data was cleaned with Eq. 1.

$$\begin{gathered} \text{value} < Q1 - 1,5^{*}\ \text{IQR} \\ \text{or} \\ \text{value} > Q3 + 1,5^{*}\ \text{IQR} \\ \text{with}\quad \text{IQR} = Q3 - Q1 \end{gathered} \tag{1}$$

3.3 Data Preprocessing and Forecasting

Figure 7 shows the method: A machine learning model is divided into two batches; The first batch contains supervised data and the second refers to unsupervised data. The supervised part is divided into two sub-parts; first regression and second

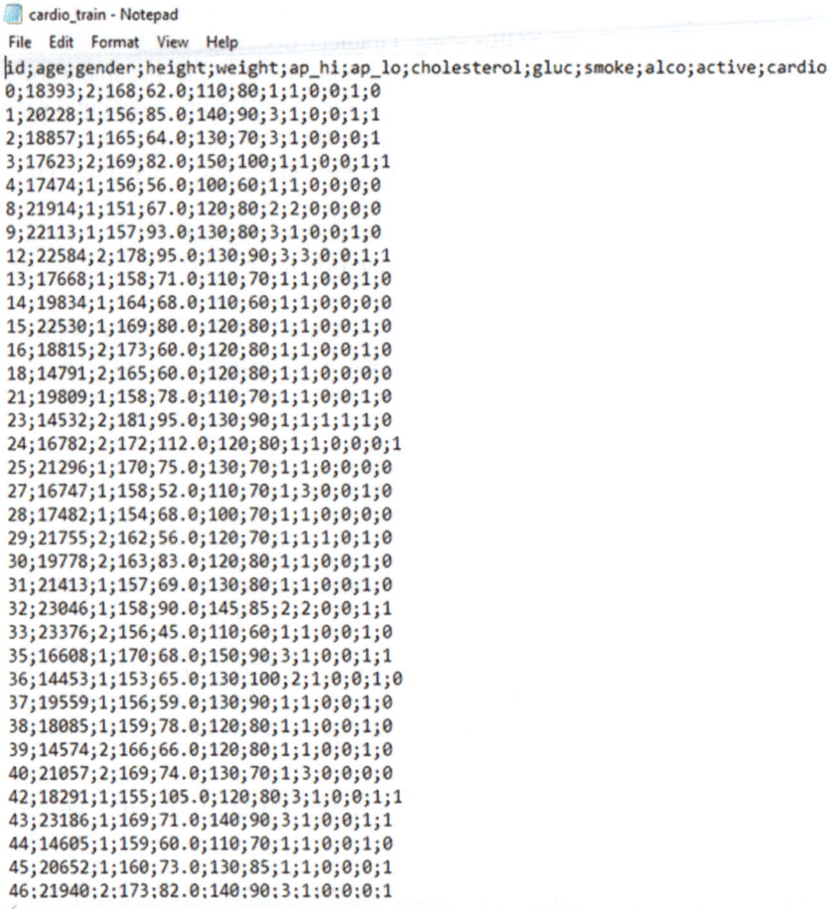

```
cardio_train - Notepad
File Edit Format View Help
id;age;gender;height;weight;ap_hi;ap_lo;cholesterol;gluc;smoke;alco;active;cardio
0;18393;2;168;62.0;110;80;1;1;0;0;1;0
1;20228;1;156;85.0;140;90;3;1;0;0;1;1
2;18857;1;165;64.0;130;70;3;1;0;0;0;1
3;17623;2;169;82.0;150;100;1;1;0;0;1;1
4;17474;1;156;56.0;100;60;1;1;0;0;0;0
8;21914;1;151;67.0;120;80;2;2;0;0;0;0
9;22113;1;157;93.0;130;80;3;1;0;0;1;0
12;22584;2;178;95.0;130;90;3;3;0;0;1;1
13;17668;1;158;71.0;110;70;1;1;0;0;1;0
14;19834;1;164;68.0;110;60;1;1;0;0;0;0
15;22530;1;169;80.0;120;80;1;1;0;0;1;0
16;18815;2;173;60.0;120;80;1;1;0;0;1;0
18;14791;2;165;60.0;120;80;1;1;0;0;0;0
21;19809;1;158;78.0;110;70;1;1;0;0;1;0
23;14532;2;181;95.0;130;90;1;1;1;1;1;0
24;16782;2;172;112.0;120;80;1;1;0;0;0;1
25;21296;1;170;75.0;130;70;1;1;0;0;0;0
27;16747;1;158;52.0;110;70;1;3;0;0;1;0
28;17482;1;154;68.0;100;70;1;1;0;0;0;0
29;21755;2;162;56.0;120;70;1;1;1;0;1;0
30;19778;2;163;83.0;120;80;1;1;0;0;1;0
31;21413;1;157;69.0;130;80;1;1;0;0;1;0
32;23046;1;158;90.0;145;85;2;2;0;0;1;1
33;23376;2;156;45.0;110;60;1;1;0;0;1;0
35;16608;1;170;68.0;150;90;3;1;0;0;1;1
36;14453;1;153;65.0;130;100;2;1;0;0;1;0
37;19559;1;156;59.0;130;90;1;1;0;0;1;0
38;18085;1;159;78.0;120;80;1;1;0;0;1;0
39;14574;2;166;66.0;120;80;1;1;0;0;1;0
40;21057;2;169;74.0;130;70;1;3;0;0;0;0
42;18291;1;155;105.0;120;80;3;1;0;0;1;1
43;23186;1;169;71.0;140;90;3;1;0;0;1;1
44;14605;1;159;60.0;110;70;1;1;0;0;1;0
45;20652;1;160;73.0;130;85;1;1;0;0;0;1
46;21940;2;173;82.0;140;90;3;1;0;0;0;1
```

Fig. 4 Getting rid of

classification. Considering that we have a target variable in our base and that the type of this variable is categorical, we will perform supervised machine learning on the classification problem (Fig. 8).

4 Results

Three algorithms were chosen to train the model; Naive Bayes, J48 Decision Tree, Function Simple Logistic, Results obtained with the model J48, Naive Bayes, and Simple Logistic in ten-fold cross-validation test mode are shown in Fig. 9. Among the root relative squared error results, the Naivest Bayes model reached the highest root relative squared error rate (89%). The other results are 87% (J48) and 86% (function simple logistic).

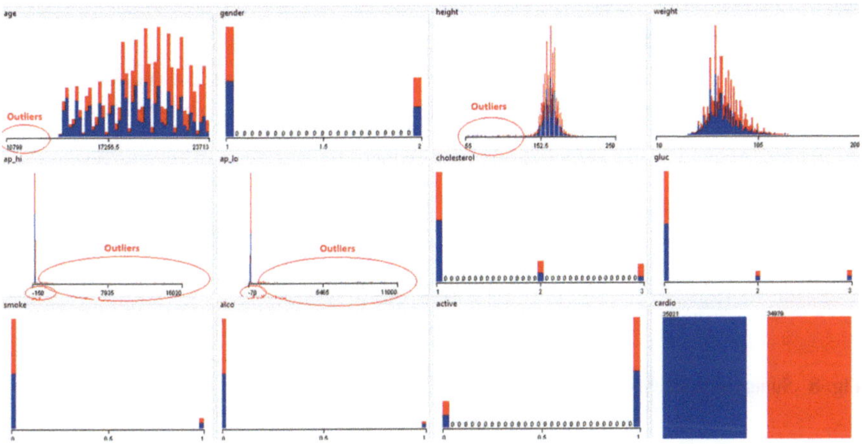

Fig. 5 Outliers

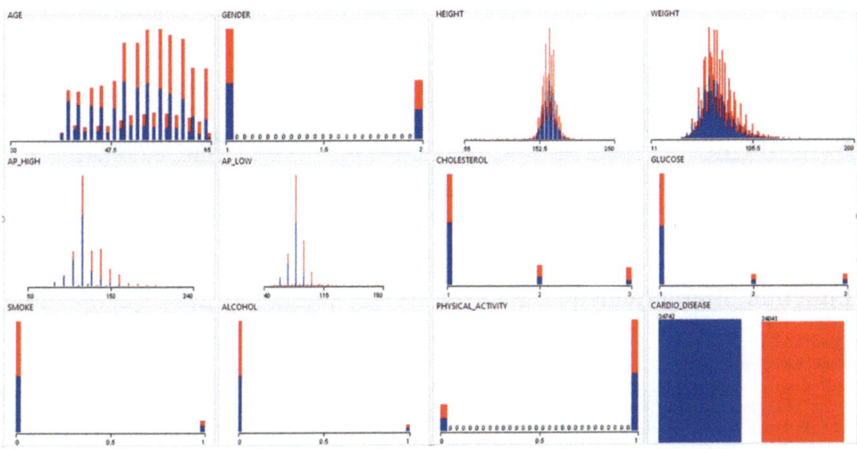

Fig. 6 Getting rid of outliers with the interquartile range (IQR) method

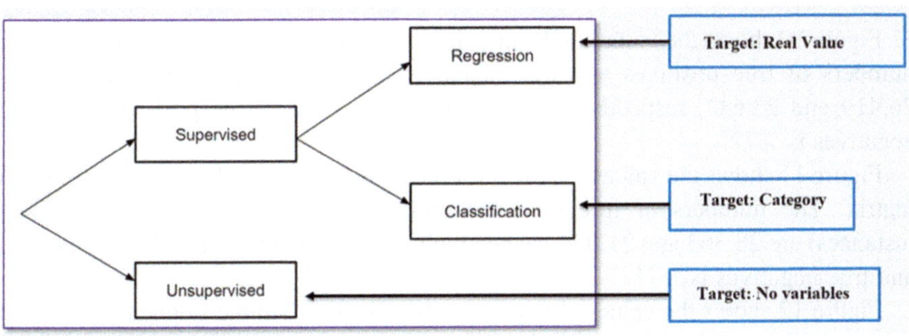

Fig. 7 Data preprocessing and forecasting

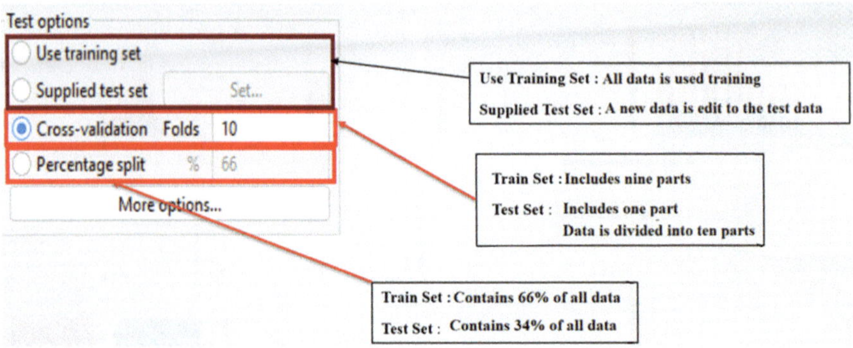

Fig. 8 Supervised machine learning

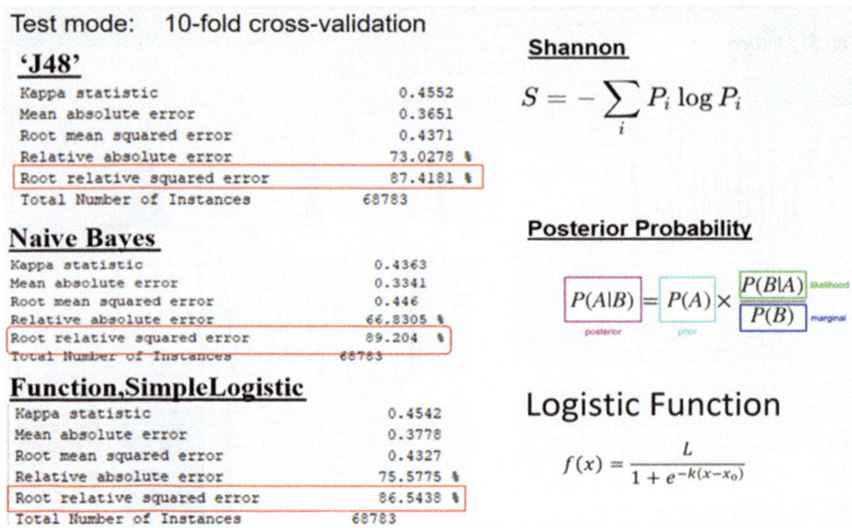

Fig. 9 10-fold cross-validation test mode

Figure 10 shows the values and confusion matrix obtained for the J48 model. The numbers of true positives and true negatives (correctly classified instances) are 26,419 and 23,641, respectively. The ratio of the sum of true positives and true negatives is %72.

Figure 11 shows the values obtained for the Naive Bayes model and the confusion matrix. The numbers of true positives and true negatives (correctly classified instances) are 28,363 and 21,074, respectively. The ratio of the sum of true positives and true negatives is %71.

Figure 12 shows the values obtained for the Function Simple Logistic model and the confusion matrix. The numbers of true positives and true negatives (correctly classified instances) are 27,324 and 22,711, respectively. The ratio of the sum of true positives and true negatives is %72.

'J48'

Correctly Classified Instances	50060	72.7796 %
Incorrectly Classified Instances	18723	27.2204 %

=== Confusion Matrix ===

a	b	<-- classified as
26419	8323	a = 0
10400	23641	b = 1

Fig. 10 J48 values

Naive Bayes

Correctly Classified Instances	49437	71.8739 %
Incorrectly Classified Instances	19346	28.1261 %

=== Confusion Matrix ===

a	b	<-- classified as
28363	6379	a = 0
12967	21074	b = 1

Fig. 11 Naive Bayes values

Function Simple Logistic

Correctly Classified Instances	50035	72.7433 %
Incorrectly Classified Instances	18748	27.2567 %

=== Confusion Matrix ===

a	b	<-- classified as
27324	7418	a = 0
11330	22711	b = 1

Fig. 12 Function simple logistic values

According to the literature review, Taşçı et al. achieved the highest result at 61% for 8500 data. Erkuş achieved 84.8% for 604 data, while Karakoyun and Hacıbeyoğlu achieved 59.72%. In another study, "Geetha et al." were able to achieve 87% for 303 data. Our study achieved 89% "root relative squared error" and 71.87% "correctly classified instances" for 70,000 recorded data.

Conflict of Interest The authors have no relevant financial or non-financial interests to disclose.

Data Availability Training and testing processes have been carried out using the cardiovascular disease dataset.

The cardiovascular disease dataset is publicly available at https://www.kaggle.com/datasets/sulianova/cardiovascular-disease-dataset?fbclid=IwAR3bpgZt5DZYJYjf4m8h9gDssOMXW7_GlAJrLqe0I3BQnnw8aH7If8ddkA. Last accessed 06/26/2022.

References

1. Who Health Organization (WHO), cardiovascular diseases, https://www.who.int/healthtopics/cardiovascular-diseases#tab=tab_1, Last accessed 2021/06/11
2. Kaba, G., Bağdatlı Kalkan, S.: Kardiyovasküler Hastalık Tahmininde Makine Öğrenmesi Sınıflandırma Algoritmalarının Karşılaştırılması. İstanbul Ticaret Üniversitesi Fen Bilimleri Dergisi. **21**(42), 183–193 (2022)
3. Who int. cardiovascular diseases [online], https://www.who.int/health-topics/cardiovascular-diseases/#tab=tab_1, Last accessed 2021/06/26
4. Uddin, M.N., Halder, R.K.: An ensemble method based multilayer dynamic system to predict cardiovascular disease using machine learning approach. Informatics Med. Unlocked. **24** (2021). https://doi.org/10.1016/j.imu.2021.100584

5. Who int. about cardiovascular diseases [online], https://www.who.int/cardiovascular_diseases/about_cvd/en, Last accessed 2021/01/26.
6. Tripoliti, E.E., Papadopoulos, T.G., Karanasiou, G.S., Naka, K.K., Fotiadis, D.I.: Heart failure: Diagnosis, severity estimation and prediction of adverse events through machine learning techniques. Comput. Struct. Biotechnol. J. **15**, 26–47 (2017)
7. Hanife, G.: Üniversite Giriş Sınavında Öğrencilerin Başarılarının Veri Madenciliği Yöntemleri İle Tahmin Edilmesi. Gazi Üniversitesi Yüksek Lisans Tezi (2012)
8. Berry, M.J.A., Linoff, G.S.: Data Mining Techniques: For Marketing, Sales, and Customer Support. Wiley Computer Pub, NewYork (1997)
9. Han, J., Kamber, M.: Data Mining: Concept and Techniques. Morgan Kaufmann Publications, San Francisco (2001)
10. Koyuncugil, A.S., Özgülbaş, N.: Surveillance Technologies and Early Warning Systems: Data Mining Applications for Risk Detection. IGI Global, USA (2010)
11. Babadağ, K.: Zeki Veri Madenciliği: Ham Veriden Altın Bilgiye Ulaşma Yöntemleri. Ind. Appl. Softw., 85–87 (2006)
12. Jacobs, P.: Data mining: What general managers need to know. Harvard Manag. Update. **4**(10), 8 (1999)
13. Alataş, B., Akın, E.: Veri Madenciliğinde Yeni Yaklaşımlar. Ya/Em-2004- Yöneylem Araştırması/Endüstri Mühendisliği XXIV Ulusal Kongresi, 15–18 Haziran, Gaziantep-Adana (2004)
14. Fayyad, U.M., Piatetsky-Shapiro, G., Smyth, P.: From Data Mining to Knowledge Discovery: An Overview, pp. 1–30. AKDDM, AAAI/MIT Press (1996)
15. Bırtıl, F.S.: Kız Meslek Lisesi Öğrencilerinin Akademik Başarısızlık Nedenlerinin Veri Madenciliği Tekniği İle Analizi. Afyon Kocatepe Üniversitesi Fen Bilimleri Enstitüsü, Yüksek Lisans Tezi (2011)
16. Fayyad, U., Gregory, P., Smyth, P.: From data mining to knowledge discovery in databases. AI Mag., 37–54 (1996)
17. Rotondo, A., Quilligan, F.: Evolution paths for knowledge discovery and data mining process models. SN Comput. Sci. **1**, 109 (2020)
18. Amal, A.M., Enas, M.H.: A Review: Data Mining Techniques and Its Applications. IJCSMA (2022)
19. Oğuzlar, A.: Veri önişleme. Erciyes Üniv. İktisadi ve İdari Bilimler Fakültesi Dergisi. **21**, 67–76 (2003)
20. Özkan, Y.: Veri Madenciliği Yöntemleri. Papatya Yayıncılık Eğitim, İstanbul (2008)
21. Amin, S., Mahmoud, A., Amir, T., Anca, D.J.: A comparative study on online machine learning techniques for network traffic streams analysis. Comput. Netw. **207** (2022)
22. Jaouja, M., Gilbert, G., Hungilo, P.: Comparison of machine learning models in prediction of cardiovascular disease using health record data. In: 2019 International Conference on Informatics, Multimedia, Cyber and Information System (ICIMCIS)., 978-1-7281-2930 (2019)
23. Mohammed, N.U., Rajib, K.H.: An ensemble method based multilayer dynamic system to predict cardiovascular disease using machine learning approach. Informatics Med. Unlocked. (2019)
24. Taşçı, M.E., Şamlı, R.: Veri Madenciliği İle Kalp Hastalığı Teşhisi, pp. 88–95. Avrupa Bilim ve Teknoloji Dergisi (2020)
25. Weka, https://weka.sourceforge.io/doc.dev/weka/classifiers/functions/SimpleLogistic.html, 2022/06/05
26. Erkuş, S.: Veri madenciliği yöntemleri ile kardiyovasküler hastalık tahmini yapılması (Yüksek Lisans Tezi). Bahçeşehir Üniversitesi Fen Bilimleri Enstitüsü, İstanbul (2015)
27. Çilhoroz, İ.A., Çilhoroz, Y.: Kardiyovasküler hastalıklara bağlı ölümleri etkileyen faktörlerin belirlemesi: OECD ülkeleri üzerinde bir araştırma. Acıbadem Üniversitesi Sağlık Bilimleri Dergisi. **12**(2), 340–345 (2021)

28. Kim, J.O., Jeong, Y.S., Kim, J.H., Lee, J.W., Park, D., Kim, H.S.: Machine learning based cardiovascular disease prediction model: A cohort study on the Korean national health insurance service health screening database. Diagnostics. **11**(6), 943 (2021)
29. Karakoyun, M., Hacibeyoğlu, M.: Biyomedikal Veri Kümeleri İle Makine Öğrenmesi Sınıflandırma Algoritmalarının İstatistiksel Olarak Karşılaştırılması. Dokuz Eylül Üniversitesi Mühendislik Fakültesi Fen ve Mühendislik Dergisi. **16**(48), 30–42 (2014)
30. Geetha, D.A., Prasada, R.B.S., Vidya, S.K.: A method of cardiovascular disease prediction using machine learning. Int. J. Eng. Res. Technol. **9**(5), 243–246 (2021)
31. Dataset, https://www.kaggle.com/datasets/sulianova/cardiovascular-disease-dataset?fbclid=IwAR3bpgZt5DZYJYjf4m8h9gDssOMXW7_GlAJrLqe0I3BQnnw8aH7If8ddkA, last accessed 06/26/2022
32. Landwehr, N., Hall, M., Frank, E.: Logistic model trees. Mach. Learn. **59**(1/2), 161–205 (2005)
33. Akaike, H.: Information theory and an extension of the maximum likelihood principle. In: Second International Symposium on Information Theory, pp. 267–281 (1973)
34. Sumner, M., Frank, E., Hall, M.: Speeding up logistic model tree induction. In: Jorge, A.M., Torgo, L., Brazdil, P., Camacho, R., Gama, J. (eds.) Knowledge Discovery in Databases: PKDD 2005. Lecture Notes in Computer Science, vol. 3721, p. 2005. Springer, Berlin/Heidelberg

Forecasting the Number of Passengers in Rail System by Deep Learning Algorithms

Aslı Asutay and Onur Uğurlu

1 Introduction

Transportation has been one of the most crucial factors influencing the selection and development of settlement centers. Rail systems, which offer a fast, safe, and comfortable transportation alternative, form the body of public transport in cities. Various urban rail transport solutions, such as commuter lines, tram networks, and metro systems, are the backbone of urban public transport. Especially in large metropolises, large-scale investments in these rail systems are admitted to provide a more effective and efficient transportation experience. The rapidly increasing population density has increased the need for public transportation. This need has further increased the weight of rail systems in terms of sustainability and efficiency of urban transportation.

With the spread of rail systems, smart transportation technologies have also developed. Especially in recent years, smart cards used in urban rail systems enable passengers to save time by facilitating payment transactions. These cards both provide a user-friendly experience and reduce the use of cash, allowing transportation systems to operate more regularly. In developed cities, the number of transactions made with passenger cards used in rail systems reaches millions every day. The increasing number of stored data makes it more possible to make predictions in public transportation. Passenger forecasting is an important part of transportation

A. Asutay (✉)
Smart Systems Engineering 35665, Izmir Bakırçay University, Izmir, Turkey
e-mail: aasliasutay@gmail.com

O. Uğurlu
Faculty of Engineering and Architecture, Department of Fundamental Sciences 35665, Izmir Bakırçay University, Izmir, Turkey
e-mail: onur.ugurlu@bakircay.edu.tr

M. N. Seyman (ed.), *2nd International Congress of Electrical and Computer Engineering*, EAI/Springer Innovations in Communication and Computing,
https://doi.org/10.1007/978-3-031-52760-9_3

networks. Forecast results can be used to help manage transportation systems, such as operations planning and station congestion management [1].

In this study, two deep learning algorithms (long-short term memory (LSTM) and recurrent neural networks (RNN)) were used on rail system data, which are taken from the New York City subway system to estimate the number of passengers [2]. The data collected through smart card systems, such as OMNY or MetroCard, were analyzed and organized in detail using time series. The effectiveness of the prediction models was assessed with different error metrics, such as root mean squared error (RMSE), mean absolute error (MAE), and coefficient of determination (R^2).

The following parts of this chapter are structured as follows: The second section provides a review of the relevant literature. The third section then outlines the deep learning algorithms utilized in this study, and the error metrics are given. The fourth section shows the results of the prediction model. Lastly, the final section presents overall assessments and outlines prospects for future research.

2 Literature Review

Several researchers have used deep learning and machine learning algorithms for passenger forecasting. Some of these works are given in this section.

Toque et al. used random forest and LSTM algorithms for long- and short-term passenger prediction over the 2-year (2014–2015) data collected on the Paris metro [3]. Zhu et al. developed a prediction model on Qingdao Metro data using deep learning and support vector machine algorithms together [4]. Gallo et al. carried out passenger estimation using artificial neural networks on the data collected from the Naples subway [5]. Atay et al. employed artificial neural networks to estimate the passenger and freight demand of the Istanbul Airport in the coming years by using past period data [6]. Cakır and Tosun aimed to estimate the number of railway passengers and compared the predictive performance of multivariate regression analysis and artificial neural networks. They reported that artificial neural networks have a better prediction success than multivariate regression analysis on the related data set [7]. Lin and Tian used random forest and LSTM models to predict short-term metro passenger flow using metro card data. They reported that using these two algorithms together increases the prediction success [8]. Yang et al. estimated the passenger volume on the Beijing airport subway line using deep learning algorithms using collected smart card data [9]. Li et al. carried out a short-time passenger estimation on the data collected on the Shanghai subway in 2015 using LSTM, support vector regression, and autoregressive integrated moving average (ARIMA) [10]. Dursun and Toraman estimated the number of passengers at Elazig Airport using Vanilla LSTM and autoregressive models [11]. Nagaraj et al. used deep learning methods, such as LSTM, RNN, and greedy layered algorithm (GLA), to predict passenger flow from Karnataka State Road Transport Corporation (KSRTC) [12]. Ma et al. presented a deep learning architecture for transit passenger prediction between transport-integrated centers in urban concentration regions and developed a transformative prediction model using LSTM networks to embed historical data using two data sets from the Beijing–Tianjin–Hebei urban concentration region in China [13]. Yifan Sun addressed the problem of those who prefer to go to work during rush hour, putting pressure on the traffic system, and compared LSTM and CNN-LSTM deep learning models to estimate the short-term passenger flow of subway stations. They concluded that the CNN-LSTM model could predict the daily passenger flow more accurately [14]. Ghandeharioun et al. studied different time units and their effects on different architectural configurations. They also analyzed 22 models representing five different architectural configurations, showing how changing the layer configurations within each architecture affected the results. The chapter's findings revealed that LSTM structures performed best for short-term time-series prediction, but more complex architectures did not significantly improve results [15]. In their research, Yao et al. utilized an advanced, LSTM model enhanced by Ensemble Empirical Mode Decomposition (EEMD) to forecast the inbound passenger flow of a metro system. Their findings indicated that the newly trained model exhibited superior predictive capabilities, especially during periods of high passenger demand, outperforming a standalone LSTM model [16].

3 Deep Learning Algorithms

Deep learning algorithms are powerful machine learning methods developed based on artificial neural networks. Thanks to their multi-layered nature, these algorithms can capture complex relationships between data. Especially on large data sets, deep learning algorithms are used successfully in many application areas, such as data mining, image processing, natural language processing, and time series forecasting.

In this study, we used RNN and LSTM algorithms, which have recently gained significant importance. The selection of these algorithms to analyze the data set was based on their inherent capacity to retain crucial information, learn from examples, and generalize. In addition, these deep learning algorithms need fewer underlying assumptions than conventional statistical techniques.

3.1 Recurrent Neural Networks

RNN emerged as a result of types of artificial neural networks developed by David Rumelhart, James McClelland, and others in the late 1980s. The influence of RNNs, however, has become more pronounced with advanced variants such as the LSTM introduced in 1993 by Sepp Hochreiter and Jürgen Schmidhuber. Over time, with the rise of deep learning from the mid-2010s, RNNs became widely used, especially in fields such as natural language processing, time series, and sequential data analysis.

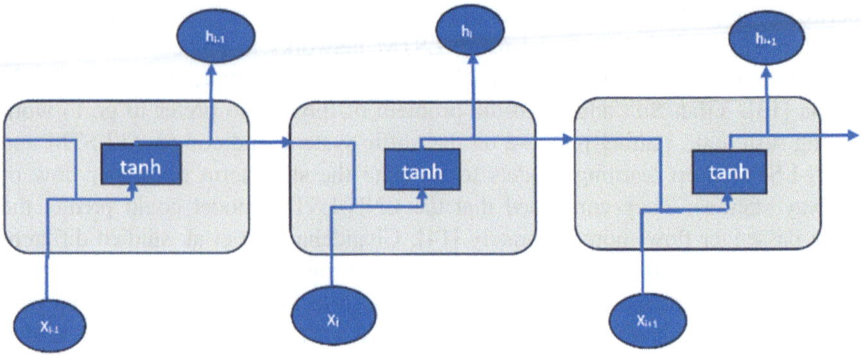

Fig. 1 RNN architecture

RNN is a type of artificial neural network designed for processing sequential data. Its basic working principle is to combine the output of the previous step with the input from the current step to produce the output of the current step. In this way, relationships and patterns over time can be captured by preserving the influence of past information [17].

The basic component of RNNs is the "cell." The cell takes the input from the current step and the output from the previous step and uses this information to generate a new state information. This new status information is passed to the next step, and this process is repeated through successive steps.

Figure 1 gives an illustration of a simple RNN architecture. In this figure, the RNN architecture consists of three layers: input layer "X," hidden layer "h," and output layer. The hidden layer consists of a tangent hyperbolic (tanh) activation function. The activation function helps the neural network learn nonlinear relationships.

The basic working principle of the RNN is as follows: At each time step, the RNN receives an input and produces an output. It also receives and updates the hidden status from the previous time step. The hidden state contains information that the RNN has learned from previous inputs and allows it to understand the context of the sequential data.

3.2 Long-Short Term Memory

LSTM is an artificial neural network architecture proposed by German researcher Sepp Hochreiter and his collaborator Jürgen Schmidhuber in 1997 [18]. LSTM is specifically designed to handle long-term dependencies encountered during the processing of sequential data and to ensure that previous sequential data are stored in memory. LSTM is widely used, especially for processing time series data, text data, and other sequential data types.

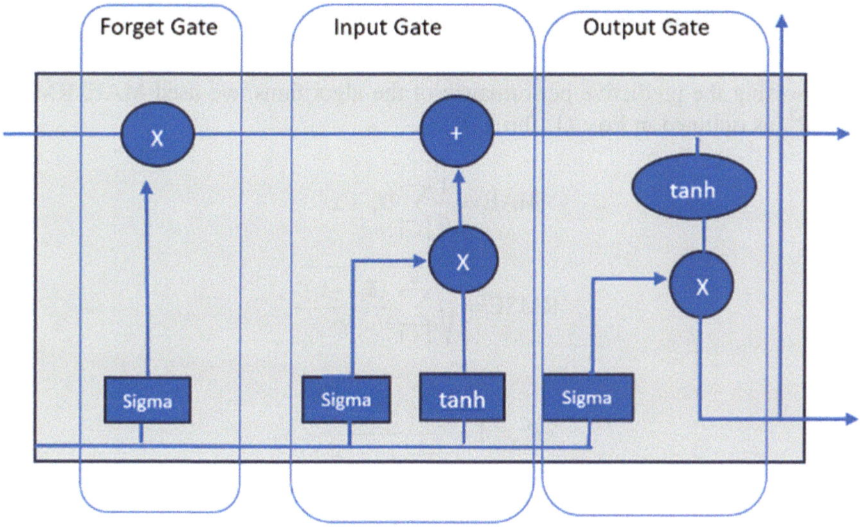

Fig. 2 LSTM cell

LSTM stores the historical information of the input data more effectively than its previous state. The LSTM cell is equipped with a gate mechanism that decides what information to keep and what information to discard.

Figure 2 gives a diagram of a basic LSTM cell. The LSTM cell is an RNN unit designed to solve the problem of slope burst or slope loss when dealing with long-time sequences in backward propagating neural networks. The LSTM cell consists of three gates: the forget gate, the entry gate, and the exit gate. These gates perform different functions to update cell state and hidden state. The forget gate decides what information from the past cell state will be forgotten. This gate is denoted by an "X" symbol and uses a sigmoid function. The sigmoid function produces a value between 0 and 1. 0 means completely forgetting, and 1 means completely remembering. The forget gate is multiplied by the previous hidden state and the current input, and the result is multiplied by the past cell state. The input gate decides what information from the current input is added to the new cell state. This gate is indicated by a "+" symbol and performs two steps. In step one, it determines which values to update to the new cell state using a sigmoid function. In the second step, it calculates the values to be added to the new cell state using a tangent hyperbolic (tanh) function. The input gate is multiplied by the previous hidden state and the current input, and the result is multiplied by the values obtained from the tanh function. The exit door decides what the new hidden state will be. This gate is indicated by a "tanh" symbol and performs two steps. In the first step, it determines which parts of the new hidden state to output using a sigmoid function. In the second step, it processes the new cell state with a tangent hyperbolic (tanh) function and multiplies the result with the values obtained from the sigmoid function.

3.3 Statistical Analysis

In assessing the predictive performance of the algorithms, we used MAE, RMSE, and R^2, as outlined in Eqs. (1) through (3).

$$\text{MAE} = \frac{1}{n}\sum_{i=1}|\overline{x}_i - x_i| \tag{1}$$

$$\text{RMSE} = \sqrt{\sum_{i=1}^{n}\frac{(\overline{x}_i - x_i)^2}{n}} \tag{2}$$

$$R^2 = 1 - \frac{\sum_{i=1}^{n}(x_i - \overline{x}_i)^2}{\sum_{i=1}^{n}(x_i - \widehat{x})^2} \tag{3}$$

In the given formula, $\overline{x}_i$ represents the predicted value for the ith sample, x_i is the actual value of the ith sample, and $\widehat{x}$ represents the mean of all the x values. The variable "n" represents the total number of samples.

4 Application

This section presents the predictive outcomes of the developed model on real-world data. For algorithm execution, Python 3.9 programming languages were chosen along with relevant packages. Python's versatility aids in efficiently addressing artificial intelligence challenges through its comprehensive package offerings.

Within the scope of this study, it is stated that the model developed using real-world data obtained from the New York Transportation Administration (MTA) will make hourly forecasts. In this manner, we employed deep learning algorithms since they capture the patterns in the data set and make highly accurate predictions in time series.

4.1 Data Set

The data set used is taken from the New York Transportation Administration [2]. The dataset includes the number of passengers, payment method, number of transits, and geographical location of each station per hour. The data set used is taken from the New York Transportation Administration [2]. The data set includes the number of passengers, payment method, number of transits, and geographical location of each station per hour. In the study, we consider three stations.

station_complex_id	transit_timestamp	ridership
A010	05/09/2023 01:00:00 AM	76
A010	12/18/2022 01:00:00 PM	1004
A010	07/26/2022 12:00:00 PM	795
A010	04/08/2023 05:00:00 PM	1591
A010	04/11/2022 11:00:00 PM	443
...	...	...
R627	03/23/2022 12:00:00 AM	9
R627	12/26/2022 07:00:00 AM	155
R627	12/23/2022 12:00:00 AM	129
R627	06/06/2022 01:00:00 AM	12
R627	10/22/2022 06:00:00 AM	148

Fig. 3 The data set used in the study

Table 1 Model parameters

	Algorithm	
Parameter	LSTM	RNN
Cell count	50	50
Activation function	Relu	Tanh
Optimization function	Adam	Adam
Loss function	MSE	MSE
Epoch	65	50

The data set used in the study includes the observed values for the years 2022–2023, which contains 37,554 rows in total. The sample data set used in the study is given below (Fig. 3):

Table 1 shows passenger numbers for different stations and dates. The table has three columns: "station_complex_id," "transit_timestamp," and "ridership." The "station_complex_id" column contains the unique identification numbers of the stations. The "transit_timestamp" column shows the date and time the data was collected. The format of this column is "month/day/year hour: minute AM/PM." For example, "1/1/2023 12:00 AM" refers to midnight on the first day of January. The "ridership" column shows the number of passengers registered at a particular station and time.

4.2 Prediction Performance Analysis of Deep Learning Algorithms

In the developed prediction model, hourly passenger numbers of each stop were modeled with time series. Then, 80% of the data set was used as a training set, and 20% was reserved as a test set. This splitting process is important so that the machine learning model can evaluate its performance in training on real-world data.

Table 2 Test data error values of deep learning algorithms

Error metrics	A010		R627		N111	
	LSTM	RNN	LSTM	RNN	LSTM	RNN
MAE	187.56	0.022	48.37	0.023	55.77	0.022
RMSE	297.9	0.030	84.66	0.034	99.81	0.033
R^2	0.81	0.973	0.67	0.964	0.66	0.958

LSTM starts with an LSTM layer equipped with parameters such as cell number and activation function determined in the model. This number of cells is determined via the "units" parameter. The cell count is set to 50, and the "relu" (Rectified Linear Unit) activation function was used in LSTM cells. Then, the output of the model is created by adding a single dense layer. Before training the model, the data were arranged in a three-dimensional array format. In the training phase, the model was trained with the "adam" optimization algorithm using the "mean_squared_error" loss function. After the training was completed, the model was evaluated separately on the training and test data [19].

In the data editing step of the RNN algorithm, input and output sequences are generated to perform the estimation of each time step using the previous 10-time steps. While creating the model, a SimpleRNN layer was added using the sequential structure and then terminated with a fully connected layer. The model was compiled using the "adam" optimizer for training and the mean square error loss function. The training process was carried out with a certain number of epochs and mini-group size. Similar to LSTM, the model was evaluated separately on the training and test data after the training was completed.

Table 2 gives the error metrics of the employed deep learning algorithms. For station A010, LSTM has higher MAE and RMSE values and fewer R^2 values than the RNN. The same situation is true for R627 and N111. RNN had an R^2 value above 0.95 for all stations; however, LSTM was able to reach 0.80 only for the A010 station. In general, RNN has an average of 0.97 R^2 and 0.022 MAE values, while LSTM has an average of 0.71 R^2 and 97.23 MAE values. Hence, it can be concluded that RNN performs better than the LSTM in the used data set.

Figure 4 compares the passenger number prediction of the RNN and LSTM for station A010. When this figure is examined, a clear difference can be seen between the RNN (Fig. 4a) and LSTM (Fig. 4b). These results show that the predictions of the RNN model are closer and can explain a large part of the data. The low MSE value shows that the predictions of the model are very close to the actual values, while the high R^2 value shows that the model explains the data successfully. Therefore, the RNN model performs better for station A010.

Figure 5 compares the passenger number prediction of the RNN (Fig. 5a) and LSTM (Fig. 5b) for station R627. The predictions of the LSTM model are further away from the actual values of station R627. On the other hand, the predictions of the RNN model are closer to the actual values of station R627 with lower error and better explain the data.

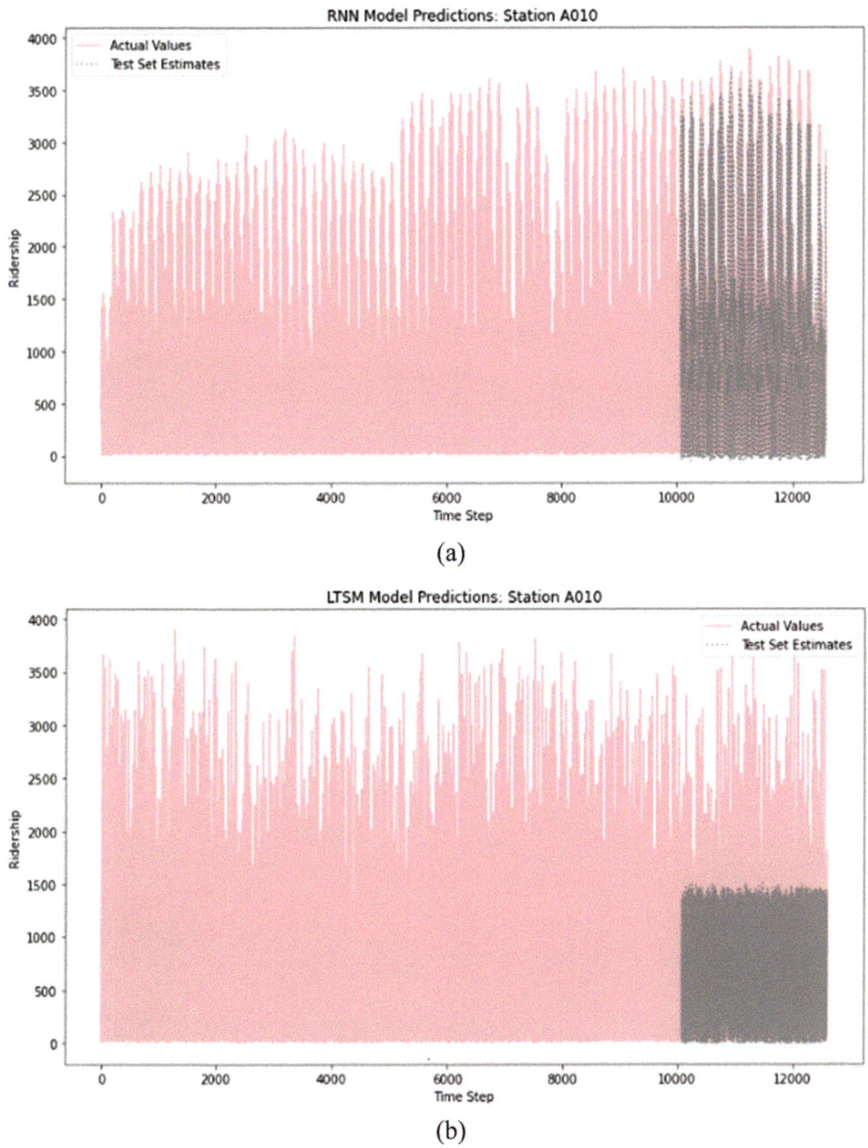

Fig. 4 Performance deep learning algorithm on A010 station: (**a**) RNN and (**b**) LSTM

Figure 6 compares the passenger number prediction of the RNN (Fig. 6a) and LSTM (Fig. 6b) for station N111. These figures show that the predictions of the LSTM deviate with a higher error rate and explain the data in a more limited way, and the predictions of the RNN model are closer to actual values and explain the data with high success. Therefore, it can be said that the RNN model provides more reliable predictions on station N111 data.

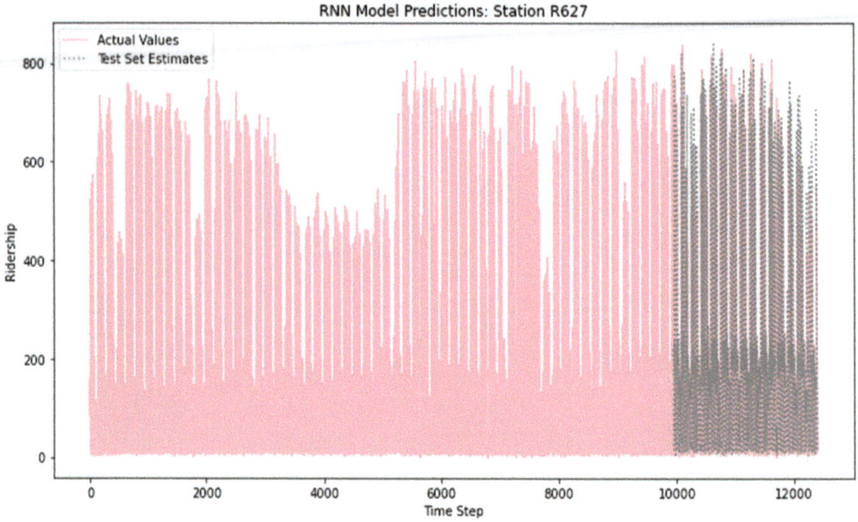

(a)

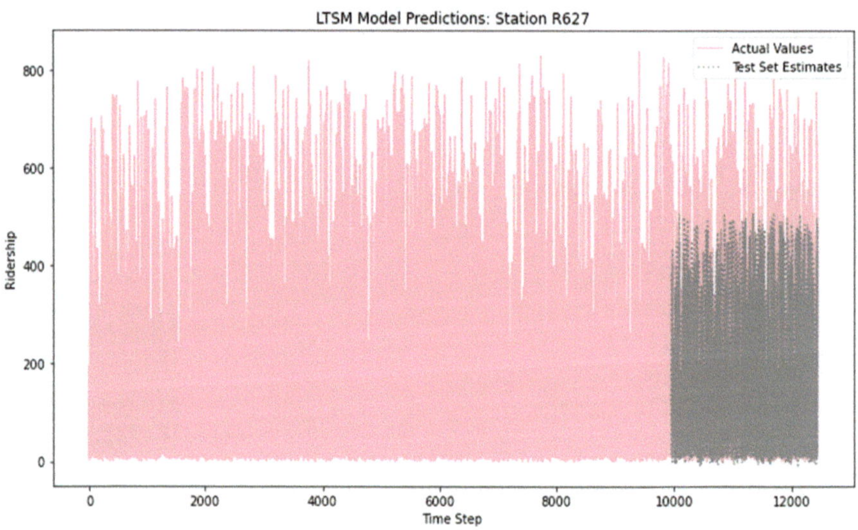

(b)

Fig. 5 Performance deep learning algorithm on R627 station: (**a**) RNN and (**b**) LSTM

5 Conclusion

Accurate prediction of time series data, particularly for passenger forecasting, assumes a key role in urban planning and the formulation of effective strategies for urban transportation, particularly in the integration and enhancement of rail

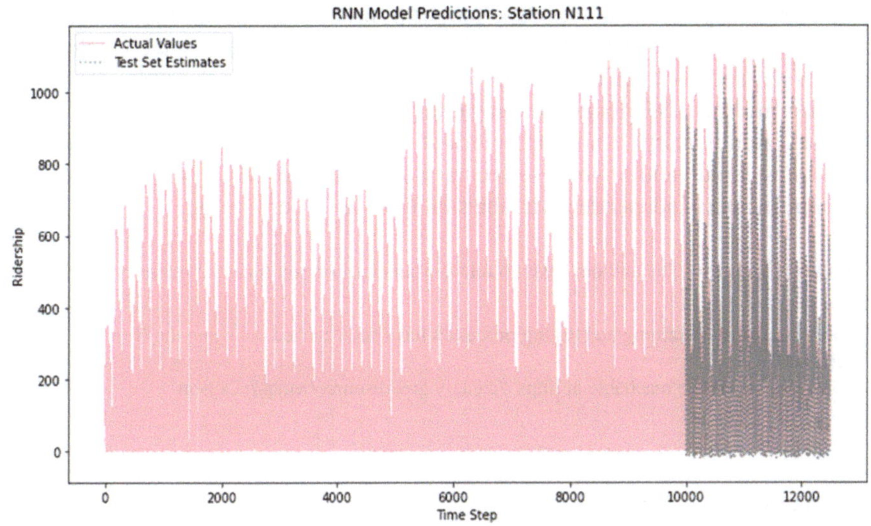

(a)

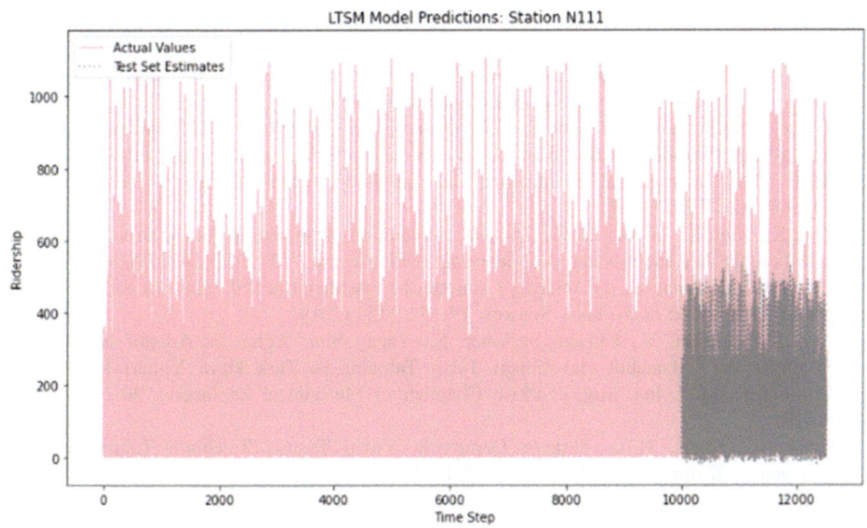

(b)

Fig. 6 Performance deep learning algorithm on N111 station: (**a**) RNN and (**b**) LSTM

systems. The primary objective of this study is to employ deep learning algorithms for precise passenger number forecasting. In order to accomplish this, we investigated the hourly forecasting accuracy of the RNN and LSTM algorithms on New York Subway data for three stations. To evaluate the prediction performance of the algorithm, we used different error metrics, such as MAE, RMSE, and R^2.

The RNN algorithm showed superior performance with R^2 values of 97%, 96%, and 96% for stations A010, R627, and N111, respectively, whereas, these values are 81%, 67%, and 66% for the LSTM. Our results indicate that the RNN algorithm outperformed the LSTM algorithm regarding predictive accuracy on the considered data set. Additionally, a notable contribution of this study is that deep learning algorithms could yield impressive accuracy without necessitating the incorporation of supplementary independent variables on time series forecasting.

Conflict of Interest The authors have no relevant financial or non-financial interests to disclose.

Data Availability Training and testing processes have been carried out using the New York State Open Data.

The dataset can be reachable at https://data.ny.gov/resource/wujg-7c2s.json.

References

1. Wei, Y., Chen, M.C.: Forecasting the short-term metro passenger flow with empirical mode decomposition and neural networks. Transp. Res. Part C Emerg. Technol. **21**(1), 148–162 (2012)
2. New York State Open Data, MTA Subway Hourly Ridership Beginning February 2020, https://data.ny.gov/resource/wujg-7c2s.json, Last accessed 2023/08/15
3. Toqué, F., Khouadjia, M., Come, E., Trepanier, M., Oukhellou, L.: Short- and long-term forecasting of multimodal transport passenger flows with machine learning methods. In: 2017 IEEE 20th International Conference on Intelligent Transportation Systems (ITSC), pp. 560–566. IEEE (2017)
4. Zhu, K., Xun, P., Li, W., Li, Z., Zhou, R.: Prediction of passenger flow in urban rail transit based on big data analysis and deep learning. IEEE Access. **7**, 142272–142279 (2017)
5. Gallo, M., De Luca, G., D'Acierno, L., Botte, M.: Artificial neural networks for forecasting passenger flows on metro lines. Sensors. **19**(15), 3424 (2019)
6. Atay, M., Eroğlu, Y., Ulusam Seçkiner, S.: Yapay Sinir Ağları ve Adaptif Nörobulanık Sistemler ile 3. İstanbul Havalimanı Talep Tahmini ve Türk Hava Yolları İç Hat Filo Optimizasyonu. J. Ind. Eng. (Turkish Chamber of Mechanical Engineers). **30**(2), 141–156 (2019)
7. Çakır, F., Tosun, H.B.: Türkiye Demiryolu Yolcu Taşıma Talebinin Tahmini. Düzce Üniversitesi Bilim ve Teknoloji Dergisi. **9**(1), 252–264 (2020)
8. Lin, S., Tian, H.: Short-term metro passenger flow prediction based on random forest And LSTM. In: 2020 IEEE 4th Information Technology, Networking, Electronic and Automation Control Conference (ITNEC), pp. 2520–2526 (2020)
9. Yang, X., Xue, Q., Ding, M., Wu, J., Gao, Z.: Short-term prediction of passenger volume for urban rail systems: A deep learning approach based on smart-card data. Int. J. Prod. Econ. **231**, 107920 (2021)
10. Li, Y., Yin, M., Zhu, K.: Short term passenger flow forecast of metro based on inbound passenger plow and deep learning. In: International Conference on Communications, Information System and Computer Engineering (CISCE), pp. 777–780 (2021)
11. Dursun, Ö.O., Toraman, S.: Uzun Kısa Vadeli Bellek Yöntemi ile Havayolu Yolcu Tahmini. J. Aviat. **5**(2), 241–248 (2021)
12. Nagaraj, N., Gururaj, H.L., Swathi, B.H., Hu, Y.C.: Passenger flow prediction in bus transportation system using deep learning. Multimed. Tools Appl. **81**(9), 12519–12542 (2022)

13. Ma, S.H., Yue, M., Chen, X.F.: Lstm-Based Transformer for Transfer Passenger Forecasting between Transportation Integrated Hubs in Urban Agglomeration. Available at SSRN 4183278 (2022)
14. Sun, Y.: Prediction of short-term passenger flow in the metro station with CNN-LSTM model. In: 2023 IEEE 3rd International Conference on Electronic Technology, Communication and Information (ICETCI), pp. 1218–1222. IEEE (2023)
15. Ghandeharioun, Z., Zendehdel Nobari, P., Wu, W.: Exploring deep learning approaches for short-term passenger demand prediction. Data Sci. Transp. **5**(3), 19 (2023)
16. Yao, Y., Jin, S., Wang, Q.: Subway short-term passenger flow prediction based on improved LSTM. In: 2023 IEEE 12th Data Driven Control and Learning Systems Conference (DDCLS), pp. 1280–1287. IEEE (2023)
17. Halyal, S., Mulangi, R.H., Harsha, M.M.: Forecasting public transit passenger demand: With neural networks using APC data. Case Stud. Transp. Policy. **10**(2), 965–975 (2022)
18. Hochreiter, S., Schmidhuber, J.: Long short-term memory. Neural Comput. **9**(8), 1735–1780 (1997)
19. Tsioulis, I.: Rapid calculation of the signal-to-noise ratio of gravitational-wave sources using artificial neural networks. Master thesis, (2023)

Viola–Jones Method for Robot Vision Purpose: A Software Technical Review

Wei Leong Khong, Ervin Gubin Moung, and Chee Siang Chong

1 Introduction

Humans possess an innate ability to detect and recognise faces effortlessly [1]. However, it is complicated and challenging for computers [2–4]. A machine learning approach, or its broader notion as Artificial Intelligence (AI), aims to train an algorithm to duplicate human activities, such as predicting the location of a face [1, 5]. Face detection, a focal point in computer vision literature [3, 6, 7], is a field dedicated to enabling computers to 'see' [8] and the foundation of many computer applications [9, 10]. Researchers have leveraged computer vision technology for diverse purposes, ranging from facial identification [11] and face recognition [7, 12] to assessing the emotional states of children with Down syndrome [13], detecting and recognising hand gestures [14], replacing outdoor lighting using robotic systems [15] and gender recognition system [16]. Recently, the availability of low-priced and high-resolution digital cameras for recording still images and video streams has pioneered the tools of digital doctoring [4]. It also motivates the development of image processing applications in medicine and healthcare, such as the video imaging technique for health status monitoring systems [17, 18], non-contact heart rate measurement techniques [19–23] and non-contact blood pressure measurement methods [24].

W. L. Khong (✉) · C. S. Chong
School of Engineering, Monash University Malaysia, Bandar Sunway, Selangor, Malaysia
e-mail: khong.weileong@monash.edu

E. G. Moung
Data Technologies and Applications (DaTA) Research Group, Faculty of Computing and Informatics, Universiti Malaysia Sabah, Kota Kinabalu, Sabah, Malaysia

M. N. Seyman (ed.), *2nd International Congress of Electrical and Computer Engineering*, EAI/Springer Innovations in Communication and Computing,
https://doi.org/10.1007/978-3-031-52760-9_4

Table 1 Summary of several studies regarding the Viola–Jones algorithm

Paper/ year	Method	DR (%)	P (%)/CT (s)	Remark(s)
[25] (2001)	VJ original paper	93.9	15FPS	Able to run in real-time
[26] (2004)	VJ's second original paper	94.1	–	Failure modes have been discussed (angle of face, backlighting, occluded eyes – Affect the accuracy).
[29] (2015)	Matlab, vision. CascadeObjectDetector	70	0.83–4.79s	Need more samples to get a better positive rate.
[30] (2016)	Cascade with histograms of oriented gradients (HOG)	94.5	–	Using HOG to construct different-granularity features.
[11] (2017)	OpenCV's VJ algorithm, with eyes and face detection	98.97	–	Haar cascade classifier (frontal face, eyes, eyes with class) is very efficient
[31] (2018)	VJ (Matlab) for multi-view face detection system	93.24	–	If the light is too dark, the number of faces detected may be incorrect
[32] (2018)	Cascaded classifier (window detection)	95.2	97.4%	Used for complementing existing 3D building models
[14] (2019)	Hand-detection, lightweight convolution neural network (CNN)	97.25	6.44–12.05 FPS	Proposed a deep learning-based architecture to detect and classify hand gestures
[33] (2020)	Revised VJ training method (extracted lesser features)	70–90	3–9%	Results showed that far distance decreases the accuracy.
[34] (2022)	Matlab train. CascadeObjectDetector is used for detecting faces with masks	100	10%	6 layers / 400 classifiers have the mentioned accuracy. It has a 70% cost reduction
[35] (2023)	Model wrapping iOS and tensor flow lite android	70–100	~120 ms	It had a 90–100% detection rate, except dark situation cases had 70%
[36] (2023)	Detecting facial U- or T-areas with Matlab vision. CascadeObjectDetector and Hough transform algorithm	100	100%	The accuracy highly relies on the position of the face. It requires the frontal face and straight neck on the image

DR detection rate, *P* precision, *CT* computation time

In 2001, Viola and Jones introduced the Viola–Jones (VJ) algorithm, the world's first real-time face detector [25, 26]. It stands out as a prominent method due to its exceptional detection speed [7, 9, 27, 28]. Researchers have extensively studied the VJ algorithm, and Table 1 summarises several notable research on the VJ method.

Generally, the VJ algorithm provides three crucial contributions enabling efficient face detection. These contributions include the integral image for rapid feature evaluation, AdaBoost for quick feature recognition and the attentional cascade for speed filtering of negative regions [37, 38]. Introduced by Freund and Schapire in

1995 [39], the AdaBoost algorithm addressed several challenges faced by earlier machine training algorithms [40]. As demonstrated in Table 1, most detectors implementing the VJ algorithm achieved a detection rate of around 90% while maintaining satisfactory timing and precision ranges. These results further validate the reliability of the VJ algorithm.

In the modern era, mobile phones become an indispensable necessity for people. It is foreseeable that detection and tracking in mobile cameras can supply significant utility [41]. Today, many visual programming environments provide robust features in processing and analysing the evaluated data [42]. MATLAB, a widely used platform for image processing due to its extensive features [43], provides vision to enable face detection using the VJ method. Apple, too, initially invented the first version of CIDetector using the VJ algorithm [44]. Nonetheless, face detection on mobile devices continues to encounter challenges in meeting real-time detection needs [45], leaving room for further study and improvement in face detection.

While not the latest technology, the Viola–Jones framework maintains its esteemed status as an effective method for face detection [34]. Numerous recent technologies have benefited from the VJ framework. By grasping the core concepts of the VJ algorithm, researchers can implement it in their works to create other astonishing technologies [46]. Therefore, this chapter aims to delve into the fundamental principles governing the practical implementation of facial detectors using the built-in functions of two renowned software platforms: 'vision.CascadeObjectDetector' of MATLAB and 'CIDetector' of iOS. These concepts are discussed in Sects. 2 and 3, respectively. Section 4 introduces the core concept of a self-trained VJ detector. Following the exposition of the general idea behind it, Sect. 5 discusses and compares the results obtained from these three detectors. Lastly, Sect. 6 draws the conclusion of this study.

2 Face Detection Using MATLAB Cascade Object Detector

In the late 1970s, Cleve Moler developed MATLAB as the software primarily intended for matrix operations. Over time, the MATLAB platform has significantly expanded in scope and become a standard tool in engineering and computer science [47]. The MATLAB software encompasses the Digital Image Processing Toolbox, offering tools and functions for processing and analysing images. These built-in tools and functions include the Image Processing Toolbox, Computer Vision System Toolbox and Neural Network Toolbox [48].

The Computer Vision System Toolbox provides several functions to detect objects in an image or a video [29]. The function 'vision.CascadeObjectDetector' is a built-in tool of MATLAB's Computer Vision System, established based on the VJ algorithm. This function allows users to develop a face detector packaged with pre-trained classifiers based on Haar-like features [29].

Users need to choose the parameters of the 'vision.CascadeObjectDetector' function before executing the detection. Six parameters are provided to suit various

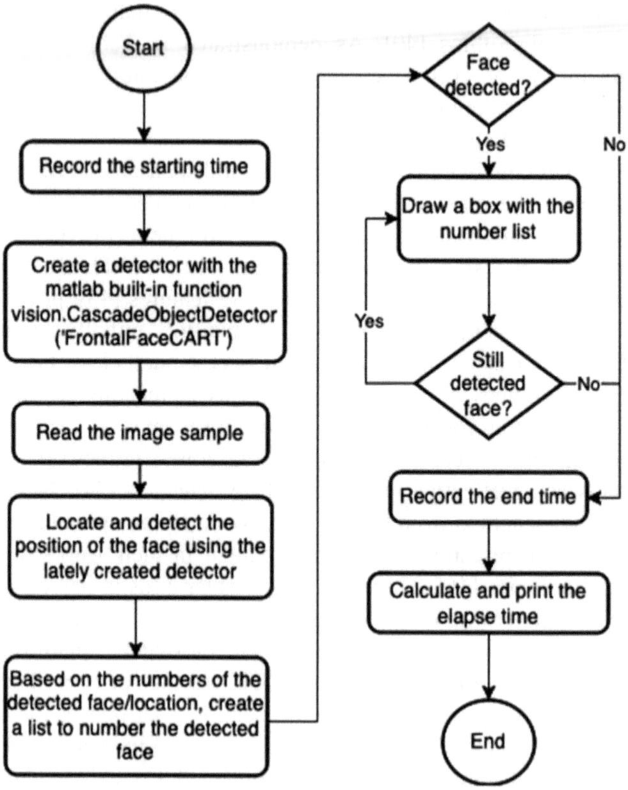

Fig. 1 The flowchart of developing the face detector using the MATLAB built-in function

detection purposes. These parameters include the choice of classification model, minimum image size, maximum image size, scale factor, merge threshold and Region of Interest specification [34]. The classification model can be chosen to detect the face, nose, eyes, mouth and upper body.

Due to the limited predetermined classifiers of the Computer Vision Toolbox, there are cases where the built-in face detector may not perform satisfactorily. Therefore, the Computer Vision Toolbox features the built-in function 'train. CascadeObjectDetector', which is competent to train a new detector using three available feature types: Haar-like, External Local Binary Pattern (LBP) and Extract Histogram of Orientation Gradients (HOG).

To assess the performance of the built-in MATLAB toolbox in face detection, default parameters of 'vision.CascadeObjectDetector' are adopted. These parameters include the classification model 'FrontalFaceCART', the minimum size '24 × 24', the maximum size as per 'the image input', the scale factor '1.0001', the merge threshold of '4' and ROI set to 'FALSE'.

Figure 1 displays a flowchart illustrating the development of a face detector using the built-in tool 'vision.CascadeObjectDetector'. Typically, the detector is created first, then the image is processed for face detection. When a face is detected, a bounding box is drawn around it. In order to study the detection efficiency, the start and end times are recorded. The detection results are discussed in Sect. 5.

3 Face Detection Using iOS CIDetector Class

Core Image, a framework Apple uses for iOS and macOS, offers around 200 built-in image filters and a wide array of valuable image processing and computer vision methods [49]. Within the Core Image framework, CIDetector is a state-of-the-art detection tool renowned for its exceptional effectiveness in analysing both still and moving images. It excels at detecting faces, Quick Response (QR) codes, rectangles and texts [50, 51]. In 2011, the first version of CIDetector, introduced by Apple in iOS 5, was developed based on the VJ detection technique [44].

When a face is detected, CIDetector delivers information about the face features, covering the positions of the eye, mouth, face angle, or whether the face is smiling. This information is crucial for tracking techniques. Figure 2 is the flowchart for developing the face detector using the built-in CIDetector class of iOS.

Generally, the user should first create a face detector and specify the options using the 'NSDictionary * Opts' and 'CIDetector *detector' codes. The NSDictionary generates an options dictionary to record the detection accuracy, where 'high accuracy' may cause longer processing time and 'low accuracy' may degrade the detection performance. CIDetector is then generated based on the mentioned detector type, such as 'detectorOfType: CIDetectorTypeFace'. The created detector is called to apply to the specified image, and the features are recorded in an array called 'NSArray * features'. Based on the computed face feature array, the characteristics of a face can be logged.

In this study, the image view is first set up and an image is loaded as illustrated in Fig. 2. Then the face detector is built up using the 'low accuracy' and 'CIDetectorTypeFace' types. When a face is detected, the characteristics of the face are logged, and boxes are drawn around the eyes to be shown as an output of the detector. The execution time of the face detection is recorded, and the performances of this built-in face detector are discussed in Sect. 5.

4 Viola–Jones Detector

The previous two sections briefly discussed the overall processes of developing a face detector using the built-in functions of MATLAB and iOS. Although these built-in functions are constructed using the VJ algorithm, the core concepts of the VJ algorithm are not presented previously. In order to review the performances of these

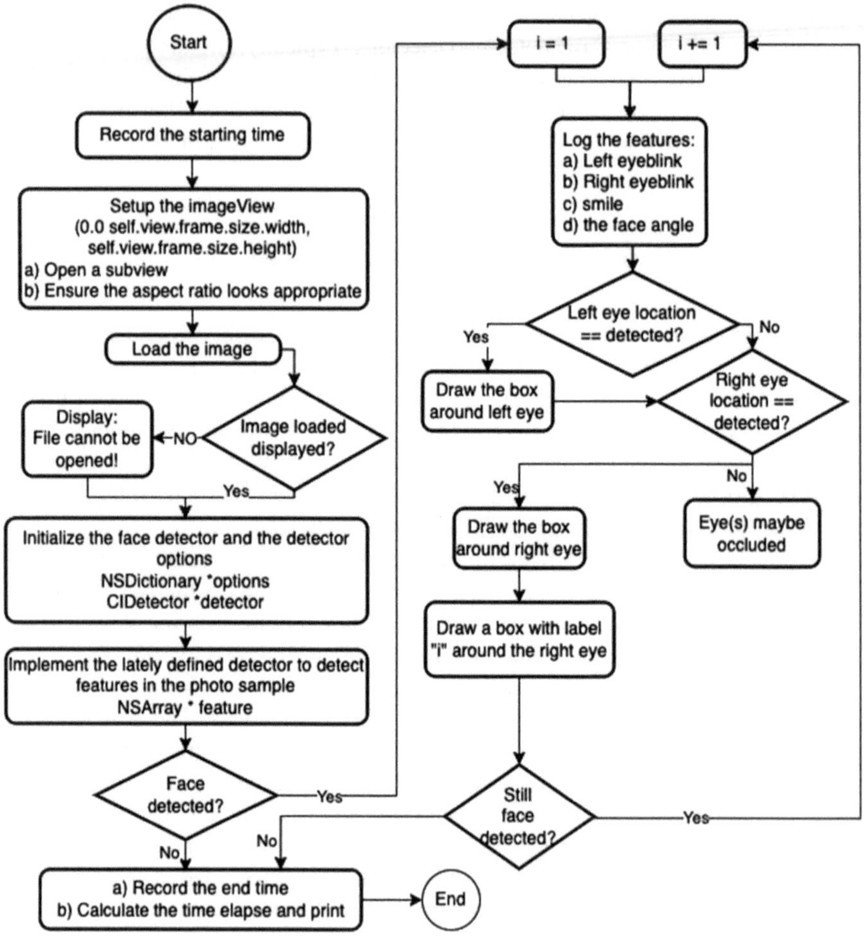

Fig. 2 The flowchart of developing the face detector using the iOS built-in function

detectors from a software perception, we have included a self-trained VJ detector in this study.

As the scope of this chapter is to review the performances of the VJ detectors, the training concept of the VJ method is not covered, and only the crucial details of the VJ detector are included. The VJ detector was trained with a sample size of 24 × 24. It has a count of 162,336 features, 34 layers and a total of 4557 committees.

Figure 3 shows how the trained cascade VJ detector is used to scan all the sub-windows and save the possible face coordinates. Usually, the trained cascade VJ detector is defined, and the image is scanned to store the pixel streams in an array. This pixel array is used to calculate the integral image array. The VJ detector divides the image sample into various sub-windows with different dimensions and coordinates. The smallest size is 24 × 24 and gradually enlarges with the scale factor.

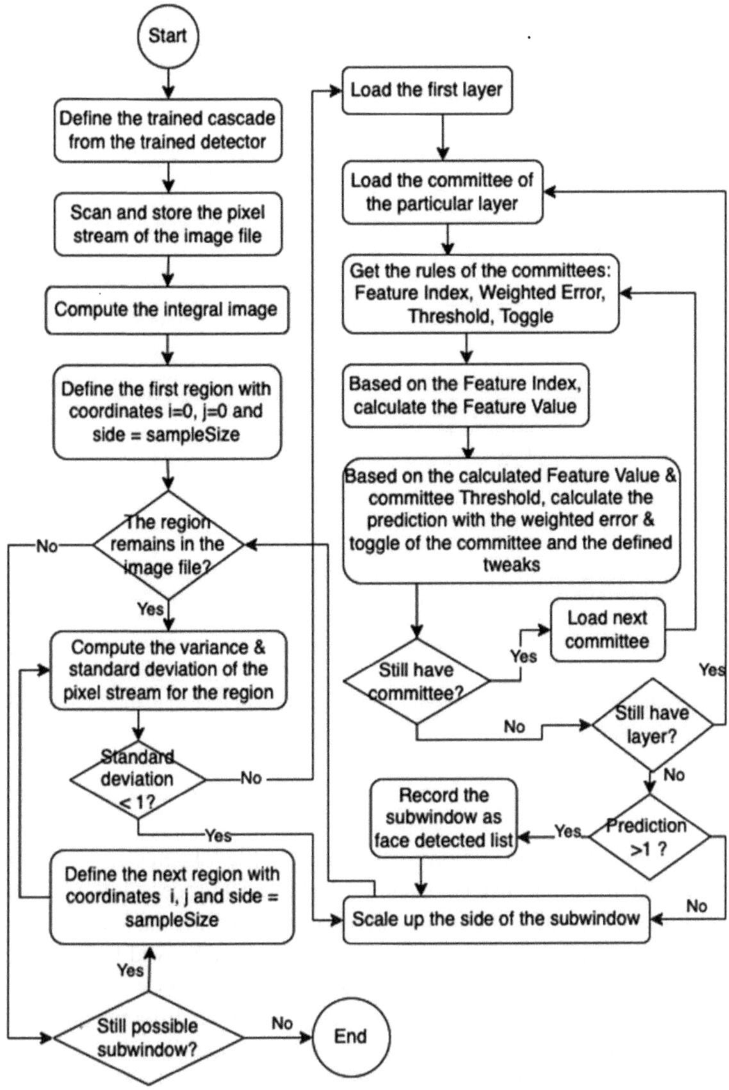

Fig. 3 The flowchart of the detail detection process for the VJ algorithm

After the integral image is acquired, the array of the sub-windows is extracted for the face detection process. As long as the sub-window remains in the image file and the variance or standard deviation of the pixel stream is more significant than one, the feature value of the integral image for the sub-window is calculated and compared to those mentioned in the rule of the committee for the particular layer. Based on the feature value and threshold of the rule, the committee decides whether the sub-window has a face. By implementing all committees of the layer, the AdaBoost algorithm calculates the prediction. If the prediction is positive, the

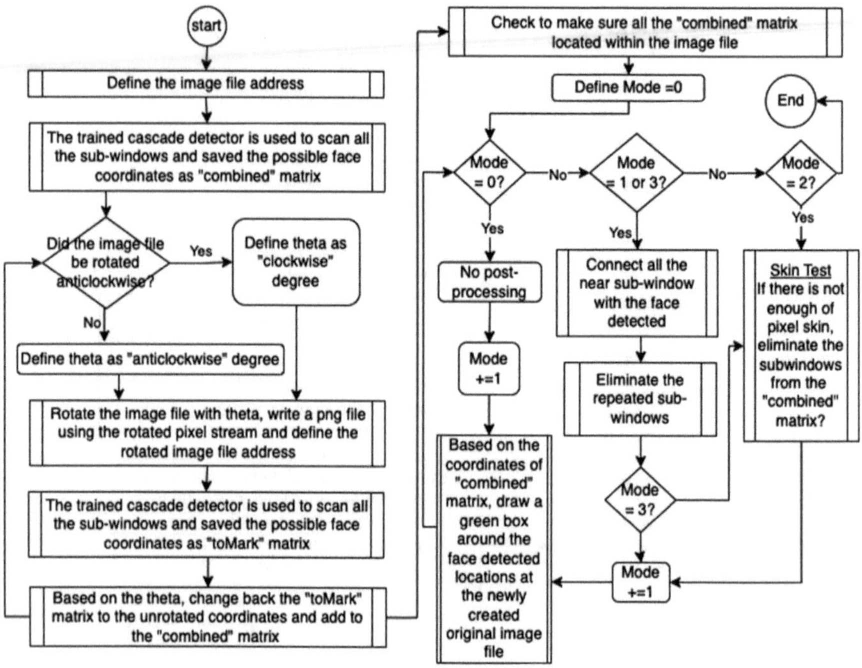

Fig. 4 The overall process of the VJ detector with several post-processing functions to enhance the face detection performance

coordinate of the sub-window is added to the matrix to store the possible face location.

In order to enhance the detection performance of the VJ detector, several pre- and post-processing functions have been added. The overall process of the VJ detector is shown in Fig. 4. Typically, the image file address is defined and sent to the trained cascade VJ detector. All the possible face coordinates are saved to a matrix named 'combined'. Then, the image file is rotated both anticlockwise and clockwise and sent again to the basic VJ algorithm to detect faces at different angles. All the possible face coordinates are added to the 'combined' matrix.

Since a skin test can quickly affirm face detection, where the pixel stream will be evaluated using the RGB rules for skin, the function can be added to enhance face detection performance. Typically, the results of the VJ detector in this study are returned as four different modes. Mode 1 is the no post-processing mode that directly draws boxes around all the detected faces. Mode 2 includes the post-processing function that filters the boxes around the same face and combines them to improve the face-detected picture's appearance. Mode 3 detects the face purely using the skin test, whereas Mode 4 detects the face based on both the VJ method and the skin test. The results obtained using this self-trained VJ detector are discussed in Sect. 5.

5 Result and Discussion

5.1 Computational Time

As discussed in Sects. 2, 3 and 4, the execution times of the developed detectors are recorded for reviewing their efficiency. Table 2 lists the computation times of the detectors, including the iOS built-in 'CIDetector' running on both iPad Pro and iPhone SE, the MATLAB built-in function 'vision.CascadeObjectDetector' and the self-trained VJ detector executed with XCode and the terminal. Twelve image samples encompassing 155 faces were processed in this study.

The third-generation iPhone SE, which uses the Apple A15 Bionic chip and a Hexa-core, 3.22 GHz CPU and the 12.9-inch sixth-generation iPad Pro, which utilises the Apple M2 chip and an eight-core CPU, were used in this study. Although the processor of the iPad Pro is more advanced than that of the iPhone SE, their computation efficiencies yielded comparable results, with differences of within 0.1 s.

When examining the performance of the MATLAB vision function 'vision. CascadeObjectDetector', we noted that it required nearly double the computational time of iOS built-in functions. Nevertheless, it is still considered an efficient detector, as it only needs half a second to analyse the detection in a 3 MB image.

To summarise the efficiency findings, the total times for processing all 12 photos were 3.54 s on the iOS built-in function and 5.41 s on the MATLAB vision built-in function. These performances are far better than those in the self-trained VJ detectors, which required 1064.594 s using the terminal and 12415.882 s using the Xcode compiler. These results also show that the self-trained detector performed significantly better when run in the terminal, requiring only one-tenth of the time compared to the Xcode compiler.

In order to review the influences of the sample sizes or face numbers on the computational time, Figs. 5 and 6 are presented. Figure 5 clearly illustrates that sample size substantially affects face detection computation time more than the number of faces. This observation is further validated by Fig. 6, which shows that the 12th sample, containing 58 faces, did not significantly increase the computation time compared to the ninth, tenth and 11th samples with 13, 18 and 18 faces, respectively. This trend holds for all cases.

5.2 Computational Time

The detection rate represents the ratio of true positives (correctly detected faces) to the total number of faces in the sample. It quantifies the detector's performance in identifying faces in the image. Conversely, precision is defined as the ratio of true positive predictions to the total positive predictions, indicating how well the detector identifies faces. These two metrics play a crucial role in assessing detector accuracy.

Table 2 The details of images and their computation time in seconds are used by the detectors

No.	Faces number	Size (kB)	Area	iOS (iPadPro)	iOS (iPhoneSE)	Matlab	VJ detector (xCode)	VJ detector (terminal)
1	3	532	960 × 540	0.232	0.304	0.385	414.475	35.415
2	4	1229	720 × 960	0.262	0.35	0.467	794.047	67.649
3	4	1318	960 × 638	0.236	0.293	0.482	837.373	72.869
4	5	1319	960 × 720	0.399	0.32	0.38	692.472	55.469
5	5	1076	960 × 720	0.252	0.264	0.363	693.094	58.162
6	6	1405	960 × 720	0.243	0.252	0.362	782.641	66.319
7	8	1271	960 × 616	0.25	0.258	0.348	789.262	62.685
8	13	2576	1440 × 1080	0.237	0.324	0.596	1791	163.741
9	13	1063	1080 × 614	0.252	0.282	0.345	555.012	47.709
10	18	3001	1080 × 1440	0.281	0.316	0.658	2194.11	188.082
11	18	3101	1080 × 1440	0.269	0.318	0.6	2125.23	182.771
12	58	1703	1949 × 403	0.295	0.264	0.428	747.166	63.723

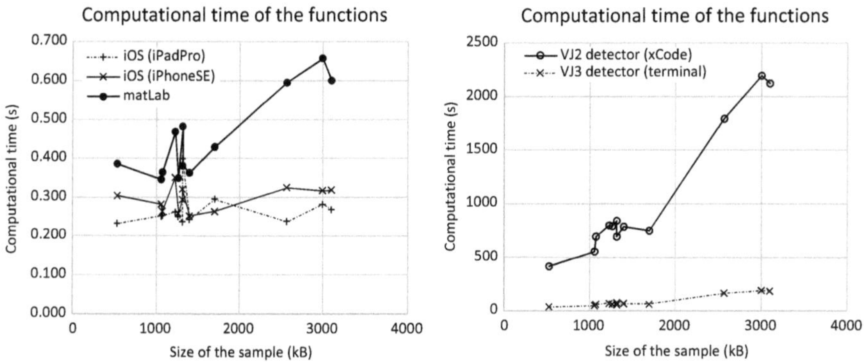

Fig. 5 The computational time used by the detectors versus the sample size

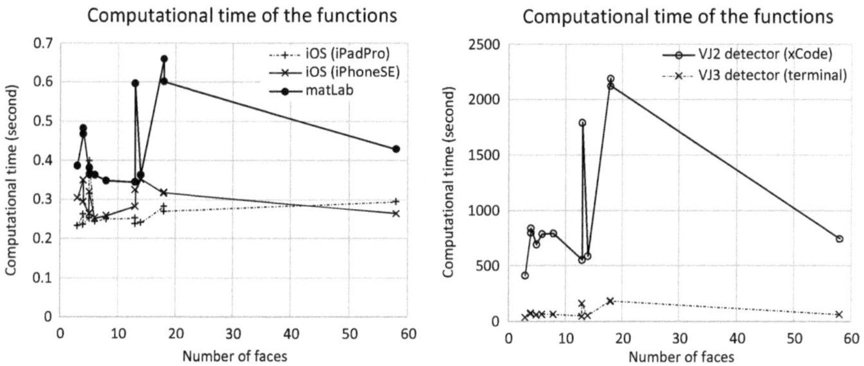

Fig. 6 The computational time used by the detectors versus the face numbers of the sample

Table 3 lists the true positive (TP), false positive (FP), detection rate (DR), and precision for the three detectors. The total number of true positives for the iOS built-in detector used on the iPad Pro was 78, leaving approximately 77 undetected. The false negatives were found in the 4th, 8th, 9th,11th, and 12th image samples. These faces were either too small or had their eyes occluded. Consequently, the detection rate dropped to 0.776, and the precision was approximately 0.833.

Even though MATLAB detected more cases of true positives, which is 154 true positives and only one false negative, the precision of the detection is much lower than that of the iOS detector. MATLAB detection has about 21 false positives, bringing the precision to 0.815 with a detection rate of 0.999. Due to the small face size in the 12th sample, the MATLAB detector had one false negative.

Same as MATLAB, the total true positives of the self-trained VJ detector were 154. However, the precision of this VJ detector is much better since it did not detect any false negatives. Thus, the self-trained VJ detector in this study had a 0.994 detection rate and precision of 1. The face that caused this VJ detector not to have a

Table 3 Output details of the detectors

No	Faces	iOS (ipadPro)				Matlab				VJ Detector			
		TP	FP	DR	Precision	TP	FP	DR	Precision	TP	FN	DR	Precision
1	3	3	0	1	1	3	2	1	0.6	3	0	1	1
2	4	4	0	1	1	4	1	1	0.8	4	0	1	1
3	4	4	0	1	1	4	4	1	0.5	4	0	1	1
4	5	3	0	0.6	1	5	1	1	0.833	5	0	1	1
5	5	5	0	1	1	5	1	1	0.833	5	0	1	1
6	6	6	0	1	1	6	3	1	0.667	6	0	1	1
7	8	8	0	1	1	8	1	1	0.889	8	0	1	1
8	13	0	0	0	0	13	3	1	0.813	13	0	1	1
9	13	10	0	0.769	1	13	0	1	1	12	0	0.923	1
10	18	17	0	0.944	1	18	1	1	0.947	18	0	1	1
11	18	18	0	1	1	18	1	1	0.947	18	0	1	1
12	58	0	0	0	0	57	3	0.983	0.95	58	0	1	1
	155	78	0	0.776	0.833	154	21	0.999	0.815	154	0	0.994	1.000

TP true positive, *FP* false positive, *DR* detection rate perfect detection rate is the one in the ninth sample, where the face had rotated to an inclined angle.

Although the VJ detector has included five degrees of clockwise and anticlockwise rotation of the image to try to detect the rotated face, it still failed to detect this face. This face was able to be detected when the rotation angle was set to 15 degrees. Viola–Jones [26] stated that the detector is unreliable for faces with more than 15 degrees in plane rotation.

5.3 *Overview of the Method*

Based on the detection rate and precision results, the performance of the self-trained VJ detector is the most reliable, although it requires more computation time than the other two. It may be a good detector option if the computation time is not the vital factor, while the detection rate and precision are the more significant considerations.

The iOS built-in function CIDetector had the best performance in computation efficiency and better precision than MATLAB. Its performance may be further improved if the faces sent to the detector could be adjusted to be larger and if the eyes are not occluded. Since the CIDetector uses the least computation time, it is the best choice if the user needs an efficient detector, and the detection rate can be slightly tolerable.

As CIDetector might fail to detect some true positives, MATLAB's built-in function 'vision.CascadeObjectDetector' may be a better choice if the user does not want to miss any true face detection. Depending on the program's purpose, the false positive may sometimes have a lesser impact than the false negative. In order to briefly illustrate the detection performance of the detectors, the outputs of the detectors are shown in Fig. 7. These notation methods can be improved accordingly.

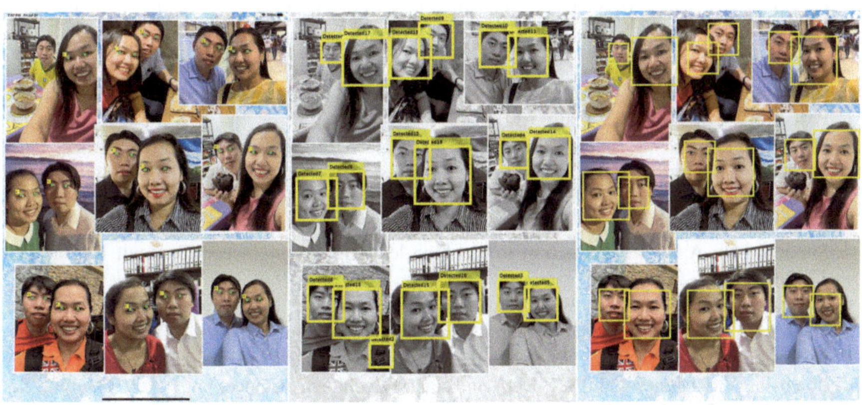

Fig. 7 The face detection output from (**a**) the iOS built-in function 'CIDetector', (**b**) the MATLAB built-in function 'vision.CascadeObjectDetector', (**c**) the self-trained VJ detector (from left to right)

6 Conclusion

The Viola–Jones (VJ) algorithm was ground-breaking when first introduced. It continues to hold a revered position as it laid the groundwork for many recent advancements in computer vision technologies. By mastering the core concepts and practical implementation methods of the VJ algorithm, researchers from various disciplines can innovate and create more powerful technology. Therefore, this chapter presents the practical implementation of the built-in functions of the MATLAB 'vision.CascadeObjectDetector' and the iOS 'CIDetector' for face detection. The core concept of the VJ detector is depicted. In this study, the total execution times for processing 12 image samples with 155 faces are 3.54 s (CIDetector), 5.41 s (vision.CascadeObjectDetector), 1064.594 s (VJ detector with terminal compiler) and 12415.882 s (VJ detector with Xcode compiler). The CIDetector had a 0.776 detection rate and 0.833 precision owing to the 77 undetected faces. Although the MATLAB detector had a 0.999 detection rate, which had only one undetected face, it had 21 false positives and caused the precision to drop to 0.815, whereas the self-trained VJ detector, which had only one false negative and zero false positives, had the most outstanding performance with a 0.994 detection rate and precision of 1. Based on these performances, the suggestions for selecting the most suitable detector for different cases are discussed.

Acknowledgments The research team would like to acknowledge the support of Common Engineering, School of Engineering, Monash University Malaysia, in making this publication successful.

References

1. Paul, T., Shammi, U.A., Ahmed, M.U., Rahman, R., Kobashi, S., Ahad, M.A.R.: A study on face detection using Viola-Jones algorithm in various backgrounds, angles and distances. Biomed. Soft Comput. Hum. Sci. **23**(1), 27–36 (2018)
2. Taloba, A.I., Sewisy, A.A., Dawood, Y.A.: Accuracy enhancement scaling factor of Viola-Jones using genetic algorithms. In: 14th International Computer Engineering Conference on Proceedings, pp. 209–212, Cairo (2018)
3. Cen, K.: Study of Viola-Jones Real Time Face Detector. Stanford University Project Page. https://web.stanford.edu/class/cs231a/prev_projects_2016/, Last accessed 2017/4/4
4. Jensen, O.H.: Implementing the Viola-Jones Face Detection Algorithm. Master Thesis, Technical University of Denmark, Denmark (2008)
5. Simeone, O.: A brief introduction to machine learning for engineers. Comput. Res. Repos. **1709**(02840), 1–231 (2018)
6. Akanksha, K.J., Singh, H.: Face detection and recognition: A review. In: 6th International Conference on Advancements in Engineering and Technology, pp. 138–140, Sangrur (2018)
7. Barnouti, N.H., Matti, W.E., Al-Dabbagh, S.S.M., Naser, M.A.S.: Face detection and recognition using Viola-Jones with PCA-LDA and square Euclidean distance. Int. J. Adv. Comput. Sci. Appl. **7**(5), 371–377 (2016)
8. Soo, S.: Object detection using Haar-Cascade classifier. In: Seminar University of Tartu, pp. 1–12, Estonia (2017)

9. Liao, S., Jain, A.K., Li, S.Z.: A fast and accurate unconstrained face detector. IEEE Trans. Pattern Anal. Mach. Intell. **38**(2), 211–223 (2016)
10. Guler, Z., Cinar, A., Ozbat, E.: A new object tracking framework for interest point based feature extraction algorithms. Electronika Ir Elektrotechnika. **26**(1), 63–71 (2020)
11. Hossen, A.M.A., Ogla, R.A.A., Ali, M.M.: Face detection by using OpenCV's Viola-Jones algorithm based on coding eyes. Iraqi J. Sci. **58**(2A), 735–745 (2017)
12. Egorov, A.D.: Algorithm for optimization of Viola-Jones object detection framework parameters. IOP Conf. Ser. J. Phys. **945**(2017), 1–5 (2017)
13. Ram, C.S.: Recognizing face emotion of down syndrome children using VJ technique. Int. J. Comput. Sci. Trends Technol. **7**(2), 93–100 (2019)
14. Mohammed, A.A.Q., Lv, J., Islam, M.S.: A deep learning-based end-to-end composite system for hand detection and gesture recognition. Sensors. **19**(5282), 1–23 (2019)
15. Slivnitsin, P., Bachurin, A., Mylnikov, L.: Robotic system position control algorithm based on target object recognition. In: 8th international conference on applied innovations in IT on proceedings, pp. 87–94, Germany (2020)
16. Hassan, B.A., Dawood, F.A.A.: Facial image detection based on the Viola-Jones algorithm for gender recognition. Int. J. Nonlinear Anal. Appl. **14**(2023), 1593–1599 (2023)
17. Khong, W.L., Rao, N.S.V.K., Mariappan, M.: National Instruments LabVIEW and video imaging technique for health status monitoring. J. Fundam. Appl. Sci. **9**(3S), 858–886 (2017)
18. Mariappan, M., Nadarajan, M., Porle, R.R., Parimon, N., Khong, W.L.: Towards real-time visual biometric authentication using human face for healthcare telepresence mobile robots. J. Telecommun. Electr. Comput. Eng. **8**(11), 51–56 (2016)
19. Huang, R., Su, W., Zhang, S., Qin, W.: Non-contact method of heart rate measurement based on facial tracking. J. Comput. Commun. **7**(2019), 17–28 (2019)
20. Khong, W.L., Rao, N.S.V.K., Mariappan, M., Nadarajan, M.: Analysis of heart beat rate through video imaging techniques. J. Telecommun. Electr. Comput. Eng. **8**(11), 69–74 (2016)
21. Khong, W.L., Mariappan, M., Rao, N.S.V.K.: National Instruments LabVIEW biomedical toolkit for measuring heart beat rate and ECG LEAD II features. IOP Conf. Ser. Mater. Sci. Eng. **705**(1), 1–7 (2019)
22. Khong, W.L., Mariappan, M.: The evolution of heart beat rate measurement techniques from contact based photoplethysmography to non-contact based photoplethysmography imaging. In: IEEE International Circuits and Systems Symposium on Proceedings, pp. 1–4. IEEE, Kuantan (2019)
23. Khong, W.L., Mariappan, M., Chong, C.S.: Contact and non-contact heat beat rate measurement techniques: Challenges and issues. Pertanika J. Sci. Technol. **29**(3), 1707–1732 (2021)
24. Khong, W.L., Rao, N.S.V.K., Mariappan, M.: Blood pressure measurements using non-contact video image techniques. In: 2nd IEEE international conference on automatic control and intelligent systems on proceedings, pp. 35–40. IEEE, Sabah (2017)
25. Viola, P., Jones, M.: Rapid object detection using a boosted cascade of simple features. In: Conference on computer vision and pattern recognition on proceedings, pp. 1–9, Kauai (2001)
26. Viola, P., Jones, M.J.: Robust real-time face detection. Int. J. Comput. Vis. **57**(2), 137–154 (2004)
27. Wang, Y.Q.: An analysis of the Viola-Jones face detection algorithm. Image Process. Line. **4**(2014), 128–148 (2014)
28. Zarkasi, A., Nurmaini, S., Setiawan, D., Kuswandi, A., Siswanti, S.D.: Implementation of facial feature extraction using Viola-Jones methos for mobile robot system. J. Phys. Conf. Ser. **1500**(2020), 1–7 (2020)
29. Alionte, E., Lazar, C.: A practical implementation of face detection by using Matlab Cascade object detector. In: 19th International Conference on System Theory, Control and Computing on Proceedings, pp. 785–790, Romania (2015)
30. Yang, H., Wamg, X.A.: Cascade classifier for face detection. J. Algorithm Comput. Technol. **10**(3), 187–197 (2016)

31. Winarno, E., Hadikurniawati, W., Nirwato, A.A., Abdullah, D.: Multi-view faces detection using Viola-Jones method. J. Phys. Conf. Ser. **1114**(2018), 1–8 (2018)
32. Neuhausen, M., Konig, M.: Improved window detection in Façade images. In: Mutis, I., Hartmann, T. (eds.) Advances in Informatics and Computing in Civil and Construction Engineering 2019, pp. 537–543. Springer (2018)
33. Tavallali, P., Yazdi, M., Khosravi, M.R.: A systematic training procedure for Viola-Jones face detector in heterogeneous computing architecture. J. Grid Comput. **18**(2020), 847–862 (2020)
34. Hindash, A., Alshehhi, K., Altamimi, A., Alshehhi, H., Mohammed, M., Alshemeili, S., Aljewari, Y.H.K.: People counting and temperature recording using low-cost AI MATLAB solution. In: International Conference on Advances in Science and Engineering Technology on Proceedings, pp. 1–6, Dubai (2022)
35. Tran, D.T., Ly, T.N.: To Wrap, or Not to Wrap: Examining the Distinctions Between Model Implementations of Face Recognition on Mobile Devices in an Automatic Attendance System, pp. 1–20. Vietnam National University, Vietnam (2023)
36. Indriyani, Giriantari, I.A.D., Sudarma, M., Widyantara, I.M.O.: An efficient segmentation of U-area and T-area on facial images by using matlab with hough transform and Viola-Jones algorithm base. In: 2nd Multidisciplinary International Conference on proceedings, pp. 1–17, Indonesia (2023)
37. Alling, A., Powers, N., Soyata, T.: Face recognition: A tutorial on computational aspects. In: Deka, G.C., Siddesh, G.M., Srinivasa, K.G., Patnaik, L.M. (eds.) Emerging Research Surrounding Power Consumption and Performance Issues in Utility Computing, pp. 405–425. IGI Global (2016)
38. Datta, A.K., Datta, M., Banerjee, P.K.: Face Detection and Recognition Theory and Practice. CRC Press, New York (2016)
39. Freund, Y., Schapire, R.A.: A decision-theoretic generalization of online learning and an application to boosting. J. Comput. Syst. Sci. **55**(1), 119–139 (1997)
40. Freund, Y., Schapire, R.A.: A short introduction to boosting. J. Jpn. Soc. Artif. Intell. **14**(5), 771–780 (1999)
41. Kishore, G., Gnanasundar, G., Harikrishnan, S.: A survey on object detection using deep learning techniques. Int. Res. J. Eng. Technol. **6**(2), 2140–2143 (2019)
42. Chong, C.S., Rao, N.S.V.K., Mariappan, M., Khong, W.L.: The validation of virtual impact tests using LabVIEW instrumentation techniques. In: Control Engineering in Robotics and Industrial Automation: Malaysian Society for Automatic Control Engineers (MACE) Technical Series 2018, vol. 371, pp. 209–237 (2021)
43. Nagane, U.P., Mulani, A.O.: Moving object detection and tracking using Matlab. J. Sci. Technol. **6**(01), 63–66 (2021)
44. Jayakody, J., Jayatilake, N.: Comparison analysis and data retrieval to identify the associated people in social media by image processing. In: 2nd International Conference on Advanced Research in Computing on Proceedings, pp. 137–141. IEEE, Sri Lanka (2022)
45. Huang, J., Shang, Y., Chen, H.: Improved Viola-Jones face detection algorithm based on HoloLens. EURASIP J. Image Video Process. **41**(2019), 1–11 (2019)
46. Ghosh, M., Sarkar, T., Chokhani, D., Dey, A.: Face detection and extraction using Viola-Jones algorithm. In: Mitra, M., Nasipuri, M., Kanjilal, M. (eds.) Proceedings of 3rd ICCACCS 2020, LNEE, vol. 786, pp. 93–107. Springer, Singapore (2022)
47. Cuevas, E., Luque, A., Escobar, H.: Computational Methods with Matlab. Springer, Cham (2024)
48. Pradeep, A., Asrorov, M., Quronboyeva, M.: Advancement of sign language recognition through technology using python and OpenCV. In: 7th International Multi-Topic ICT Conference on Proceedings, pp. 1–7. IEEE, Pakistan (2023)
49. Marques, O.: Image Processing and Computer Vision in iOS. Springer, Cham (2020)

50. Kalantarian, H., Washington, P., Schwartz, J., Daniels, J., Haber, N., Wall, D.P.: Guess what? Towards understanding autism from structured video using facial affect. J. Healthc. Informatics Res. **3**(43), 43–66 (2019)
51. Li, P., Guo, Y., Luo, Y., Wang, X., Wang, Z., Liu, X.: Graph neural networks based memory inefficiency detection using selective sampling. In: International Conference for High Performance Computing, Networking, Storage and Analysis on Proceedings, pp. 1–14. IEEE, USA (2022)

Comparing ChatGPT Responses with Clinical Practice Guidelines for Diagnosis, Prevention, and Treatment of Diabetes

Melike Sah and Kadime Gogebakan

1 Introduction

On November 30, 2022, Open AI released the AI chatbot, ChatGPT [1]. Since then, ChatGPT has attracted great interest from researchers in healthcare and different communities. ChatGPT is a language model based on natural language processing (NLP) using deep learning. Recently, NLP has been advanced significantly with the introduction of transformer [2] models. Transformers are deep-learning models that work with multiple attention heads and feedforward layers. ChatGPT utilizes multiple layers of transformer blocks called the GPT-3.5 model that helps to learn contextual data from text. The GPT-3.5 model was trained on large text corpora from Web resources and it contains more than 175 billion parameters. Recently, ChatGPT-4 was released; in addition to text inputs, GPT-4 accepts multi-model inputs of image and text.

Comparing to Web search, AI chatbots are becoming popular. Although Web search can be utilized for accessing Web sources and information, AI chatbots such as ChatGPT have the following advantages: (a) ChatGPT is AI based, therefore, it understands the meaning of the queries to an extent. (b) In addition, ChatGPT is a conversational tool that is more engaging and merges data from various sources for users to communicate in a more human manner. (c) Finally, using chatbot APIs, applications can also integrate intelligence for better human computer interaction.

M. Sah (✉)
Computer Engineering Department, Cyprus International University, Nicosia, Turkey
e-mail: msah@ciu.edu.tr

K. Gogebakan
Directorate of Information Technologies, Istanbul Technical University, Famagusta, Turkey
e-mail: kadime.gogebakan@itu.edu.tr

M. N. Seyman (ed.), *2nd International Congress of Electrical and Computer Engineering*, EAI/Springer Innovations in Communication and Computing,
https://doi.org/10.1007/978-3-031-52760-9_5

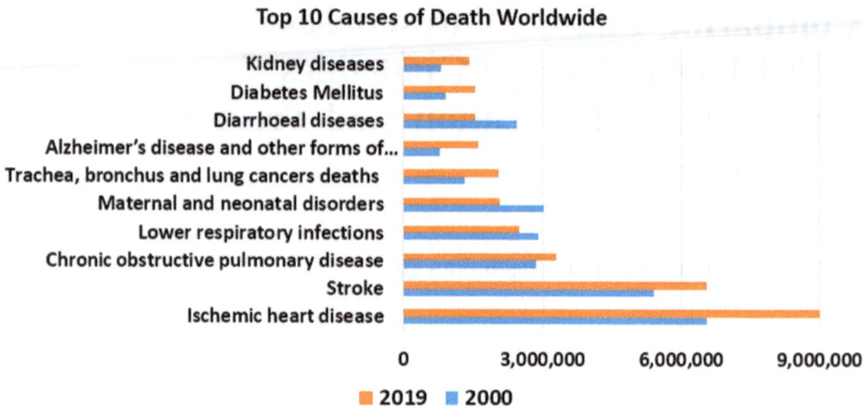

Fig. 1 The top 10 causes of death worldwide between 2000 and 2019 [8]

Recently, people started to use AI chatbots and Web searches together in order to reach out for information they need. Therefore, for disease diagnosis and prevention, ChatGPT is a potential tool for people in disadvantageous countries or people with low incomes. In this work, we investigate the potential use of ChatGPT for diagnosis, prevention, and treatment of diabetes mellitus (diabetes) education. If diabetes cannot be prevented in the early stages, it may pose many risks in terms of human health and the result may bring great burdens to the states.

Diabetes is actually a chronic non-communicable disease caused by the metabolic syndrome. According to the 2021 report of the International Diabetes Federation [3]:

- 537 million adults aged 20–79 years have diabetes mellitus, which means that 1 out of every 10 people in the world has diabetes. If necessary precautions are not taken, these numbers will grow exponentially in the coming years.
- 75% of people living with diabetes live in low- and middle-income countries. It also shows that people in these countries do not have access to proper medical services, which is the motivation of this study.
- Diabetes was the cause of death of 6.7 million people in 2021.
- Diabetes, which has increased rapidly in the last 15 years, has caused 966 billion dollars of health expenditures [3].

Furthermore, according to the analysis of GHDx data [4], diabetes is among the top 10 causes of death worldwide, as shown in Fig. 1. Therefore, early diagnosis and appropriate treatment plans are essential [5, 6], which is the motivation of this study. In this work, we investigated whether ChatGPT can be a potential tool for diabetes education. We collected data about diabetes from the well-known clinical practice guidelines of the Clinical Practice Guideline for Diabetes Management in Chronic Kidney Disease (KDIGO) [7] and the International Diabetes Federation (IDF) [3]. Then, the key questions related to the diagnosis, prevention, and treatment of diabetes are identified that were utilized for the assessment of ChatGPT regarding diabetes education. In this work, we compare how well ChatGPT answers to diabetes-related questions match with the accepted clinical practice guidelines for diabetes. To the best of our knowledge, none of the previous works compare answers of ChatGPT to clinical practice guidelines for diabetes and it is the contribution of our work. Our contributions can be summarized as follows:

- Key questions related to the diagnosis, prevention, and treatment of diabetes are identified and then answers to these questions are extracted from clinical practice guidelines of KDIGO and the International Diabetes Federation.
- Although there are works that use AI chatbots for diabetes education, in this work, for the first time, answers given by ChatGPT to diabetes-related questions are compared with the well-known clinical practice guidelines of KDIGO and the International Diabetes Federation.
- By comparing ChatGPT answers to diabetes-related questions based on the clinical practice guidelines, we aim to provide a quantitative evaluation of ChatGPT answers that is the novelty of the proposed work.
- Furthermore, we discuss potential ethical and security issues when using ChatGPT for disease diagnosis, prevention, and treatment.

The rest of the chapter is organized as follows: Sect. 2 is related work. Section 3 discusses diabetes symptoms and the potential of ChatGPT for diagnosis, prevention, and treatment. Section 4 compares the answers of ChatGPT to the diabetes-related questions and discusses how relevant are the answers regarding to KDIGO and IDF clinical practice guidelines. Section 5 discusses the potential issues such as security, ethical issues when using ChatGPT. Finally, Sect. 6 reveals conclusions and future work.

2 Related Work

Recently, other researchers also investigated the potential of ChatGPT and other AI chatbots in healthcare [9] and for the diagnosis of diabetes disease. Seetharaman [10] reviewed the role of ChatGPT in medical education in general. Arslan [11] explored ChatGPT's potential for obesity treatment by analyzing ChatGPT answers. Hasnain [12] investigated ChatGPT applications for controlling monkeypox. Praveen and Vajrobol [13] investigated healthcare workers' perception of ChatGPT. Vaishya et al. [14] explored the answers of ChatGPT to different health-related questions. Vaishya [15] investigated the potential of ChatGPT for healthcare and research in general. Zhang and Chen [16] analyzed the potential use of ChatGPT for colorectal surgery. Sharma and Sharma [17] investigated the potential of ChatGPT usage by mariners in Maritime Health. Naim et al. [18] developed a chatbot for diabetes classification using a combination of various classifiers for diabetes prediction. Mash et al. [19] developed a WhatsApp chatbot for diabetes information and education during the COVID-19 pandemic. Please note that [18, 19] did not use ChatGPT. In this work, we investigate the potential use of ChatGPT for diabetes education. We identified key questions related to diagnosis, prevention, and treatment of diabetes, and then compared the results of ChatGPT with the accepted clinical practice guidelines of KDIGO and IDF. To the best of our knowledge, none of the previous works compare ChatGPT answers with clinical practice guidelines, which is the novelty of our work.

3 Diabetes Symptoms and Potential Use of ChatGPT for Diagnosis, Prevention, and Treatment

The symptoms of diabetes, which do not cause any complaints in the early stages, are dryness in the mouth and accordingly the desire to drink water, the desire to overeat and the feeling of hunger, frequent urination, weight loss, blurred vision, fatigue and weakness, non-healing wounds, and itching [5]. Even if there is no one of these complaints, early diagnosis is of vital importance and diagnosis can be made by looking at the results of at least two blood tests done at different times. With the early diagnosis of diabetes, the disease can be prevented, the quality of life of the patient can be increased, and the burden of the countries in the field of health can be reduced [8]. Moreover, many complications and diseases caused by diabetes are preventable. In the treatment of diabetes, along with drugs, the patient's lifestyle and diet are also very important factors. As a result of delayed or improperly applied treatments, patients have to use diabetes drugs for the rest of their lives and continue their nutrition and social activities accordingly. Along with their daily lives, they have a very busy schedule as a result of diabetes.

An AI chatbot, such as ChatGPT, can assist patients with usage of drugs, nutrition diets, physical activities, and social activities that can be applied to diabetes patients. In addition, if a person thinks that s/he has diabetes, s/he can use ChatGPT to investigate early symptoms before visiting a health specialist. Specifically, ChatGPT can be utilized to find out diabetes symptoms, possible prevention mechanisms, and treatment plans. As long as their condition does not worsen, diabetic patients continue their treatment from home and ChatGPT can be a potential solution to support the situation of diabetic patients. To assess the feasibility of ChatGPT for diabetes support, next, we outline the key questions as well as compare the answers of ChatGPT with the accepted clinical practice guidelines of KDIGO and IDF.

4 Comparing Answers of ChatGPT with Clinical Guidelines of KDIGO and IDF

In Table 1, we summarize the list of key diabetes questions regarding diagnosis, prevention, and treatment. In terms of the relevance of ChatGPT answers to clinical practice guidelines, we used a four-scale relevance score; *Highly relevant*; ChatGPT answer matches the clinical practice guidelines very well. *Relevant*; ChatGPT

Table 1 Comparing ChatGPT answers with KDIGO [7] and IDF [3] clinical practice guidelines

Question	Answer – *KDIGO/ **IDF	ChatGPT answer	Relevance of answers
Causes of diabetes?	**Physical inactivity, obesity and Westernized foods	Genetics, lifestyle factors, insulin resistance, autoimmune disorders, pancreatic disease or injury, and medications	Somewhat relevant
Symptoms of diabetes?	**Dehydration, stomach pain, nausea and vomiting can lead to deep and rapid breathing, flushed face, dry skin and mouth, fruity breath odor, fast and weak pulse, and low blood pressure	ChatGPT recognized the key symptoms of diabetes as Frequent urination, excessive thirst, fatigue, increased hunger, autoimmune disorders, and so forth	Somewhat relevant
How to prevent diabetes?	**The IDF recommends Performing at least 30–45 min of physical activity at least 3–5 days a week Prevention of type 2 diabetes includes a healthy lifestyle: a balanced diet, regular exercise, and avoidance of sugary and processed foods. Maintaining a healthy weight is very important as obesity increases the risk	ChatGPT acknowledged that maintaining a healthy weight, engaging in regular physical activity, eating a healthy diet, quitting smoking, limiting alcohol intake, managing stress, and getting regular checkups can help to manage risk factors for diabetes	Highly relevant
Difference of type1 and type 2 diabetes?	**Type 1 diabetes generally occurs in childhood and cannot be prevented. People with type 1 diabetes can sustain their lives with continuous use of insulin, education, support, and constant control of blood sugar. Type 2 diabetes occurs at later ages and it is possible to prevent it. Medication treatment and a healthy lifestyle can be controlled by adopting a lifestyle. Knowing the difference between type 1 and type 2 diabetes is important since diverse types of diabetes require different treatments	ChatGPT answered that Type 1 diabetes is an autoimmune disorder and that genetics and environmental factors likely play a role. Type 1 diabetes is often diagnosed in childhood or adolescence and is not typically associated with lifestyle factors. Type 2 diabetes, is primarily caused by insulin resistance and other lifestyle factors, such as obesity, physical inactivity, and poor diet. It is more commonly diagnosed in adults, but it can also occur in children, especially as rates of childhood obesity rise	Highly relevant

(continued)

Table 1 (continued)

Question	Answer – *KDIGO/ **IDF	ChatGPT answer	Relevance of answers
What are the tests that can be used to diagnose diabetes?	**Currently, WHO and IDF recommend the two-hour oral glucose tolerance test (OGTT) for the detection of Impaired Glucose Tolerance (IGT) and Impaired fasting glycemia (IFG). However, there is accumulating evidence that the one-hour OGTT is used as a more sensitive method with the capacity to detect moderate hyperglycemia at an earlier time	ChatGPT provided 10 common tests that are used for the diagnosis of diabetes such as OGTT, Hemoglobin A1c (HbA1c) test, Fasting Plasma Glucose (FPG) Test, and others. It is not clear which tests are more reliable	Somewhat relevant
What diet is suitable for diabetic patients?	**Prevention of type 2 diabetes includes a healthy lifestyle: a balanced diet, avoidance of sugary, and processed foods. Maintaining a healthy weight is very important as obesity increases the risk	ChatGPT acknowledged the importance of healthy and balanced diet and provided a list of food categories, such non-starchy vegetables (broccoli, spinach, kale, tomatoes, carrots, etc.), whole grains (brown rice, etc.), lean proteins (chicken, Turkey, etc.), and other diabetes-friendly food categories	Highly relevant
What are the timing of meals and their intervals for diabetes patients?	**Eat 3 main meals a day. However, eat at least three servings of vegetables each day, including up to three servings of fresh fruit and leafy greens. Another important thing for diabetes patients to consider is timing of their meals so that their blood sugar levels can be kept consistent throughout the day	ChatGPT suggested to eat consistent meals at regular intervals which is important for diabetes patients to regulate blood sugar levels. In addition, it suggested to avoid skipping meals, spacing out meals and snacks by at least 2–3 h, consider eating smaller, more frequent meals, avoiding eating too close to bedtime	Highly relevant
Are physical activities suitable for diabetes patients?	**The IDF recommends regular physical activity	ChatGPT acknowledged that physical activities are crucial for managing diabetes. It suggested different activities and advised to talk to a health provider for personalized plans	Highly relevant

(continued)

Table 1 (continued)

Question	Answer – *KDIGO/ **IDF	ChatGPT answer	Relevance of answers
What are social activities for diabetes patients?	**Social conditions of the person are also among the causes of diabetes. Stress is one of the major causes of diabetes, so it is necessary to socialize	ChatGPT emphasized that social activities can be very beneficial for diabetes patients since they can provide emotional support and reduce stress	Highly relevant
Which medications are used for treating diabetes?	*Metformin, Sulfonylureas, and Sodium-Glucose Co-Transporters Type 2 (SGLT-2) inhibitors	ChatGPT provided the list of commonly used drugs for treating diabetes such as Metformin, Sulfonylureas, meglitinides, Alpha-Glucosidase Inhibitors, and other inhibitors. Security issues will be discussed below	Relevant
Which drug is used in the first-line treatment of type 2 diabetes?	*Metformin	ChatGPT also answered as Metformin	Highly relevant
What is the purpose of drug therapy for type 2 diabetes?	*Lowering and managing glucose level.	ChatGPT answered to help manage and control blood sugar levels effectively	Highly relevant
Does the use of multiple drugs in the treatment of type 2 diabetes cause drug-drug interaction?	*Drug interactions are more likely to occur in diabetic patients who take multiple medications for diabetic complications	ChatGPT responded that multiple drugs in the treatment of Type 2 diabetes can potentially cause drug-drug interactions	Highly relevant
What are the complications of type 2 diabetes?	**Heart disease and stroke, nerve damage, foot problems, vision loss and blindness, miscarriage and stillbirth, kidney problems and others	ChatGPT responded as cardiovascular problems, kidney disease, eye problems, nerve damage, foot problems, and other complications	Highly relevant

response matches the clinical practice guidelines, but there is some missing information. *Somewhat relevant*; some of the ChatGPT answers are matching to the clinical practice guidelines, but in general answers are not in line. *Not relevant*; ChatGPT answers do not match the clinical practice guidelines.

Table 1 illustrates that some answers given by ChatGPT are not relevant such as the answer given to the "tests used to diagnose diabetes." According to the WHO and IDF, two-hour or one-hour OGTT tests are more reliable for detecting diabetes. On the other hand, ChatGPT provided a list of tests and it is not clear that which one is the most suitable and reliable. In addition, answers of ChatGPT to the symptoms and causes of diabetes partially match the clinical practice guidelines. Therefore, some answers may not reflect the current guidelines. On the other hand, many diabetes-related questions are answered by ChatGPT with the highest relevance such as drug–drug interactions, physical/social activities, and others, as shown in Table 1. Special attention should also be given to the drug-related questions. ChatGPT provides detailed information about drugs and their usage. Although ChatGPT warns about consulting a health specialist regarding drugs, providing such detailed information to a non-specialist may have some consequences as we discuss in the next section.

5 Concerns Regarding the Usage of ChatGPT

Although ChatGPT can answer diabetes-related questions, it has disadvantages:

- Accuracy of ChatGPT is a problem since it relies on the existing relevant trained data. Integrating specialist doctors in order to prevent incorrect information is crucial.
- Suggestions may be biased since ChatGPT relies on its text corpora for suggestions. Again, integrating specialist doctors might be the solution.
- Patient privacy and confidentiality should be secured.
- There might be limited engagement due to a person's trust in an AI chatbot.
- ChatGPT has no emotional connection with a patient. Thus, answers to disease questions might emotionally affect the people.
- There are possible side effects of inaccurate medical advice (i.e., drug suggestions and medical guidelines) by ChatGPT and patients that may have a worsening effect on their health. Integrating specialist doctors is the solution.
- There are possible potential risks due to wrong diagnosis, wrong guidance, or misleading information.
- It is not clear when updates to medical data are applied by ChatGPT. Thus, users should be aware of this information.

6 Conclusions and Future Work

In this work, we investigate the potential of ChatGPT for diabetes education. Diabetes mellitus is one of the fastest-growing diseases around the world. If it can be detected in early stages, with simple lifestyle changes, many diabetes cases can be prevented. In cases of delayed or improper treatments, on the other hand, patients have to use diabetes drugs for the rest of their lives. In this work, the key diabetes-related questions and their answers are collected from KDIGO and the International Diabetes Federation clinical guidelines. Then diabetes-related questions are asked to ChatGPT and answers are compared to assess how they are aligned with the clinical practice guidelines. This comparison illustrated that ChatGPT can answer many diabetes-related questions with high relevancy according to the clinical practice guidelines, especially for diet, physical/social activities, and prevention plans. ChatGPT can help with the early diagnosis of diabetes, by providing awareness to the people. With simple lifestyle changes, the disease can be prevented. As a result, the quality of life of the patient can be increased and the burden of the countries in the field of health can be reduced. On the other hand, ChatGPT partially answered some of the questions related to causes, symptoms, tests, and drugs. Therefore, some answers may not reflect the current clinical practice guidelines for diabetes. Finally, we outlined possible risks while using ChatGPT for diabetes education such as ethical/security issues, and some potential risks associated with inaccurate, and misleading medical advice that may have negative consequences on a person's health. If these potential risks can be alleviated, ChatGPT can be utilized as a potential tool for diabetes education. In future, we plan to develop a specialized diabetes mobile application based on ChatGPT API in order to create awareness, as well as, help arrange treatment plans for diabetes patients.

References

1. OpenAI ChatGPT, https://chat.openai.com/, Last accessed 7/9/2023
2. Vaswani, A., Shazeer, N., Parmar, N., Uszkoreit, J., Jones, L., Gomez, A.N., Kaiser, L., Polosukhin, I.: Attention is All You Need. 2017. arXiv.org. https://arxiv.org/abs/1706.03762
3. International Diabetes Federation: IDF Diabetes Atlas, 10th edn. International Diabetes Federation, Brussels (2022)
4. Global Burden of Disease Study 2019 (GBD 2019) Results. Seattle, United States: Institute for Health Metrics and Evaluation/(IHME), 2020. Available from http://ghdx.healthdata.org/gbd-results-tool, Last accessed 7/9/2023
5. Diabetes: Retrieved from World Health Organization, https://www.who.int/news-room/fact-sheets/detail/diabetes, Last accessed at 7/9/2023
6. Diagnosis: Retrieved from American Diabetes Association, https://diabetes.org/diabetes/a1c/diagnosis, Last accessed at 7/9/2023
7. KDIGO: Clinical Practice Guideline for Diabetes Management in Chronic Kidney Disease, Available at https://kdigo.org/wp-content/uploads/2022/10/KDIGO-2022-Clinical-Practice-Guideline-for-Diabetes-Management-in-CKD.pdf, Last accessed at 7/9/2023
8. Göğebakan, K., Şah, M.: A review of recent advances for preventing, diagnosis and treatment of diabetes mellitus using semantic Web. In: 2021 3rd International Congress on Human-Computer Interaction, Optimization and Robotic Applications (HORA), pp. 1–6, Ankara (2021)
9. Biswas, S.S.: Role of chat GPT in public health. Ann. Biomed. Eng. **51**, 868–869 (2023)
10. Seetharaman, R.: Revolutionizing medical education: Can ChatGPT boost subjective learning and expression? J. Med. Syst. **47**, 61 (2023)
11. Arslan, S.: Exploring the potential of chat GPT in personalized obesity treatment. Ann. Biomed. Eng. (2023)
12. Hasnain, M.: ChatGPT applications and challenges in controlling monkey pox in Pakistan. Ann. Biomed. Eng. (2023)
13. Praveen, S.V., Vajrobol, V.: Understanding the perceptions of healthcare researchers regarding ChatGPT: A study based on bidirectional encoder representation from transformers (BERT) sentiment analysis and topic modeling. Ann. Biomed. Eng. (2023)
14. Vaishya, R., Misra, A., Vaish, A.: ChatGPT: Is this version good for healthcare and research? Diabetes Metab. Syndr. Clin. Res. Rev. **17**(4) (2023)

15. Grünebaum, A., Chervenak, J., Pollet, S.L., Katz, A., Chervenak, F.A.: The exciting potential for ChatGPT in obstetrics and gynecology. Am. J. Obstet. Gynecol. (2023)
16. Li, W., Zhang, Y., Chen, F.: ChatGPT in colorectal surgery: A promising tool or a passing fad? Ann. Biomed. Eng. (2023)
17. Sharma, M., Sharma, S.: Transforming maritime health with ChatGPT-powered healthcare services for mariners. Ann. Biomed. Eng. **51**, 1123–1125 (2023)
18. Naim, I., Singh, A.R., Sen, A., Sharma, A., Mishra, D.: Healthcare CHATBOT for Diabetic Patients Using Classification. Lecture Notes in Networks and Systems, vol. 425. Springer, Singapore (2022)
19. Mash, R., Schouw, D., Fischer, A.E.: Evaluating the implementation of the GREAT4Diabetes WhatsApp Chatbot to educate people with type 2 diabetes during the COVID-19 pandemic: Convergent mixed methods study JMIR. Diabetes. **7**(2), e37882 (2022)

ML-Based Optimized Route Planner for Safe and Green Virtual Bike Lane Navigation

Nourhan Hazem Hegazy, Ahmed Amgad Mazhr, and Hassan Soubra

1 Introduction

Cycling is one of the simplest ways of being physically active and, in addition, it is one of the most sustainable means of transport [1]. Furthermore, reducing cars on the road and replacing them with bikes reduces air pollution which results in the death of half a million people every year as per the World Health Organization (WHO), and causes severe health problems such as heart failure and many more [2].

Since Cairo is ranked as one of the top 10 polluted cities due to high traffic [3–5], actions are needed, e.g., the ones presented in [6, 7].

It has been shown that funding safe cycling infrastructure can play a critical role in general health, mitigating climate change, and improving the environment, according to a new WHO publication [8]. The problem with old cities such as Cairo is that it would be almost impossible to create bike lanes after all the infrastructure of the city has been built, so a viable solution is to implement a virtual lane to make sure that cyclists are navigated through the safest green path possible. The platform used in this study is a smart autonomous E-bike prototype presented in [9–16]

In this chapter, we aim to use machine learning to optimize the route planner system described in [10] made to promote safe cycling that navigates the riders in the best possible safest green path to their desired destination according to different criteria that also ensure the fulfillment of the rider. However, populating the system

N. H. Hegazy (✉) · A. A. Mazhr
German University in Cairo, Cairo, Egypt
e-mail: nourrhan.hegazy@gmail.com; ahmedmazhar20@gmail.com

H. Soubra
Ecole Centrale d'Electronique-ECE, Lyon, France
e-mail: hsoubra@ece.fr

M. N. Seyman (ed.), *2nd International Congress of Electrical and Computer Engineering*, EAI/Springer Innovations in Communication and Computing,
https://doi.org/10.1007/978-3-031-52760-9_6

with various Application Programming Interface (API) calls makes the system very slow, therefore, we aim to minimize the use of API calls by using machine learning techniques to utilize data from past experiences.

The rest of the chapter is structured as follows: Sect. 2 provides the existing related works in the literature. Section 3 presents an overview of our optimized route planner proposed. While Sect. 4 demonstrates the system advancements compared to the previously proposed one, Sect. 5 displays the system evaluation and results. Finally, Sect. 6 concludes the paper by summarizing the key findings.

2 Literature Review

2.1 Background Study

In 2023, Seoudi et al. [10] proposed a state-of-the-art idea whose objective is to make cycling more convenient in the streets of Cairo through a path-planning algorithm dedicated to biking purposes. The paper pointed out the gap in considering the safety of the rider when deciding the optimal route and thus it opted to provide a multi-criterion route planner established in Cairo that provides the best possible path that maintains the comfort, health, and safety of the rider considering the absence of bike lanes and traffic laws for cyclists. The system considers five criteria as input that contribute to the output:

1. Thermal comfort.
2. Rider's effort.
3. Exposure to air pollution.
4. Weather.
5. Safety:
    (a) Speed limit.
    (b) Crossroads.

These are extracted through Open Street Maps library and Google Maps API in Python to get the best path by considering the map as a network that deals with the street segments as "edges" and the nodes are represented using "longitudes" and "latitudes." The thermal comfort is the number of green areas in the area buffer conducted around the street which is then transformed into the "Ultra Violet (UV) cost." The rider's effort is acquired by extracting the elevation of each road edge from Google Maps API and thus its cost is calculated as "Elevation cost." As for the pollution, it is divided into sub-elements that are combined to result in the final pollution value. The first element is the traffic and as mentioned before by Weltbank [4], Cairo has one of the highest traffic rates. Needless to add, the more traffic the higher the pollution as a result of the gas emissions from the currently available means of transport. This equation states that every extra minute of traffic results in a 10% increase in the pollution value. The second element that affects the final pollution value is the green areas, and every 5% increase in green areas in the area buffer results in a 0.5 decrease in the pollution level.

People who cycle frequently will decide whether to ride their bikes at the current moment or simply rely on other means of transport all will be dependent on the current weather conditions. Since the weather plays the most important role in the decision to cycle, it is added to the equation as a way to assign the weight to the other mentioned criteria. Each of the criteria is affected by the weather element that relates to it. The four weather elements are:

1. Wind speed.
2. Air Quality Index.
3. Ultraviolet index.
4. Visibility.

The solution is presented by applying Eq. 1 to alternative paths. W1, W2, W3, and W4 represent four different weather elements, while C1, C2, C3, and C4 represent the corresponding criterion that is affected and weighted by the specified weather element.

$$C = W1 * C1 + W2 * C2 + W3 * C3 + W4 * C4 \tag{1}$$

2.2 *Machine Learning in Route Planners*

Using machine learning has been a point of interest to further improve the available route planning applications by having prediction features for certain properties of the road.

In 2020, Shah et al. [17] addressed the problem that the present cyclist route planning applications do not consider safety factors like traffic, rain, or visibility into account when providing cycling routes. They explored a solution to their case study on Auckland City where the traffic and weather data are acquired using Google, Bing, and Wunderground APIs. After comparing the three applications for route planning, they decided to go with Google Maps as it was found to outperform the others in all their aspects. Google Maps contains many cyclist-friendly features such as elevation information and cycle path support for routes and is very efficient at finding the shortest path. However, it does not take the safety of the cyclist and weather conditions into consideration. The authors were concerned with collecting data according to real-time constraints to make the system more accurate and safer for the cyclist and found that the most accurate method of real-time data collection is through the use of floating car data: Utilizing the phones of road users to provide an easily obtainable, accurate source of real-time data. They also considered looking into weather prediction but unfortunately, they found that this can be very inaccurate and is difficult to use in their application. Shah et al. added a machine learning algorithm that can be used by users to plan their future trips and ensure that they are traveling during safe times. It predicts the traffic levels on common routes in advance to ensure their safety. The information can be shown in the application in a graph representation; it plots the level of congestion against the time of day.

In 2018, Valatka [18] proposed a model that "quantitatively evaluates recurrent neural network capabilities to generate optimal routes in real road networks." They fetched the real-time data of the roads from Open Street Maps which were then trained on 123,354 optimal routes for 10 epochs with a batch size of 64. Afterwards, traveling time was calculated for each edge by dividing the edge length and the edge speed limit attributes. It was shown that optimal routes were generated with the Dijkstra shortest path algorithm from the graph. The model was presented with 84.2% accuracy.

In 2017, Said et al. [19] discussed a green adaptive transportation decision system for choosing the best transportation route calculated for different means of transport (train, metro, and bus) to reach a certain destination at time t. The system is based on the following parameters: carbon dioxide emissions, ticket tariff, connection time, and comfortability feedback. It classifies the collected route traces and distinguishes between them to select the best route to reach a certain destination at time *t*. Their system is made of two main machine learning layers: Q-learning, which is a reinforcement learning technique, and the second layer is a support vector machine that is a supervised learning technique. The Q-learning layer nominates the best route at time t among the proposed routes, while the support vector machine layer is responsible for building the prediction of the best transport route, classifies the data into two classes, and differentiates between classes where the first class includes the best route(s) and the second one contains all other available routes. Eventually, it is expected to find a linear/non-linear separator between these two classes and calculate the accuracy of the entire model.

In 2021, Czuba et al. [20] presented machine learning and deep learning algorithms for solving Vehicle Routing Problems (VRP) and dynamic VRP related to delivering products to customers. They further explained the vehicle as an undirected graph in Eq. 2

$$G=(V,E); V=v_0, v_1, \ldots, v_n \tag{2}$$

where V is a set of vertices that represents customers.

$$E=\left(v_i, v_j\right); v_i, v_j \in v, i<j \tag{3}$$

Equation 3 is the set of edges that correspond to the connection between customers. Each edge has an associated cost (time or length between customers, or customer and depot) that is represented by weight C_{ij}. Additionally, each customer has associated demand di which must be fulfilled by a single vehicle. Their implementation architecture is composed of three main layers. The first layer is a neural network that consists of two recurrent neural networks: encoder and decoder. The encoder network reads the input sequence and stores knowledge about it in a vector representation while the decoder takes encoded information and converts it back to the output sequence. Another layer is the graph neural network where the network assigns to each vertex a multi-dimensional embedding and performs message-passing tasks from the embedding to their adjacencies. Vertices aggregate incoming messages to be passed through Recurrent Neural Networks (RNN) and after many message-passing interactions, the model is asked to decide whether a specific route with cost $< C$ exists. The last layer is a reinforcement learning one where the state of the environment is seen as a partial solution of the instance, each node features, and the action is seen as a choice of the next customer to visit. The reward is the negative route length and the overall policy corresponds to a strategy given by the neural network.

2.3 Literature Review Conclusion and Research Gap

It is clear that safety factors such as traffic and pollution play an important role in the cycling experience and that bike lanes are an essential factor in encouraging people to use their bikes as a main means of transport. It is concluded that machine learning is widely used in different ways to enhance route planning systems by providing an efficient, fast, and simple way of retrieving the path based on the specified criteria.

To our knowledge, applying machine learning to all five safety criteria mentioned in [10] has never been implemented. All the mentioned papers have applied machine learning to some of our desired criteria but none of them presented an approach for them all. In our implementation, we worked on filling this gap by providing a new accurate machine learning approach that outputs the path according to all the criteria mentioned in [10] in a fast and efficient way. Added to that, the rider would have the option to view the path several hours or several days in advance if the rider wishes to plan the trip.

3 Our Enhanced Routing System Proposed

The system proposed uses machine learning techniques for the prediction of the road properties several hours and several days in advance. It analyzes all possible paths and predicts the value for each criterion on each one of the lists of possible paths to finally determine the optimal route for the rider which takes into account multiple criteria without the need for the entire API calls. Dealing with a multi-criterion route planner system, the proposed system can be divided into three machine learning sub-systems one for each criterion that provides a fast yet accurate response without having to wait several minutes on the APIs for the path to be displayed.

3.1 Problem Definition and Solution Formulation

In [10], multiple problems were detected. The most evident one is the system's tedious response time due to the use of multiple API calls sequentially to extract the data. In addition, the system is incapable of learning from the past experiences, therefore, failing to identify common road edges between different start- and end-points. To optimize the system's performance, machine learning techniques were incorporated that will minimize API usage by providing predictions based on the previous real-time values. Furthermore, they will easily interpolate between the edges to predict values for never-seen road segments while also learning the newly presented edges.

3.2 Dataset Formation

Provided that the system incorporates previously unrecorded factors in Cairo, it had to be done from scratch. In the previous system, real-time data of each of the criteria were collected for each road segment; however, there is a drawback: assigning the path according to the location and desired destination could take up to 15 min. The system outputs the set of criteria mentioned in [10] per edge of the road network graph. To address this, we automated the previous system's code on different locations, every hour of the day, and every day of the week to obtain the dataset of real-time values for all hours of the day. The code was automated for 7 days on 10 distinct locations with their corresponding edges that make up each path to each location using a Python script hosted on Amazon Web Services (AWS) and stored the data collected. The dataset size is 63,771 rows that consist of 12 features per edge in each path of the possible paths; namely, month, day, hour of day, edge length, start location of each edge in the path, end destination location of each edge in the path, pollution, amount of green areas, road elevation, traffic, number of intersections, and shadow present. These 12 features make up our 4 criteria mentioned in [10]: thermal comfort, rider's effort, exposure to pollution, and safety. The fifth criterion, weather elements, is extracted without machine learning; therefore, it was not included in the dataset. The model was trained on 80% of the dataset and tested on the remaining 20%.

3.3 Model Development

Feature Extraction: The road segments, the crossroads, and the weather elements are extracted in the following manner:

Edge Extraction
The user inputs the start location and the desired destination, which are then passed to Open Street Maps library in Python. The library transforms the input addresses into longitudes and latitudes, representing the coordinates of the locations. Afterwards, the system extracts all possible road edges used to construct the path to reach the destination.

Crossroad Extraction
Similar to the edge extraction, intersections are calculated by Open Street Maps API in Python by reviewing all the edges extracted and checking if any two or more edges share the same nodes. In the presence of a shared node, a crossroad is identified, and its dimension, known as the "n-way crossroad," is based on how many edges share the same node. Subsequently, each edge is assigned an intersection cost, symbolizing a safety factor according to the dimension "n" of the n-way crossroads. This criterion is significantly important in Cairo as proper traffic laws are not applied leading to dangerous crossroad circumstances for riders.

Weather
Weather is extracted per edge according to the requested day and hour using Open Weather Map API that outputs four weather elements, namely, ultraviolet index, air quality index, visibility, and wind speed. Each weather element is assigned a weight, which is used to weigh its corresponding road property into the final equation represented in Eq. 4.

$$\begin{aligned}\text{FinalWeight} = \text{index} * \text{uv cost} + \text{windspeed} * \text{elevationcost} + \text{visibility}\\ * \text{intersection cost} + \text{airquality index} * \text{pollution cost}\end{aligned} \tag{4}$$

Machine Learning Models: Several machine learning models have been created and evaluated for each of the remaining criteria, namely, pollution, thermal comfort, and Rider's effort. In this section, we will examine every model created for each of the criteria.

Pollution Model
The objective of this model is to provide the system with a list of all possible edges from the start location to reach the desired destination. The model predicts the pollution cost for each edge that will contribute to the weighted equation according to the time and day of the user's request. Furthermore, several models have been developed and tested: Transformer Neural Network (TNN), sequential neural network (NN), Recurrent Neural Network (RNN), and long short-term memory (LSTM), all of which showed inaccurate results that will be illustrated in the System Evaluation and Results section.

Given that we are implementing a machine learning model that has an output dependent on time intervals, the time series forecasting ML model demonstrated promising results in [21] as it is designed to find the correlation of features concerning the timestamp.

According to Bontempi et al. [21], a time series is a collection of past observations made at regular intervals of an observable variable y. Time series can be a single-step or multi-step model depending on the desired output being one or more time steps ahead. Similarly, there are also univariant and multi-variant time series models. A univariate time series is a collection of measurements of a single variable across time, whereas a multi-variant time series is a series of measurements of numerous variables over time [22].

We are dealing with a multi-variant time series model as we need to forecast the pollution according to multiple inputs which are: the longitude and latitude of both the start point and the end point, hour, and day. In addition, we have implemented it to be a multi-step time series to add the pleasant feature of being able to predict multiple hours. Time series models typically use a neural network embedding and, in our implementation, LSTM is used.

A Random Forest model is a vector sampled randomly, with the same distribution across all of the trees in the forest, which combines various tree predictors to make each tree dependent [23]. Random Forest was also implemented and tested and has shown promising results as will be illustrated in the System Evaluation and Results Section.

Thermal Comfort and Rider's Effort Models

Thermal comfort and Rider's effort are physical road properties that are constant for each edge, therefore, time series is not considered as one of the possible models since these properties do not depend on time. The use of machine learning in these models is for its interpolation property when dealing with never-seen data, namely, never-seen locations. Several neural network models such as transformer, sequential, RNN, and LSTM neural networks have been implemented and tested and both Thermal Comfort and Rider's effort models yield the same results that will be shown in the System Evaluation and Results section.

eXtreme Gradient Boosting (XGBoost) is presented in [24] as the gradient boosting framework used by the decision-tree-based ensemble machine learning method. This classical machine learning model has also been examined for both Thermal Comfort and Rider's Effort.

Integration: After all, three machine learning models have been successfully implemented, and web extractions are completed. All the sub-models are integrated to form our system. In this subsection, we will discuss the flow of integrating all sub model's inputs and outputs.

After the user inputs the start location and destination, the system extracts all possible edges from the start point to the endpoint. All these edges are simultaneously inputted to the three machine Learning models to make predictions for the three features: pollution, thermal comfort, and rider's effort. Intersections and weather elements are extracted and then all the criteria are combined and represented as a weight for every single edge. Afterwards, Dijkstra's shortest path algorithm is applied leading us to the output of the safest green path.

4 System Comparison

In this section, comparisons of the previous route planner system and our proposed system are presented.

In the previous system, it took 386.63 s (Fig. 1) to output the path which is 6.44 min. Whereas, in our machine learning system, it only took 38.37 s (Fig. 2) which is almost 10 times faster. In addition, our system can predict several hours and several days in advance by choosing the day and hour from drop-down lists found in the user interface as shown in Figs. 3 and 4.

Figure 5 presents a comparison of the chosen path between the original system and the machine learning system. The paths are not always exactly the same but they are extremely similar. This is because Google Maps API provides only one Free API key with a limited number of extraction requests per API key, which resulted in limited data collection. Moreover, most of the edges are similar and sometimes even the same in both systems which makes us to conclude that the system performs well in comparison with the original slow system. This proves that our system is one with high precision compared to the original system. Added to that, the system has a fine feature of predicting the path in advance according to the future values for riders who wish to plan their trip.

```
Execution Time:  386.63610009999684  Sec
PS C:\Users\Nourhan Hegazy>
```

Fig. 1 Previous system execution time

```
3/3 [==============================] - 1s 8ms/step
Execution Time:  38.37378673699999  Seconds
```

Fig. 2 Our new improved execution time

Fig. 3 User interface input

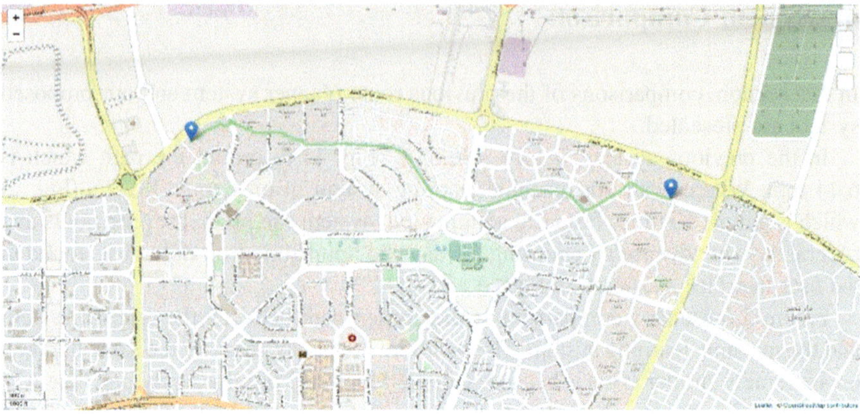

Fig. 4 User interface map view

(a) Original Path 1

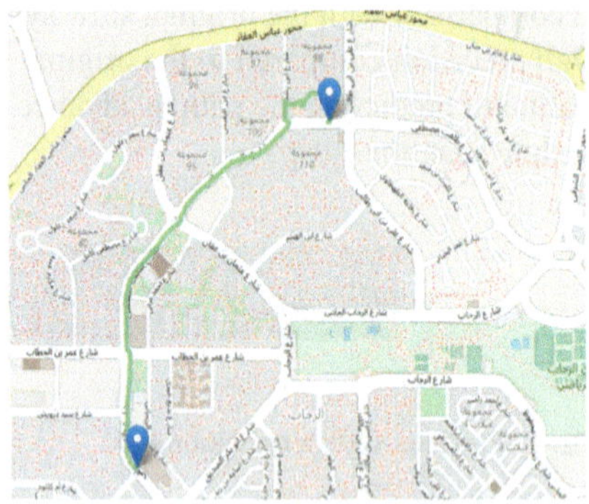

(b) Our System's Path 1

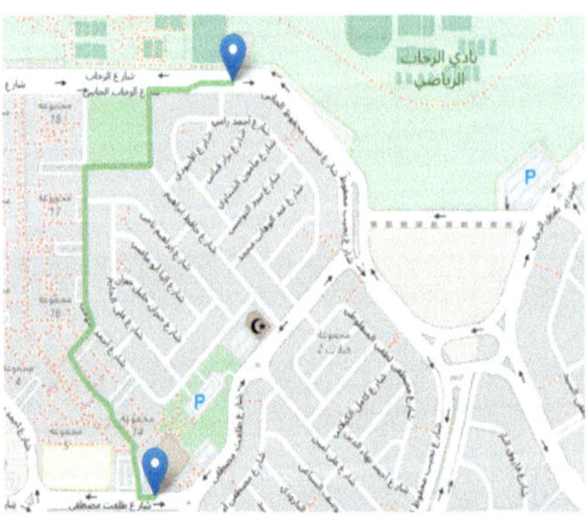

(c) Original Path 2

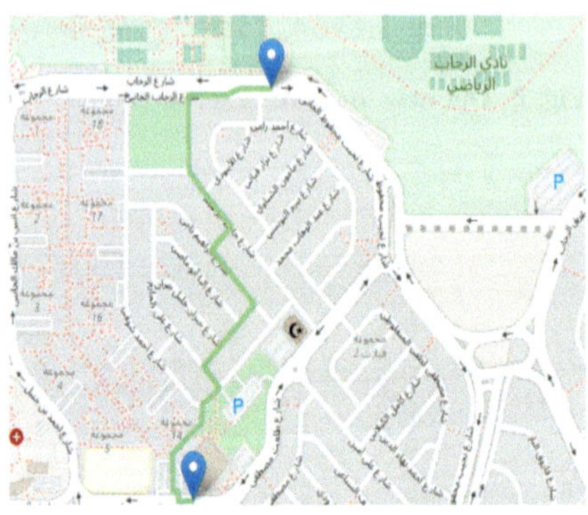

(d) Our System's Path 2

Fig. 5 Original and our system's paths comparisons

5 System Evaluation and Results

In this section, an overview of the training and testing results will be explained. The testing was based on a real-time network analysis of the city of Cairo. The training phase is made of several locations and destinations over the city and each one of these locations' criteria was extracted at every hour of the day and every day of the week. The system was tested on both locations that are known for the model from the training phase, as well as on those that were never seen by the model.

In this section, the numerical results of the models' losses will be shown and discussed. The loss functions are mean squared error (MSE), mean absolute percentage error (MAPE), and mean absolute error (MAE).

5.1 Pollution Models Losses

Table 1 displays the training losses presented by the pollution models, while Table 2 displays the models' losses for the testing phase.

The Random Forest model is presented with the accuracy displayed in Fig. 6.

Table 1 Pollution models' training losses

Loss function	Sequential NN	Transformer	NN	Recurrent NN	LSTM	Time series
MSE	249.37	401.69		402.17	9711.5	6.41
MAPE	66878868.00	3836811.00		3628990.75	291335040.00	7832480.00

Table 2 Pollution models' testing losses

Loss function	Sequential NN	Transformer NN	Recurrent NN	LSTM	Time series
MSE	251.45	409.31	406.85	9571.57	28.37
MAPE	64281728.00	3182014.25	3079775.25	317329120.20	9.46
MAE	9.43	9.57	9.39	8.26	0.75

```
score = model.score(x_test, y_test) *100
score

<ipython-input-11-3d3c8295157a>:3: DataConversionWarning:
  model.fit(x_train, y_train)
99.9729551030226
```

Fig. 6 Random Forest accuracy

5.2 *Thermal Comfort and Rider's Effort Models Losses*

Table 3 displays the training losses presented by the thermal comfort and rider's effort models while Table 4 displays the models' losses for the testing phase.

The XGBoost model is presented with the accuracy displayed in Fig. 7.

5.3 *Results Conclusion*

The time series model was proven to be optimal with the lowest MAE having a value of 0.75 as shown in Table 2. In addition, the Random Forest model resulted in a 99.97% accuracy as shown in Fig. 6. If opted to use time series forecasting alone, the model would have high accuracy when needed to predict a known location to the model but unfortunately would predict poorly when it comes to never-seen locations because previous pollution values are not available for the model to predict accordingly. As for if Random Forest is used alone, the system would interpolate genuinely adequate when it comes to predicting new locations, which gives it an advantage over time series forecasting in this specific feature. It would also adequately predict known locations; however, as the dataset gets larger, Random Forest tends to overfit data with much noise. In this case, time series forecasting would be a more optimal option. Since we are interested in delivering a universal system, both approaches are used together in our system according to whether the edge is known to the model or not.

As for the thermal comfort and rider's effort, the neural network models' loss function values shown in Tables 3 and 4 show the discrepancy between the actual

Table 3 Thermal comfort and rider's effort models' training losses

Loss Function	Sequential NN	Transformer NN	Recurrent NN	LSTM
MSE	10413.23	1715.43	1699.86	51217868.00
MAPE	3165813504.00	12898707.00	49802436.00	180934082560.00

Table 4 Thermal comfort and rider's effort models training losses

Loss Function	Sequential NN	Transformer NN	Recurrent NN	LSTM
MSE	10300.00	1793.84	1698.80	48238888.00
MAPE	3159980288.00	13240838.00	49800916.00	192033521664.00
MAE	17.25	18.81	22.05	17.33

```
score = model.score(X_test, y_test)*100
score

99.1980565860829
```

Fig. 7 XGBoost accuracy

real-world path and the predicted path, concluding that these particular neural network models are unable to correctly find the correlation between the data points. The XGBoost model on the other hand is the only model that was able to identify the data correlations for the thermal comfort and rider's effort resulting in 99.19% accuracy shown in Fig. 7.

6 Conclusion

In this paper, the implementation of a multi-criterion routing system using machine learning has been discussed. It has aimed to further enhance a previously proposed system intended for the streets of Cairo. We have compared various machine learning models and techniques to examine which is the best fit for our specified criteria. The criteria predicted on each edge by the machine learning models are passed through an equation that weighs the criteria according to their corresponding weather elements on the requested day and hour to output the safest green path which is essentially the path having the lowest weights. Comparisons of the system before and after applying the machine learning layers show that the proposed machine learning system is indeed highly accurate and almost ten times faster, and the plan in advance feature makes it more user-satisfactory.

Google Maps API was used which provides only one Free API key with a limited number of extraction requests per API key, which resulted in limited data collection. For further improvement, additional API keys are required to gather data properly so that the models used in our route planner system can predict with higher accuracy.

The system proposed in this chapter paves the way for further enhancements: the rerouting algorithm could be applied to the system proposed to make it more interactive and updated according to the user's movements. In addition, an augmented reality feature could be implemented for better visualization and more safety for the rider because the rider would be viewing the path to follow along with the real-time events happening on the street to prevent having an accident due to looking at the screen and keeping eyes off the road.

References

1. Kwiatkowski, M.A.: Urban cycling as an indicator of socio-economic innovation and sustainable transport. Quaestiones Geographicae. **37**(4), 23–32 (2018)
2. Hoek, G., Brunekreef, B., Fischer, P., van Wijnen, J.: The association between air pollution and heart failure, arrhythmia, embolism, thrombosis, and other cardio- vascular causes of death in a time series study. Epidemiology, 355–357 (2001)
3. Cheng, Z., Luo, L., Wang, S., Wang, Y., Sharma, S., Shimadera, H., Wang, X., Bressi, M., de Miranda, R.M., Jiang, J., et al.: Status and characteristics of ambient pm2. 5 pollutions in global megacities. Environ. Int. **89**, 212–221 (2016)
4. Weltbank: Cairo Traffic Congestion Study: Executive Note (2014)

5. Khalil, Y., Zaky, M.O., El Hayani, M., Soubra, H.: Real life pollution measurement of cairo. In: ICORES, pp. 222–230 (2022)
6. Aboulyousr, Y., Azab, F., Soubra, H., Zaky, M.: Priority-based pollution management its. In: International Conference on Computing, Intelligence and Data Analytics, pp. 1–16. Springer (2022)
7. Zaky, M.O., Soubra, H.: An intelligent transportation system for air and noise pollution management in cities. In: VEHITS, pp. 333–340 (2021)
8. Cycling, W.: Walking can help reduce physical inactivity and air pollution, save lives and mitigate climate change (2022)
9. Elhusseiny, N., Sabry, M., Soubra, H.: B2x multiprotocol secure communication system for smart autonomous bikes. In: 2023 IEEE Conference on Power Electronics and Renewable Energy (CPERE), pp. 1–6. IEEE (2023)
10. Seoudi, M.S., Mesabah, I., Sabry, M., Soubra, H.: Virtual bike lanes for smart, safe, and green navigation. In: 2023 IEEE Conference on Power Electronics and Renewable Energy (CPERE), pp. 1–6. IEEE (2023)
11. Khalifa, H.H., Sabry, M., Soubra, H.: Visual path odometry for smart autonomous e-bikes. In: 2023 IEEE Conference on Power Electronics and Renewable Energy (CPERE), pp. 1–6. IEEE (2023)
12. Halim, C.E., Sabry, M., Soubra, H.: Smart bike automatic autonomy adaptation for rider assistance. In: 2023 IEEE Conference on Power Electronics and Renewable Energy (CPERE), pp. 1–6. IEEE (2023)
13. Shams, A., Sabry, M., Soubra, H., Salem, M.A.M.: Online reconfigurable convolu- tional neural network for real-time applications. In: 2022 18th International Com- puter Engineering Conference (ICENCO). vol. 1, pp. 31–36. IEEE (2022)
14. Sabry, N., Abobkr, M., ElHayani, M., Soubra, H.: A cyber-security prototype mod- ule for smart bikes. In: 2021 16th International Conference on Computer Engineer- ing and Systems (ICCES). pp. 1–5. IEEE (2021)
15. Abdelrahman, A., Youssef, R., ElHayani, M., Soubra, H.: B2x communication sys- tem for smart autonomous bikes. In: 2021 16th International Conference on Com- puter Engineering and Systems (ICCES). pp. 1–6. IEEE (2021)
16. Elnemr, M., Soubra, H., Sabry, M.: Smart autonomous bike hardware safety met- rics. In: International Conference on Computing, Intelligence and Data Analytics. pp. 132–146. Springer (2022)
17. Shah, M., Liu, T., Chauhan, S., Qi, L., Zhang, X.: Cyclesafe: Safe route planning for urban cyclists. In: Cloud Computing, Smart Grid and Innovative Frontiers in Telecommunications: 9th EAI International Conference, CloudComp 2019, and 4th EAI International Conference, SmartGIFT 2019, Beijing, China, December 4-5, 2019, and December 21-22, 2019 9. pp. 312–327. Springer (2020)
18. Valatka, L.: Recurrent neural network models for route planning in road networks (2018)
19. Said, A.M., Abd-Elrahman, E., Afifi, H.: A comparative study on machine learn- ing algorithms for green context-aware intelligent transportation systems. In: 2017 International Conference on Electrical and Computing Technologies and Applica- tions (ICECTA). pp. 1–5. IEEE (2017)
20. Czuba, P., Pierzchala, D.: Machine learning methods for solving vehicle routing problems. Proceedings of the 36th International Business Information Management Association (IBIMA), Granada, Spain pp. 4–5 (2021)
21. Bontempi, G., Ben Taieb, S., Le Borgne, Y.A.: Machine learning strategies for time series forecasting. Business Intelligence: Second European Summer School, eBISS 2012, Brussels, Belgium, July 15-21, 2012, Tutorial Lectures 2 pp. 62–77 (2013)
22. Brownlee, J.: Deep learning for time series forecasting: predict the future with MLPs, CNNs and LSTMs in Python. Machine Learning Mastery (2018)
23. Breiman, L.: Random forests. Machine learning 45, 5–32 (2001)
24. Morde, V., Setty, V.: Xgboost algorithm: long may she reign! towards data science; 2019 (2020)

Pre-Trained Variational Autoencoder Approaches for Generating 3D Objects from 2D Images

Zafer Serin, Uğur Yüzgeç, and Cihan Karakuzu

1 Introduction

Systems created by humans today are equipped with the ability to think, make decisions, generate ideas, and express emotions [1]. While some studies aim to classify and distinguish objects, others aim to develop freedom of choice and decision-making ability in the face of events and situations. This concept, first introduced in 1956 and known as "artificial intelligence," is gaining popularity and being applied in various fields [2]. Artificial intelligence is a multifaceted concept comprising various sub-fields, including machine learning, object recognition, natural language processing, autonomous systems, advanced data mining, and computational intelligence. Machine learning's objective is to decipher data and discover undiscovered relationships [3]. Although it draws on data analysis and feature extraction processes by humans, machine learning adopts a less human-dependent approach. Many learning algorithms are employed in the field of artificial intelligence. Some of these are decision trees, support vector machines, random forests, and Bayesian models. However, artificial neural networks are widely favored for their capability to learn from extensive datasets and manage complexity [4].

Artificial intelligence, machine learning, and deep learning fields have made significant strides in various sectors and are continuously being developed and adapted for previously untapped or challenging domains [5]. One such domain is imaging methods and computer-aided modeling, which have faced inherent

Z. Serin (✉)
Bilecik Seyh Edebali University, Pazaryeri Vocational School, Bilecik, Turkey
e-mail: zafer.serin@bilecik.edu.tr

U. Yüzgeç · C. Karakuzu
Bilecik Seyh Edebali University, Computer Engineering Department, Bilecik, Turkey
e-mail: ugur.yuzgec@bilecik.edu.tr; cihan.karakuzu@bilecik.edu.tr

M. N. Seyman (ed.), *2nd International Congress of Electrical and Computer Engineering*, EAI/Springer Innovations in Communication and Computing,
https://doi.org/10.1007/978-3-031-52760-9_7

challenges since their inception. Ongoing studies strive to overcome these difficulties. One of the most significant challenges is the difficulty of representing 3D objects on a 2D computer screen. Many software solutions have been developed to address this issue and are widely used today. With the gaming industry surpassing both the music and film sectors [6], the demand for 3D objects is also increasing. The demand for 3D objects is rapidly rising, not only in video games but also in fields including movies, virtual reality, augmented reality, architectural design, industrial design, and metaverses. As artificial intelligence advances and the need for 3D objects increases, it is becoming increasingly necessary to incorporate and tailor artificial intelligence and related technologies in the realm of 3D modeling.

The presentation of 3D objects in a computerized setting can be classified into three sub-categories: voxel, point cloud, and polygon [7]. Voxel, among them, can be viewed as an extension of a 3D pixel, i.e., a volumetric pixel. Simply, adding x and y-axis pixels in a 2D environment is equivalent to including a z-axis and thus converting a point with x and y coordinates into a cube. The point cloud approach involves presenting data obtained from sensors in a 3D environment using laser scanning, 3D cameras, and similar tools. Meanwhile, the polygon method represents object surfaces with triangles, which are 2D surfaces with three points in 3D space. These triangles are combined to form more complex surfaces. These techniques are all used to represent 3D objects. Voxel-based object representation has been employed in the present study due to its appropriateness for implementation in artificial neural networks.

With the rise in demand for 3D items and the advancement of artificial intelligence, certain models are being developed for 3D object generation. The ability to generate a 3D object from a single 2D image has various implications. Such an output can serve as an initial representation for those involved in 3D modeling, as well as being applicable in architectural modeling, video gaming, and movie production. In this study, the 3D-VAE-GAN models, which combine the GAN [8] and VAE [9] models, are employed with a voxel-based representation to address the problem. The VAE model's encoder part maps the input data to the latent space and can restore this mapped data to its initial state, preserving information about the original data. In a typical 3D-VAE-GAN model, the encoder can operate as a CNN model. In this study, we evaluate feature extraction and mapping to the latent space using pre-trained encoder networks. We compare different pre-trained networks (DenseNet121 [10], EfficientNetB0 [11], RegNet16 [12], and ResNet18 [13]) to analyze their performances. Additionally, we present the results obtained when using a standard encoder network without the fully connected layer and with the fully connected layer.

1.1 Related Works

In this subsection, some studies in the literature related to the subject of this study are discussed. 3D-GAN, developed in 2016 by Jiajun Wu and his team, is regarded as a crucial advancement in 3D object manufacturing. This artificial intelligence model employs both GAN and volumetric convolutional neural network structures to create entirely new objects from pre-existing datasets. Additionally, the novel 3D-VAE-GAN model is proposed for the creation of 3D objects using 2D images. Both models utilize voxels as their bases. Based on the evaluations conducted on the ModelNet [14] dataset, it is evident that these models deliver exceptional accuracy rates.

In 2017, Edward J. Smith and David Meger presented a study on the 3D-GAN model. Their work employed gradient penalization and the Wasserstein distance function with the objective of generating successful results across not only single classes but also multiclass generations. The direct object generation work is referred to as 3D-IWGAN, while 3D-VAE-IWGAN pertains to the generation of 3D objects from 2D images. These voxel-based models underwent testing on the IKEA dataset [15] and produced good precision outcomes [16].

In 2018, Jianwen Xie, Zilong Zheng, and colleagues proposed the 3D DescriptorNet, which is a model for generating 3D objects that includes a deep convolutional energy-based neural network. This model is based on voxelized data and uses a probability density function signal to operate. The probabilistic distribution is defined using Markov Chain Monte Carlo, and the 3D object pattern is then derived from this probabilistic distribution. It is a versatile approach to addressing issues such as object recovery and object resolution enhancement. The approach was tested on reconstruction and classification problems using the ModelNet10 dataset, yielding successful outcomes [17].

In 2019, Hui Huang, Zhijie Wu, et al. introduced SAGNet, a model that can learn the relationships between objects and their parts, as well as the connections between parts of different objects, in a structure-sensitive manner. The model incorporates an autoencoder that can learn the geometry and structure of objects and be evaluated using a voxel-based approach with a set of semantically meaningful parts. The reconstruction results obtained by the model were highly successful [18].

In 2020, Yanran Guan, Tansin Jahan, and Oliver van Kaick proposed a 3D-GAE model for generating 3D objects using the generalized autoencoder (GAE) technique. This model is characterized by fast training and high generalization capability. 3D-GAE employs a combination of a standard autoencoder loss function and a unique loss function. It calculates the generated objects' similarity using the Chamfer Distance method and quantifies successful outcomes. The tests on the COSEG [19–21] and ModelNet 40 datasets yielded an effective baseline [22].

In 2020, Rundi Wu, Yixin Zhuang, and colleagues put forward PQ-NET for 3D object generation. This model is founded on the sequential generation of object parts to create an object. This method is akin to constructing a sentence, where, just as the position of words is determined during sentence construction, a specific sequence must be followed and produced sequentially when generating objects. Based on a part-based analysis, PQ-NET employs PartNet, a variant of the ShapeNet dataset with semantically separated objects. PQ-NET, featuring a voxel-based approach, has undergone assessment under the IoU metric for reconstruction and has attained favorable outcomes [23].

2 3D-VAE-GAN Model

In this study, the 3D-VAE-GAN models were utilized to generate 3D objects from 2D images. The voxel object representation method was chosen due to its compatibility with deep learning models, fast processing, boundary recognition, and direct usage with raw data, making it particularly advantageous [24]. The 3D-VAE-GAN architecture combines VAE and GAN models. Standard autoencoder models comprise encoder and decoder neural networks. The encoder network maps the provided input to a latent space of a specific size, while the decoder network aims to replicate the original input by retrieving data from this latent space. While autoencoders are primarily used for operations such as data compression, VAE models are employed for data generation. The main contrast between these models, which are highly alike, is that VAE utilizes a certain probability distribution for encoding data.

On the contrary, GAN is a method comprising two neural networks, a generator and a discriminator neural network, and is based on competitive learning. The generator network in the GAN models is fed input either randomly or adhering to certain norms and attempts to generate an output with specific characteristics and size. The generated output is inputted to the discriminator network, which is another part of GAN and has received prior training with real and fake data. Its task is to distinguish (classify) whether the data is real or fake. As the model's training progresses, the generator network produces more precise data, and the discriminator network becomes adept at distinguishing the generated data more accurately. Essentially, the 3D-VAE-GAN model consists of three neural networks. These are encoder, generator, and discriminator neural network models. Instead of the decoder component in the VAE model, the GAN model incorporates the generator component. The encoder part of the VAE model takes $224 \times 224 \times 3$ images as input, which are then converted into latent space vectors consisting of 200 elements. The images utilized in this study were obtained from the ShapeNet [25] image dataset and featured chairs as objects of interest. Some of these images are illustrated in Fig. 1.

These vectors, which have been mapped to the latent space, are utilized for input into the generator network with the objective of producing the 3D object in the image. The generated objects are subsequently subjected to the classification process through the discriminator network, and the performance of the model is assessed by calculating the average loss values for all three networks. Kullback–Leibler divergence is employed in this study to measure the reconstruction error. Figure 2 exhibits a standard 3D-VAE-GAN model.

The generator used includes five layers. The first four layers comprise transpose convolution, batch normalization, and ReLU layers, whereas the last layer involves a sigmoid layer directly following the transpose convolution process. The kernel size was set to 4, and a stride of 2 was used throughout all layers, with (1,1,1,1) being used for the padding value. 3D transpose convolution and batch normalization operations were used in compliance with three dimensions. A $32 \times 32 \times 32$ environment was used for the 3D depiction of voxel objects.

Fig. 1 2D sample object images from the ShapeNet dataset

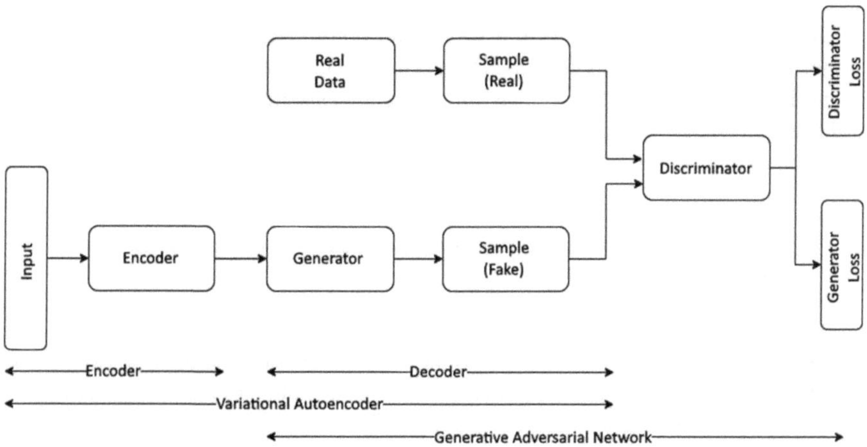

Fig. 2 3D-VAE-GAN model

The discriminator used comprises five layers. The first four layers include convolution, batch normalization, and LeakyReLU layers, while the final layer has a sigmoid layer immediately following the convolution process. For all layers, a kernel size of 4, a stride of 2, and (1,1,1,1) as the padding value were used. Convolution and batch normalization techniques were implemented in line with 3D. A $32 \times 32 \times 32$ environment was employed to create the 3D representation of voxel objects.

3 Pre-trained Network Models

Pre-trained networks can be defined as networks that have undergone pre-training on specific problems and already possess prior knowledge. Generally, these networks are pre-trained with substantial data corresponding to the problems they will address. They have the capability to incorporate the knowledge gained from previous data (known as transfer learning) while adapting to new tasks and problems. In essence, they involve two distinct training phases. Firstly, pre-trained networks undergo initial training on large datasets, followed by a second stage of training on specific, personalized problems. As pre-trained networks already possess knowledge of the problem, they achieve superior performance compared to networks trained from scratch. Additionally, pre-trained networks may possess a basic structure specific to the problem they are employed for. For instance, pre-trained networks employed for image classification may essentially involve a CNN model and units, whereas a pre-trained transformer is used in natural language processing.

DenseNet121, RegNet16, ResNet18, and EfficientNetB0 are among the pre-trained networks utilized in this study. Structural modifications were implemented in the pre-trained networks. Typically, the fully connected layers located at the end of the pre-trained networks were excised, and optimization was performed on the issue. The rationale for this customization is that such networks are generally tailored to particular problems, such as binary or multiclass classification, and output suitable values.

3.1 Densely Connected Convolutional Neural Networks 121 (DenseNet121)

DenseNet121 is a pre-trained convolutional neural network architecture extensively employed in computer vision tasks such as image classification and object recognition. It is a part of the DenseNet model family, which is comprised of "Densely Connected Convolutional Networks." DenseNet121, in particular, comprises 121 layers in its architecture. The dense connectivity pattern of DenseNet effectively resolves the vanishing gradient issue in deep neural networks. In this architecture, each layer is linked to every other layer in a feed-forward method. The dense connectivity enables enhanced feature reuse and information flow between layers, allowing for more effective training of deeper networks while using fewer parameters in comparison to conventional CNN structures such as VGG or ResNet. In Fig. 3, the structure of a standard DenseNet121 is shown.

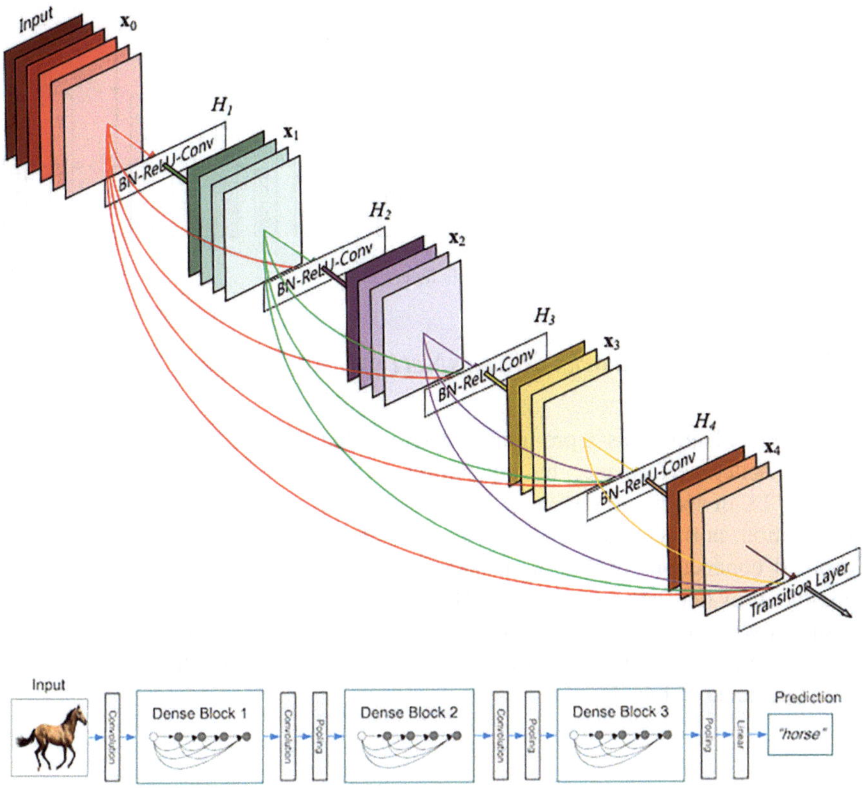

Fig. 3 Structure of a standard DenseNet121 pre-trained network [10]

3.2 *Regular Networks 16 (RegNet16)*

Regular network models optimize the structure of deep learning layers and computational resources. They are pre-trained to focus on two separate hyperparameters: the network's width, which is the number of filters (or neurons) in each layer, and the network's depth. Wider networks may yield better results, but with increasing width come higher computational costs and memory requirements. Network depth is determined by the number of layers in the network, and as depth increases, the learning capacity of the network may also increase, but at the expense of greater computational and memory demands. RegNet provides various approaches to optimizing network width and depth. When constructing RegNet models, various configurations of network width and depth are employed to obtain optimal values. As a result of these aspects, RegNet models can possess attributes such as minimal computational cost, low memory demand, and improved generalization.

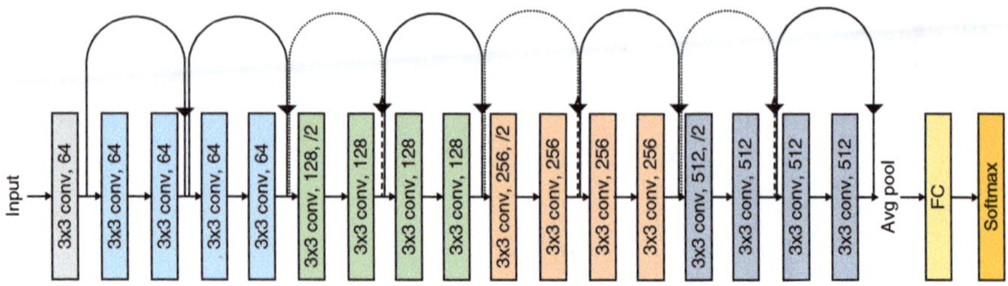

Fig. 4 Diagram of the ResNet18 model and its layers [13]

3.3 Residual Networks 18 (ResNet18)

It has been identified as a pre-trained network that facilitates deeper networks to achieve better performance by integrating the notion of residual learning. The number 18 in its name denotes the number of layers. These layers encompass convolution, normalization, and activation layers, along with residual connections. Residual connections enable input data to be directly added to the output when moving from one layer to another. This facilitates the backward propagation of gradients with greater ease and efficiency. Additionally, the network can be made deeper, thereby mitigating the risk of overfitting. The standard ResNet18 architecture is depicted in Fig. 4.

3.4 Efficient Networks B0 (EfficientNetB0)

This model, like others presented in this study, belongs to the deep learning family. The structure has designations such as B0, B1, and B2, with the number of layers typically expanding in line with the problem's complexity. Each of these adjustments is an optimized CNN model, aiming to deliver superior efficiency and performance. The primary objectives of such models are to attain heightened performance levels while reducing the number of parameters relative to larger and more intricate models. This is vital for systems with restricted computational resources and is specifically tailored for use on mobile devices. EfficientNet architectures are scalable, with Swish being the preferred activation function over ReLU in EfficientNetB0 due to its superior efficiency. While ReLU returns the input value of x if it is positive and 0 if it is negative, Swish returns a value in the format of x multiplied by sigmoid (x). Since weight sharing is employed in EfficientNetB0, it results in a reduction in the total amount of weight being used within the network. This approach scales the size and resources of every layer as per the size per channel, enabling every layer to function more effectively. Figure 5 typically depicts the EfficientNetB0 model.

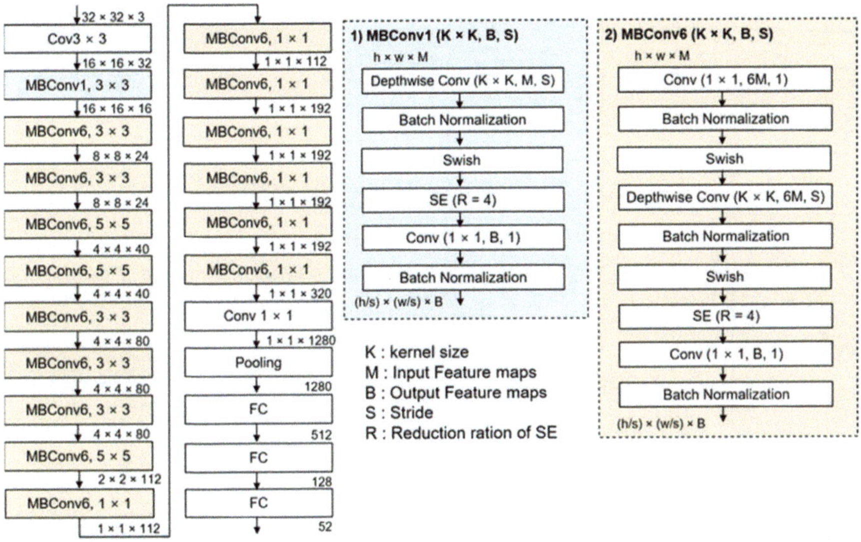

Fig. 5 The structure of the EfficientNetB0 model and the units it contains [26]

4 Results

In this study, the soft labeling method was employed to label both real and fake data. For labeling real data, a uniform distribution ranging from 0.7 to 1.2 was used, whereas a uniform distribution ranging from 0 to 0.3 was employed for fake data. The discriminator network learned at a faster pace compared to the generator network, and the training of the discriminator was halted when its accuracy surpassed 0.8 to prevent competitive behavior. The generator network and discriminator network utilized binary cross entropy as their loss function, while the encoder network utilized Kullback–Leibler divergence to gauge the likeness between the generated and real objects. The model's input consisted of 224 × 224 × 3-size images, which were converted to a 200-element latent space via the encoder network. Using the mapped data as input, the generator network created 32 × 32 × 32 3D voxel objects. The discriminator network determined whether these objects were real or fake, engaging in continuous competition in order to achieve optimization. The encoder network mapped 224 × 224 × 3 images to a 200-element latent space using pre-trained networks, which were then compared against each other. The study employed the Adam function and implemented a learning rate of 0.0025, 0.0001, and 0.001 for the generator, encoder, and discriminator networks, respectively, with a beta value of 0.5. A batch size of 32 was used. The training and evaluations were carried out on an i7-4790 CPU, GTX 980 Ti GPU, and 16 GB of RAM. The study utilized the PyTorch library and CUDA technologies [27].

Figure 6 shows the generator network loss, discriminator network accuracy and loss, Kullback–Leibler divergence, and total iteration time obtained as a result of

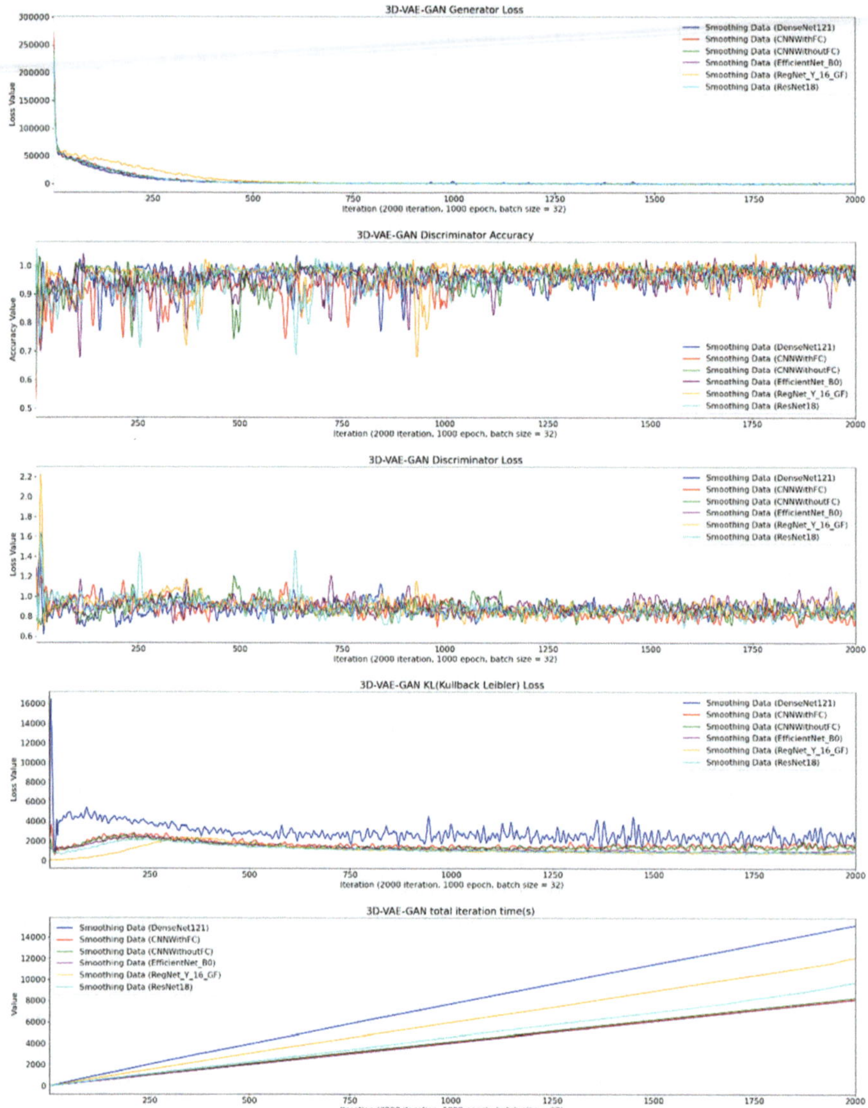

Fig. 6 Effect of pre-trained networks on performance metrics in encoder networks

training the encoder network with pre-trained networks. According to the obtained results, the loss of the generator network swiftly converges to zero. It is evident that there is a distinct disruption in the generator network loss for the RegNet16 model, unlike the other networks, up to 500 iterations. The analysis of discriminator network accuracy and loss values reveals similar outcomes. Upon examining the running times of the algorithms, it is noticeable that DenseNet121 operates the slowest, followed by RegNet16 and then ResNet18. EfficientNetB0, the CNN with and without a fully connected layer, was faster than the other models and had similar running times among themselves. When assessing the Kullback–Leibler divergence value, which measures the similarity between the generated and real objects, it is evident that DenseNet121 and a CNN that lacks a fully connected layer obtain relatively lower results compared to other algorithms. ResNet18, RegNet16, and EfficientNetB0 achieve similar values. It is seen that the models start with a low Kullback–Leibler divergence value, which increases over time and then decreases steadily. This may be due to reasons such as the complexity of the model, learning speed, number of iterations, and data distribution.

Table 1 shows the changes in Kullback–Leibler divergence values of the models used throughout the epochs. To reach a more precise conclusion, Table 2 is given, which contains the mean, maximum, minimum, and standard deviations of the Kullback–Leibler values at the end of 1000 epochs. As can be seen in Table 2, RegNet16 gave the lowest value on the mean. Based on the mean values, RegNet16, ResNet18, and EfficientNetB0 can be ranked, respectively. However, if Fig. 6 is examined again, it will be seen that the order is exactly the opposite in terms of iteration time.

In this instance, if improved results for the problem are desired and time is not important to us, we should choose RegNet16; if time is very important to us and we want acceptable results, we should choose EfficientNetB0; and if we want an above-average value for both time and performance, we should choose ResNet18. It is advantageous to use these three pre-trained networks due to their superior performance as compared to other networks. As RegNet16 is considered superior in terms of performance, Fig. 7 displays the test results of the 3D-VAE-GAN model with the RegNet16 encoder. The upper part presents the 2D images provided as input, while the lower part showcases the 3D object outputs of 3D-VAE-GAN with RegNet16 as voxels.

5 Conclusions

Based on the findings of the study, an investigation into the application of 3D-VAE-GAN models in 3D object generation from 2D images was conducted. Specifically, the examination focused on utilizing several pre-trained convolutional neural networks (CNNs) as potential encoder networks in the VAE component. The examined pre-trained network models are DenseNet121, EfficientNetB0, RegNet16, and ResNet18, a standard CNN model with fully connected layers and a CNN model without fully connected layers. Feature extraction was executed within pre-trained network models, and the extracted features were then assigned to the latent space. The generator and discriminator networks of the GAN model were trained using the binary cross-entropy loss function, while the encoder network of the VAE model was trained using the Kullback–Leibler divergence to enhance the training process. Importantly, our methodology differs from traditional VAE models as we replace the

Table 1 Changes in the Kullback–Leibler divergence value during epochs

Epoch	Model name	Kullback–Leibler divergence value
1	DenseNet121	3215.43
	ResNet18	1662.01
	RegNet16	126.45
	EfficientNetB0	1367.98
	CNN with fully connected	924.78
	CNN without fully connected	2287.20
200	DenseNet121	2747.81
	ResNet18	1701.45
	RegNet16	2318.09
	EfficientNetB0	1936.44
	CNN with fully connected	2680.55
	CNN without fully connected	1959.02
400	DenseNet121	2403.31
	ResNet18	1253.69
	RegNet16	1106.81
	EfficientNetB0	1430.30
	CNN with fully connected	1536.77
	CNN without fully connected	1255.60
600	DenseNet121	2035.97
	ResNet18	1044.08
	RegNet16	986.19
	EfficientNetB0	1089.31
	CNN with fully connected	1561.37
	CNN without fully connected	1726.35
800	DenseNet121	3638.48
	ResNet18	950.78
	RegNet16	841.22
	EfficientNetB0	977.22
	CNN with fully connected	1766.02
	CNN without fully connected	1295.69
1000	DenseNet121	2534.85
	ResNet18	903.51
	RegNet16	909.86
	EfficientNetB0	924.78
	CNN with fully connected	1588.52
	CNN without fully connected	1772.49

decoder part of the VAE model with the generator network of the GAN model. We illustrate the duration of each iteration of the various algorithms and methods, along with the overall iteration time, using visual representations. Our focus during the training and testing phases is directed toward the category of "chair" contained within the ShapeNet dataset. As for the representation of 3D objects, we opted for a

Table 2 Mean, maximum, minimum, and standard deviation values of the Kullback–Leibler divergence value over 1000 epochs

Model name	Mean	Maximum	Minimum	Std. Dev.
DenseNet121	2893.807	30179.875	1079.060	1261.083
ResNet18	1219.067	2219.813	624.192	359.986
RegNet16	1129.660	2429.420	110.130	478.198
EfficientNetB0	1352.815	2579.661	876.164	424.084
CNN with fully connected	1696.749	4437.990	552.489	447.603
CNN without fully connected	1538.489	3579.641	679.022	396.773

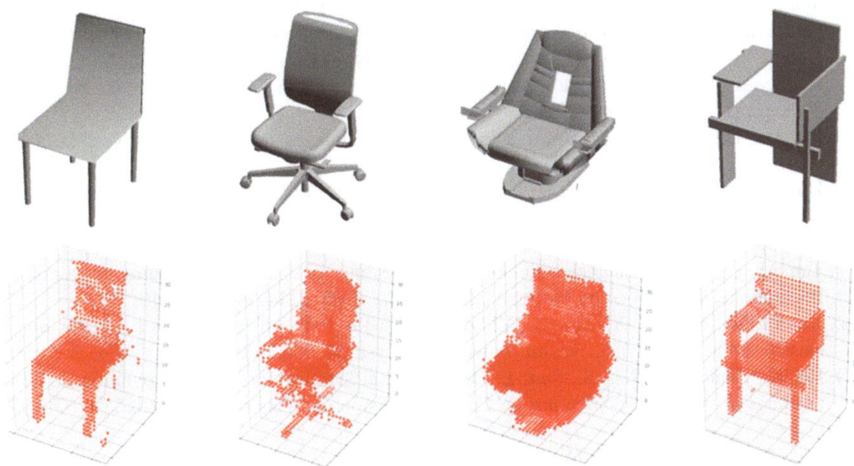

Fig. 7 Test results of the 3D-VAE-GAN model with the RegNet16 encoder

voxel-based approach, which is well-suited for neural networks and deep learning processes. An indirect labeling technique is preferred over a direct one in determining an object's veracity. To ensure training stability, we impose constraints on the parser network to match the speed of the generator network. Our methodology uses a single 2D image as an input to create 3D objects. Based on our test outcomes, the pre-trained RegNet16 network exceeds other techniques, exhibiting exceptional efficiency in our setting.

The study indicates that utilizing a pre-trained network for the encoder network produces positive results. Future studies could investigate intervention not only in the encoder network but also in the GAN model's component. The generator and discriminator networks within the GAN structure may produce positive effects that, when combined with the findings of this study, enable the creation of successful models in both the VAE and GAN components. This may result in the production of overall more effective 3D-VAE-GAN models. Further research could explore diverse objects beyond chairs, as well as experimenting with point cloud and polygon approaches instead of voxel techniques. The use of multiple and cleaner images as input may also enhance performance.

References

1. Tai, M.C.-T.: The impact of artificial intelligence on human society and bioethics. Tzu-Chi Med. J. **32**, 339 (2020)
2. Moor, J.: The Dartmouth College artificial intelligence conference: the next fifty years. AI Mag. **27**, 87–87 (2006)
3. Russell, S.J.: Artificial intelligence a modern approach. Pearson Education, Inc., London (2010)
4. Murphy, K.P.: Machine learning: a probabilistic perspective. MIT Press, Cambridge (2012)
5. Ongsulee, P.: Artificial intelligence, machine learning and deep learning. In: 2017 15th International Conference on ICT and Knowledge Engineering (ICT&KE), pp. 1–6. IEEE (2017)
6. Seo, Y., Dolan, R., Buchanan-Oliver, M.: Playing games: advancing research on online and mobile gaming consumption. Internet Res. **29**, 289–292 (2019)
7. Funkhouser, T.: Overview of 3d object representations. Princeton University, Princeton (2003)
8. Goodfellow, I., Pouget-Abadie, J., Mirza, M., Xu, B., Warde-Farley, D., Ozair, S., Courville, A., Bengio, Y.: Generative adversarial nets. In: Advances in Neural Information Processing Systems, p. 27, New York (2014)
9. Kingma, D.P., Welling, M.: Auto-encoding variational bayes. arXiv preprint arXiv:1312.6114 (2013)
10. Huang, G., Liu, Z., Van Der Maaten, L., Weinberger, K.Q.: Densely connected convolutional networks. In: Proceedings of the IEEE Conference on Computer Vision and Pattern Recognition, pp. 4700–4708 (2017)
11. Tan, M., Le, Q.: Efficientnet: rethinking model scaling for convolutional neural networks. In: International Conference on Machine Learning, pp. 6105–6114. PMLR (2019)
12. Xu, J., Pan, Y., Pan, X., Hoi, S., Yi, Z., Xu, Z.: RegNet: self-regulated network for image classification. IEEE Trans. Neural Netw. Learn. Syst. **34**, 9562 (2022)
13. He, K., Zhang, X., Ren, S., Sun, J.: Deep residual learning for image recognition. In: Proceedings of the IEEE Conference on Computer Vision and Pattern Recognition, pp. 770–778 (2016)
14. Wu, Z., Song, S., Khosla, A., Yu, F., Zhang, L., Tang, X., Xiao, J.: 3d shapenets: a deep representation for volumetric shapes. In: Proceedings of the IEEE Conference on Computer Vision and Pattern Recognition, pp. 1912–1920 (2015)
15. Lim, J.J., Pirsiavash, H., Torralba, A.: Parsing ikea objects: fine pose estimation. In: Proceedings of the IEEE International Conference on Computer Vision, pp. 2992–2999 (2013)
16. Wu, J., Zhang, C., Xue, T., Freeman, B., Tenenbaum, J.: Learning a probabilistic latent space of object shapes via 3d generative-adversarial modeling. In: Advances in Neural Information Processing Systems, p. 29. MIT, Cambridge (2016)
17. Xie, J., Zheng, Z., Gao, R., Wang, W., Zhu, S.-C., Wu, Y.N.: Learning descriptor networks for 3d shape synthesis and analysis. In: Proceedings of the IEEE Conference on Computer Vision and Pattern Recognition, pp. 8629–8638 (2018)
18. Wu, Z., Wang, X., Lin, D., Lischinski, D., Cohen-Or, D., Huang, H.: Sagnet: structure-aware generative network for 3d-shape modeling. ACM Trans. Graph. **38**, 1–14 (2019)
19. Wang, Y., Asafi, S., Van Kaick, O., Zhang, H., Cohen-Or, D., Chen, B.: Active co-analysis of a set of shapes. ACM Trans. Graph. **31**, 1–10 (2012)
20. Sidi, O., Van Kaick, O., Kleiman, Y., Zhang, H., Cohen-Or, D.: Unsupervised co-segmentation of a set of shapes via descriptor-space spectral clustering. In: Proceedings of the 2011 SIGGRAPH Asia Conference, pp. 1–10 (2011)
21. Van Kaick, O., Tagliasacchi, A., Sidi, O., Zhang, H., Cohen-Or, D., Wolf, L., Hamarneh, G.: Prior knowledge for part correspondence. In: Computer Graphics Forum, pp. 553–562. Wiley Online Library, Hoboken (2011)
22. Guan, Y., Jahan, T., van Kaick, O.: Generalized autoencoder for volumetric shape generation. In: Proceedings of the IEEE/CVF Conference on Computer Vision and Pattern Recognition Workshops, pp. 268–269 (2020)

23. Wu, R., Zhuang, Y., Xu, K., Zhang, H., Chen, B.: Pq-net: a generative part seq2seq network for 3d shapes. In: Proceedings of the IEEE/CVF Conference on Computer Vision and Pattern Recognition, pp. 829–838 (2020)
24. Deng, J., Shi, S., Li, P., Zhou, W., Zhang, Y., Li, H.: Voxel r-cnn: towards high performance voxel-based 3d object detection. In: Proceedings of the AAAI Conference on Artificial Intelligence, pp. 1201–1209 (2021)
25. Chang, A.X., Funkhouser, T., Guibas, L., Hanrahan, P., Huang, Q., Li, Z., Savarese, S., Savva, M., Song, S., Su, H.: Shapenet: an information-rich 3d model repository. arXiv preprint arXiv:1512.03012. (2015)
26. Gang, S., Fabrice, N., Chung, D., Lee, J.: Character recognition of components mounted on printed circuit board using deep learning. Sensors. **21**, 2921 (2021)
27. Nickolls, J., Buck, I., Garland, M., Skadron, K.: Scalable parallel programming with cuda: is cuda the parallel programming model that application developers have been waiting for? Queue. **6**, 40–53 (2008)

Enhancing Ground Vehicle Route Planning with Multi-Drone Integration

Murat Bakirci and Muhammed Mirac Özer

1 Introduction

With the rapid advancement of technology, unmanned aerial vehicles (UAVs) have assumed pivotal roles across various domains [1]. These versatile aircraft find utility in an array of applications, spanning from military operations [2] to energy infrastructure management [3], agriculture [4], and disaster response [5]. Their potential for enhancing efficiency and effectiveness is continually under exploration across diverse sectors [6]. While UAVs were initially deployed primarily for military purposes like intelligence gathering, border surveillance, and enemy tracking, their sphere of influence has extended significantly into civilian domains [7–10]. In the energy sector, UAVs undertake vital tasks, including fault detection and gas measurements, contributing to the safety and security of energy infrastructure [11]. In agriculture, UAVs offer invaluable support to farmers by aiding in tasks such as monitoring agricultural regions, analyzing productivity, and detecting diseases, leveraging their data collection capabilities [12]. Furthermore, UAVs are making significant contributions to fields like cartography [13], archaeological site documentation [14], and forestry management [15]. Their ability to provide high-resolution images and data leads to faster and more precise outcomes [16, 17]. UAVs have proven to be instrumental in emergency response and search-and-rescue operations [18], offering accessibility to otherwise hard-to-reach areas and thus holding the potential to save human lives.

The two-stage iterative solution approach introduced in this study for addressing the simultaneous disaster response challenge involving ground vehicles and UAVs

M. Bakirci · M. M. Özer (✉)
Unmanned/Intelligent Systems Lab, Faculty of Aeronautics and Astronautics, Tarsus University, Mersin, Turkey
e-mail: muratbakirci@tarsus.edu.tr; muhammed_mirac@tarsus.edu.tr

M. N. Seyman (ed.), *2nd International Congress of Electrical and Computer Engineering*, EAI/Springer Innovations in Communication and Computing,
https://doi.org/10.1007/978-3-031-52760-9_8

distinguishes itself from prior methods by its capacity to handle larger-scale scenarios and its adaptability to multiple UAV scenarios. Its most significant contribution lies in addressing a previously overlooked aspect of the field: multiple UAV scenarios. The primary objective of this extended approach is to minimize intervention times in disaster response scenarios. Additionally, the creation of a hybrid genetic algorithm, incorporating machine learning and function prediction, presents a novel perspective for solving the concurrent intervention problem concerning ground vehicles and UAVs. This algorithm's flexibility further enhances its utility in managing multiple UAV scenarios. Offering a pioneering contribution to the literature, this definitive solution approach for the simultaneous intervention of ground vehicles and multiple UAVs in disaster-stricken areas is the first to prioritize intervention time reduction. Its development not only holds the potential to apply the problem to more realistic scenarios but also seeks to address a literature gap and lay the groundwork for future studies in the field.

2 Related Work

The Vehicle Routing Problem (VRP) has been extensively studied over time, giving rise to a wide array of variants in the literature [19, 20]. This problem has evolved considering multiple factors, including the number of vehicles, vehicle diversity, capacity, customer service hours, service types, and number of warehouses, and has embraced new dimensions with advancements in technology [21–23]. The ongoing technological progress in UAV systems has prompted a fresh perspective on routing challenges resulting from the collaboration between ground vehicles and UAVs [24]. This perspective has led to the emergence of the "Drone Assisted Vehicle Routing Problem," which primarily hinges on the service capabilities of UAVs [25]. In this scenario, where both types of vehicles possess the capacity to serve customers, it is essential to efficiently determine which customer is served by which vehicle. The response process in this problem involves the simultaneous operation of both ground vehicles and UAVs, emphasizing the paramount importance of coordinating these two types of tools. This collaborative operation of ground vehicles and UAVs is being explored to enhance the efficiency of disaster response processes. Consequently, there is a growing body of research, particularly in the "Traveling Salesman Problem with Drone" category, addressing the Drone Assisted Vehicle Routing Problem in the literature [26, 27]. These studies aim to develop effective solution approaches that ensure the seamless coordination of these tools, thus holding significance in both academic and industrial contexts.

Reference [28] explores studies centered on the utilization of UAVs for environmental surveillance and cartography. It discusses the optimization of UAV operations for these tasks through Informative Path Planning (IPP), Route Planning (PR), and Autonomous Exploration of Environments. The aim of this comprehensive review is to elucidate the challenges and objectives in this domain, shedding light on current approaches. It serves as a valuable resource for those interested in the field of data collection and monitoring. In [29], attention is directed toward the parallel drone planning Traveling Salesman Problem, a crucial component of UAV applications in urban logistics. This problem aims to minimize delivery times by distributing deliveries between vehicles and drones. The study introduces a two-step heuristic solution, addressing the client string as a two-criteria shortest path problem and employing one coding step followed by dynamic programming. Reference [30] explores the amalgamation of trucks and drones for cargo delivery. Specifically, it focuses on the "Traveling Salesman Problem with m Drones" (TSP-mD), where a truck collaborates with multiple drones rather than a single one. The study adapts and compares the Adaptive Large Neighborhood Search (ALNS) heuristic with the Greedy Randomized Adaptive Search Procedure (GRASP) to address this challenge. Experimental results demonstrate that the combination of a truck with multiple drones can yield more efficient solutions with GRASP, while ALNS outperforms GRASP in this context. Differing from [29], this research ventures into more complex and large-scale problem-solving techniques, incorporating hybrid genetic algorithms and machine learning, offering an advantage in addressing larger and more intricate logistics challenges.

These investigations delve into the integration of UAVs with carrier vehicles and explore how this collaboration can effectively address routing problems. This novel perspective underscores the adaptability of routing problems to complex, real-world scenarios and emphasizes the value it brings to the table. This study introduces a novel approach that employs a hybrid genetic algorithm enriched by machine learning and function estimation to optimize the utilization of UAVs in ground vehicle route planning for disaster response missions. This approach stands out as a unique contribution to the existing literature. It harmonizes exact and heuristic solution methods, resulting in a substantial reduction in disaster response times. This work exhibits remarkable potential, particularly in the context of larger and more intricate intervention challenges.

3 Methodology

The main objective of the investigation into the benefits and significance of methodologies for solving the Vehicle Routing Problem with Drones (VRPD) was to develop strategies for situations that involve the coordinated operation of a ground vehicle with a single drone, following a two-stage problem-solving approach. However, as these methods have proven to be primarily effective for small-scale issues, the need to incorporate heuristic techniques becomes apparent when dealing with medium- and large-scale challenges. Consequently, the comprehensive solution devised in the initial phase encompasses the simultaneous deployment of ground vehicles with multiple drones, creating a more intricate scenario. This phase showcases the optimization of both drones and carrier vehicles and contributes to exploring more realistic scenarios that capture the complexities of response processes. In the second stage, the hybrid genetic algorithm approach is adapted to scenarios involving multiple drones. This step underscores the versatility of the hybrid genetic algorithm and introduces the analysis of previously unexplored scenarios.

The Vehicle Routing Problem with Multi-Drone (VRP-mD) represents a highly practical scenario, emphasizing the capacity of response ground vehicles to carry and operate multiple drones simultaneously. An operational assessment of this scenario underscores the necessity to place the problem within a more realistic context. The study primarily centers around VRP-mD, delving into the simultaneous operation of multiple drones. In the context of VRP-mD, the vehicle fleet comprises two distinct vehicle types, ground vehicles and drones, with the total fleet size denoted as $a + 1$. In this context, a signifies the maximum number of drones that ground vehicles can transport simultaneously. VRP-mD encapsulates a scenario in which drones have the flexibility to operate either jointly with ground vehicles or autonomously. It is essential for drones to recharge their batteries before each flight and prepare for the next intervention. While a separated ground vehicle can attend to one or more disaster areas in this timeframe, other drones accompanying the ground vehicle may also detach concurrently.

Essentially, an iterative exact solution algorithm initially developed for VRPD with drones has been adapted for solving a complex response problem, where a single ground vehicle operates concurrently with multiple drones. A mathematical model has been formulated to optimize the routing of multiple drones alongside the ground vehicle. In the multi-drone scenario, the objective function seeks to minimize the arrival time of the last vehicle reaching the ground station. When the ground vehicle intends to depart from a node, it must await the arrival of all drones that have reached the same node for rendezvous. Consequently, the objective function has been articulated as depicted in Eq. (1) to minimize the earliest time required for the ground vehicle's departure from the ground station.

$$\min\left[t\max_{b+1}^{0}\right] \tag{1}$$

where b represents the number of intervention zones within the response network. For the tour assignments in the problem to be considered feasible, certain conditions must be satisfied. The first of these conditions is presented in Eq. (2).

$$\sum_{e=1}^{a}\sum_{z\in C_r} l_{xz} n_z^e = 1, \quad x \in C_r \tag{2}$$

In this context, the vehicle index is denoted as e, the drone tour index as z, and the iteration index as r. The set of intervention regions assigned to the drone in iteration number r is represented as C_r. Additionally, l signifies the condition that in tour r, if drone x intervenes in the disaster area, it is assigned the value 1; otherwise, it is 0. Similarly, n_z^e represents the condition that, in tour z, if drone e is assigned, it is 1; if not assigned, it is 0. As Eq. (3) demonstrates, in the initial stage, each disaster area can only be assigned to one drone. Additionally, any node in the ground vehicle's route can be chosen only once as the starting point for each drone tour. These constraints govern drone assignments and node selections, ensuring that each disaster response zone is assigned to a single drone, and each drone tour originates at a unique node.

$$\sum_{z\in C_r} k_{xz} n_z^e \leq 1, \quad x \in R \cup \{0\} \backslash C_r, e = 1, 2, \ldots, a \tag{3}$$

In this context, we represent the set of intervention regions as R. Additionally, k_{xz} indicates whether, in tour e, a drone's route starts with node x (assigned value of 1) or not (assigned value of 0). It is worth noting that the same node can serve as the starting, ending, or intermediate point for the routes of multiple drones. Conversely, each node along the ground vehicle's route can be chosen as the termination point for at most one drone's tour. These constraints govern the paths drones follow between nodes and their allocation. Consequently, these rules ensure that each node can fulfill only one role within a specific drone tour, as illustrated in Eq. (4).

$$\sum_{z\in C_r} m_{xz} n_z^e \leq 1, \quad x \in R \cup \{b+1\} \backslash C_r, e = 1, 2, \ldots, a \tag{4}$$

where m_{xz} is used to indicate whether, in tour e, the route concludes at node x (assigned value of 1) or not (assigned value of 0). Another vital constraint, as defined in Eq. (5), is imposed to ensure viable tour assignments. This constraint restricts each drone from commencing a new tour before finishing an ongoing one. For instance, if a drone tour begins at location y and ends at location x, it is not permissible to select another drone tour that initiates or concludes at any node z situated between these two positions. This regulation governs the assignment of drone tours and prevents infeasible assignments by ensuring that each drone cannot enter another tour without completing its ongoing one. This significantly enhances the logical and effective development of the intervention plan.

$$\begin{gathered}\sum_{z\in C_r} k_{c_q z} n_z^e + \sum_{z\in C_r} m_{c_q z} n_z^e \leq 2\left(1 - \sum_{z\in C_r} k_{c_x z} m_{c_y z} n_z^e\right), \\ x = 0, 1, \ldots, p_r - 1, \ \ y = x + 2, \ldots, p_r + 1, \ \ e = 1, 2, \ldots, a, \ \ q \in \underset{J}{\longrightarrow}, \\ \mathbf{J} = \{j | x < j < y\}, \ x \neq y\end{gathered} \tag{5}$$

In this equation, q represents the intervention zone on the ground vehicle's route. c_q designates the position to which intervention zone q is assigned in iteration r. p_r signifies the number of intervention zones assigned to the ground vehicle's route in that iteration. The hidden layer output matrix is denoted as J, and j stands for the number of nodes in the hidden layer. Equation (6a) is used to calculate the arrival time for the ground vehicle when it moves from a node at position x to the next node while on its route y. Conversely, Eq. (6b) computes the arrival time for the ground vehicle when it departs from a node at position x and progresses to the subsequent node within route y. During these calculations, a service time called d_s is taken into account if any flight has commenced at a node located at position x. The service times encompass the time necessary for the drone to equip itself for disaster response before initiating a new flight and for replacing the drone's battery upon completing the flight. If a ground vehicle is assigned a tour that starts at a node located at position x and concludes at a node situated at position y, the arrival time at position y is determined by adding the flight time to the earliest time the ground vehicle can depart from position x. These calculations are instrumental in coordinating the response process's timing and ensuring synergy between the drones and ground vehicles.

$$t^e_{c_{y+1}} \geq t \max{}^e_{c_y} + s_{c_{y+1},c_y}, \quad y = 0, \ldots, p_r, \quad e = 0 \tag{6a}$$

$$t^q_{c_y+1} \geq t \max{}^q_{c_y} + s_{c_{y+1},c_y} + d_s \sum_{z \in C_r} k_{c_y z} n^e_z, \quad y = 1, \ldots, p_r, q = 0, \quad e = 1, 2, \ldots, a \tag{6b}$$

Here, t^e_x denotes the position x on the route of ground vehicle number e. The arrival time at a given node is denoted as t, and the nodes 0 and b + 1 correspond to the departure from the ground station and the return to the ground station, respectively. $s_{x,\ y}$ represents the travel time of the ground vehicle between nodes x and y, while d_s signifies the service time. If a ground vehicle is assigned to a tour commencing at a node located at position x and culminating at a node situated at position y, the calculation of the arrival time at position y involves the addition of flight time to the earliest departure time possible for the ground vehicle from position x. Additionally, a specified duration d_s is allocated at the take-off node, as indicated in Eq. (7).

$$t^e_{c_y} \geq t \max{}^e_{c_x} + d_z \sum_{z \in C_r} k_{c_x z} m_{c_y z} n^e_z, \quad y = 1, \ldots, p_r + 1, \quad x < y, \quad x \cup 0, \quad e = 1, 2, \ldots, a \tag{7a}$$

$$t^e_{c_y} \geq t \max{}^e_{c_x} + (d_z + d_s) \sum_{z \in C_r} k_{c_x z} m_{c_y z} n^e_z, \quad y = 1, \ldots, p_r + 1, \quad x < y, x \cup 0, e = 1, 2, \ldots, a \tag{7b}$$

In this context, d_z signifies the completion time of round z. When transitioning from a node at position y, the ground vehicle remains stationary, awaiting the arrival of each drone scheduled to reach that node for the duration computed in Eq. (8a). Similarly, when departing from a node at position y, the ground vehicle delays its departure to accommodate each drone it is scheduled to meet at that node, with the delay duration determined as indicated in Eq. (8b).

$$t\max{}_{c_y}^{q} \geq t_{c_y}^{e},\ \ y=0,\ \ldots,p_r+1,\ \ q=0,\ \ e=1,2,\ \ldots,a \tag{8a}$$

$$t\max{}_{c_y}^{q} \geq t_{c_y}^{e} + d_d\sum_{z\in C_r} m_{c_y z} n_z^e,\ \ y=0,\ \ldots,p_r+1,\ \ q=0,\ \ e=1,2,\ \ldots,a \tag{8b}$$

The departure time of the vehicles from a node occurs after the arrival time at that node, as computed in Eq. (9a). The ground vehicle may depart from this node once the drone has waited at the node for a period of d_d following its arrival. As outlined in Eq. (9b), the departure time of vehicles from a node cannot exceed the time of arrival at that node.

$$t\max{}_{c_y}^{e} \geq t_{c_y}^{e}, y=0,\ \ldots,p_r+1,\ \ e\in E \tag{9a}$$

$$t\max{}_{c_y}^{e} \geq t_{c_y}^{e} + d_d\sum_{z\in C_r} m_{c_y z} n_z^e, y=0,\ \ldots,p_r+1,\ \ e\in E \tag{9b}$$

Here, E represents the ensemble of ground vehicles and drones. Before commencing a new flight from any node along the ground vehicle route, drones must await the arrival of the ground vehicle at that node, as determined in Eq. (10a). This stipulation ensures that drones depart from the node only after the ground vehicle has arrived, fostering coordination in the process. In the event of a rendezvous at a node located at position y, the drones wait for a duration of d_d before departing from this node. Similarly, drones cannot initiate a new flight from a node on the ground vehicle route without meeting the ground vehicle first. Consequently, this necessitates the ground vehicle to await the duration calculated in Eq. (10b) to reach the node and allocate d_d time following the rendezvous.

$$t\max{}_{c_y}^{e} \geq t_{c_y}^{q},\ \ y=0,\ \ldots,p_r+1,\ \ q=0,\ \ e=1,2,\ \ldots,a \tag{10a}$$

$$t\max{}_{0}^{e} = 0,\ \ e\in E$$

$$t\max{}_{c_y}^{e} \geq t_{c_y}^{q} + d_d\sum_{z\in C_r} m_{c_y z} n_z^e,\ \ y=0,\ \ldots,p_r+1,\ \ q=0,\ \ e=1,2,\ \ldots,a \tag{10b}$$

The vehicles depart from the warehouse simultaneously, initiating the response mission at time 0. As depicted in Eq. (11), when drones embark on a flight starting at position x and concluding at position y, they rendezvous with the ground vehicle prior to exhausting their battery power. This regulation guarantees that drones can efficiently manage their energy resources to fulfill their mission and safely regroup with the ground vehicle upon mission completion.

$$t^{q}_{c_y} - t\max{}^{e}_{c_x} \leq \mathrm{B} + \mathrm{a}\left(1 - \sum_{z \in C_r} k_{c_x z} m_{c_y z} n^{e}_{z}\right), \quad y = 1, \ldots, p_r + 1, \quad x < y, \quad x \cup 0, \quad q = 0, \quad e = 1, 2, \ldots, a \tag{11}$$

In this context, B represents the maximum flight duration a drone can maintain before its battery is depleted, denoting its battery life. When a drone, starting at node x within the same location, initiates its flight, the ground vehicle has the flexibility to await another drone at that node. Should a drone arrive at the rendezvous point prior to the ground vehicle, it must remain airborne, waiting for the ground vehicle to arrive. Therefore, the drone's flight duration should not only be shorter than its battery life but also shorter than the time from its takeoff to its rendezvous with the ground vehicle, which accounts for the duration it remains airborne. This critical constraint also restricts the travel time of the ground vehicle between the departure and merger nodes concerning the drone. In this context, Eq. (12) illustrates the variables describing the tours designated for the drones.

$$n^{e}_{z} \in \{0, 1\}, \quad z \in C_r, \quad e = 1, 2, \ldots, a \tag{12}$$

Equation (13) defines the decision variables that represent the arrival times of vehicles at each node along the ground vehicle's route.

$$t^{e}_{c_y} \geq 0, \quad y = 1, \ldots, p_r + 1, \quad e \in E \tag{13}$$

On the other hand, Eq. (14) outlines the decision variables that indicate the departure times of the vehicles from each node along the ground vehicle's route.

$$t\max{}^{e}_{c_y} \geq 0, \quad y = 1, \ldots, p_r + 1, \quad e \in E \tag{14}$$

4 Results

To facilitate the evaluation of the VRP-mD problem, this section is divided into two main parts: The initial section provides exact solutions for sample problems and presents various solution statistics. These scenarios involve the simultaneous intervention of two and three drones alongside a ground vehicle. The problems were generated with multiple target regions, uniformly distributed across a range of 0 to 10,000 meters by randomizing x and y coordinates. The calculations assume a ground vehicle speed of 11.11 m/s and a drone speed of 15.55 m/s. Each drone is assumed to have a battery life of 1200 s, and no service times are considered. The second part discusses heuristic solutions obtained through the proposed hybrid genetic algorithm. It delves into the outcomes of these heuristic solutions. Both

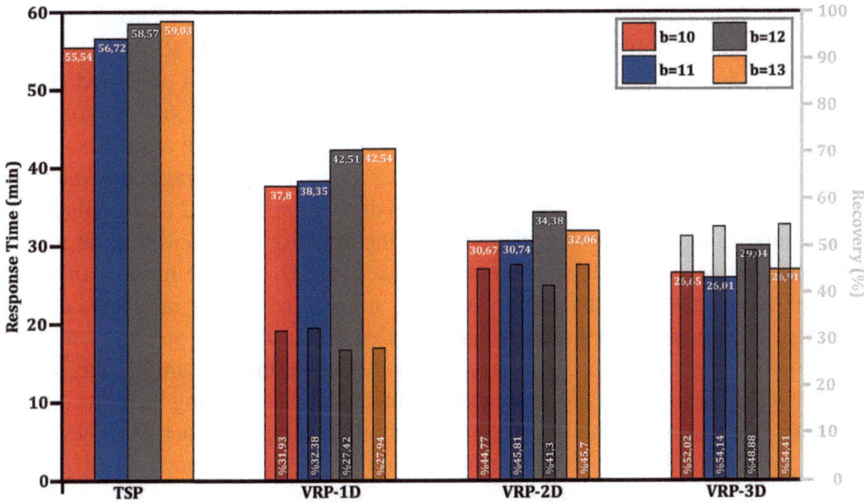

Fig. 1 Evaluation of VRP solutions based on the number of drones

parts offer a comprehensive exploration of the VRP-mD problem, allowing for a thorough assessment of the operational effectiveness of this novel intervention model across various scenarios.

To begin, we compare problem solutions involving two and three drones with VRPD solutions. These results are obtained without considering service times. Figure 1 provides an overview of the average response times for scenarios involving only the ground vehicle, the ground vehicle, and one, two, or three drones, and improvements achieved by integrating drones into the response when compared to the TSP solution. The intervention times shown in Fig. 1 represent the averages from 10 sample problems for each problem dimension. Interestingly, there is a trend that emerges as the number of drones increases, indicating that the algorithm's capacity to tackle larger problem sizes also expands. This is intriguing because an increase in the number of drones augments the problem's complexity by introducing more variables. The proposed exact solution algorithm effectively enhances the upper bounds by dividing the problem into two stages, assigning the largest drone intervention zone first, and subsequently addressing the smaller ground vehicle routes, thereby significantly reducing the number of ground vehicle routes generated. However, in scenarios involving multiple drones, it is noticeable that the algorithm's ability to solve larger problems diminishes as the number of drones increases. When examining the improvement rates achieved in response times relative to the TSP solution, it is apparent that integrating a single drone leads to an average response time reduction of 30%, while integrating two drones results in a 45% average reduction, and integrating three drones results in a 50% average reduction. It is noteworthy that there is no clear correlation between these recovery rates and the number of disaster areas. Figure 1 illustrates that the increase in the number of drones is associated with diminishing returns in terms of improved response times.

In Fig. 2, a comparison of various solution statistics for the 1-Drone Assisted Vehicle Routing Problem (VRP-1D), 2-Drone Assisted Vehicle Routing Problem (VRP-2D), and 3-Drone Assisted Routing Problem (VRP-3D) is presented. These statistics encompass parameters such as solution time, number of iterations, the count of ground vehicle routes generated, time required for route creation, and the number of assignments to drones. These values represent the averages derived from 10 sample problem solutions for each problem dimension. Figure 2 clearly illustrates that solution time, the number of iterations, the quantity of ground vehicle routes, and the time needed for their creation increase directly with the number of target regions while inversely decreasing with the number of drones involved. Consequently, the interpretation that there is a reduction in solution time and the number of iterations does not provide conclusive results due to the limited sample size.

The results depicted in Fig. 2 are closely interconnected, making it challenging to evaluate them in isolation. The two-step problem-solving approach involves iteratively improving the holistic upper bound and step-by-step generation of ground vehicle routes, starting with the shortest route. This approach effectively narrows the solution space and reduces the number of ground vehicle routes generated, resulting in a more efficient route creation process. Consequently, this significantly reduces the overall solution time. It is evident that these statistics tend to increase with the growing number of disaster areas. However, a notable observation is the reduction in these statistics with an increase in the number of drones. More drones allow for a higher maximum number of disaster zones to be assigned to drones, which, in turn, decreases the number of response zones on the ground vehicle route. This reduction in disaster zones assigned to ground vehicles significantly cuts down the number of ground vehicle routes required. Fewer ground vehicle routes contribute to reaching a solution more swiftly with fewer iterations. Therefore, despite the increased number of drones, the solution time decreases, while the problem's size that can be effectively solved expands. Conversely, the number of response zones assigned to drones increases as expected with the growing number of disaster zones and drones.

The scenarios involving only 10 regions in need of intervention yield small and somewhat unrealistic problems for the multi-drone response model. Moreover, solving problems with 50 intervention zones considerably extends the solution time compared to the single-drone version. Because of this, the numerical analyses of VRP-mD were conducted using 30 sample problems associated with areas requiring intervention. To this end, 30 sample problems were generated, featuring disaster response regions with x and y coordinates uniformly distributed between −16.0934 and +16.0934. Figure 3 displays the response times, solution times, and TSP-based solutions derived from these 30 sample problems for scenarios involving one, two, and three drones. The presented values for each problem reflect the averages obtained from 10 iterations of the respective problem.

In the purpose column, response times are displayed. The results indicate a significant increase in solution times as the number of drones increases. Specifically, single-drone solutions are accomplished within just one minute, while this time extends to one hour for two drones and up to 2.5 hours for three drones. As the number of drones grows, the rate of improvement in recovery percentages, as

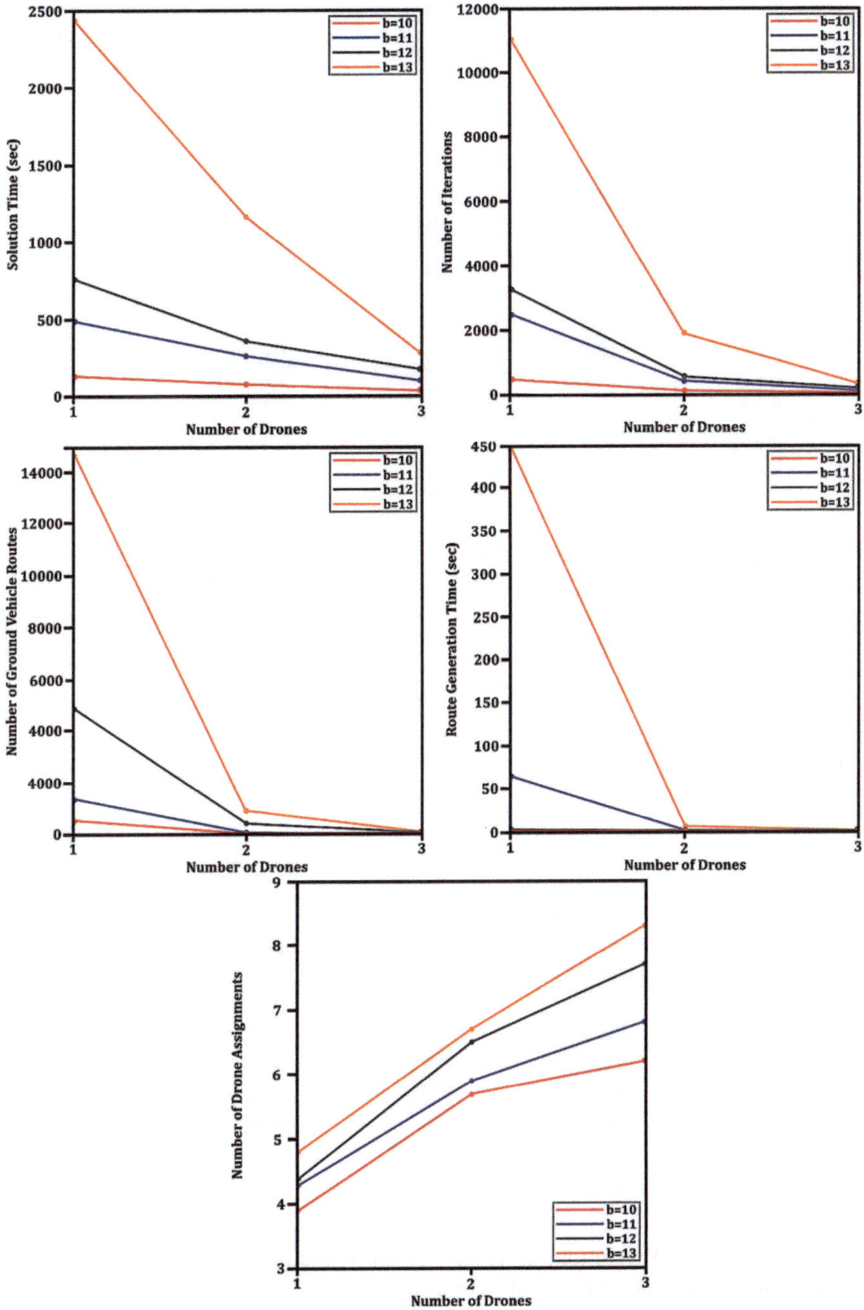

Fig. 2 Data of TSP solution for different numbers of drones

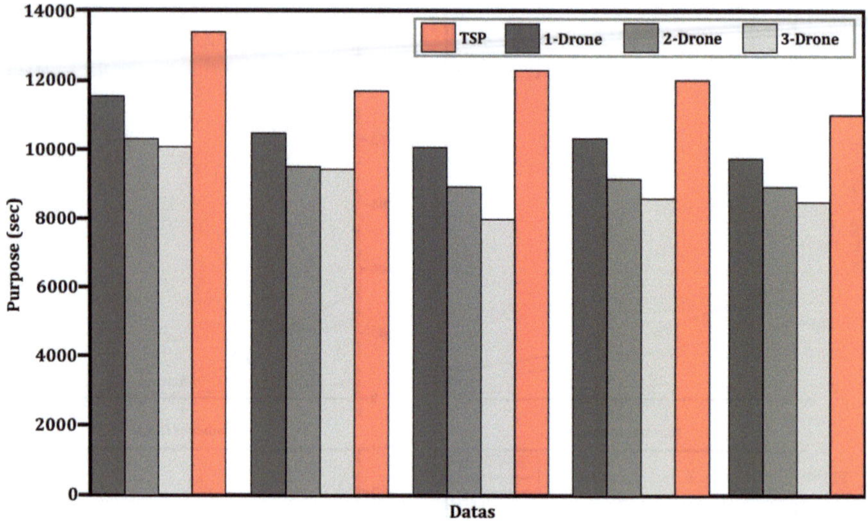

Fig. 3 Heuristic solutions with different numbers of drones

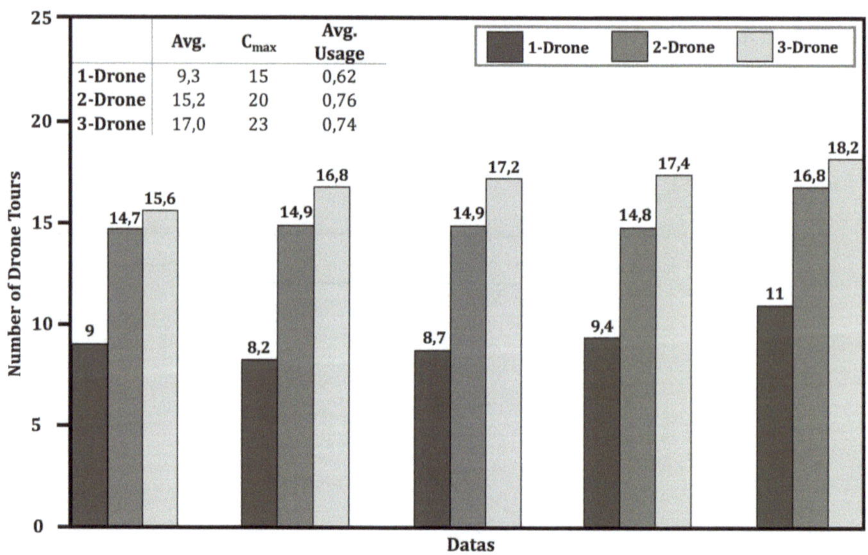

Fig. 4 The use of drones for different problems

compared to the TSP solution, decreases. Notably, the inclusion of a third drone in the response network led to substantially lower recovery rates, particularly in problems 1 and 2, compared to the two-drone scenario. A similar trend is observed in the usage rates of drones. Figure 4 illustrates the drone usage rates for problems consisting of 30 intervention zones. According to the data in Fig. 4, the drone usage rate decreased in cases involving three drones compared to those with two drones.

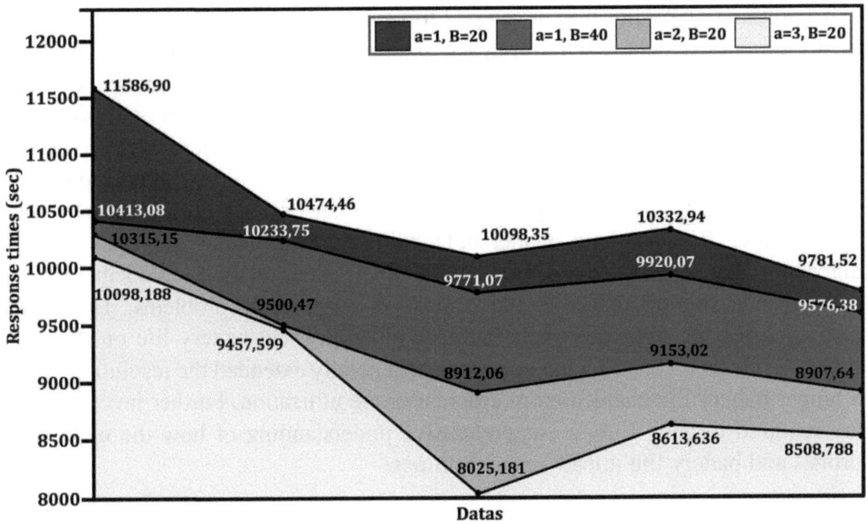

Fig. 5 Intervention times of drones depending on different numbers and battery life

This trend is believed to be linked to the slower response times and the increase in the number of drones, which, in turn, affects the waiting time of ground vehicles for the drones.

The results depicted in Fig. 4 unambiguously demonstrate that an increase in the number of drones leads to reduced drone efficiency. This prompts the question of whether the improvements achieved by increasing the number of drones can be replicated by extending the drones' battery life. Consequently, the VRPD problem for the 30 intervention areas has been resolved under the assumption that the drones possess a 2400-s battery life. Figure 5 illustrates the average response times across 10 replicates for various combinations of drone count and battery life. Figure 5 provides a more detailed view of how the combination of drone number and battery life affects response times. As indicated by this figure, in only one out of the five sample problems, the combination of a single drone with a 2400-s battery life yielded response times similar to those achieved with two drones and a 1200-s battery life. This outcome underscores that the battery life and, consequently, battery technology are just as crucial as the number of drones in the context of simultaneous interventions by ground vehicles and drones. To conduct a more comprehensive analysis, it becomes evident that further research is required, considering cost factors as well.

5 Conclusion

This study aimed to advance our understanding by addressing the intricate task of coordinating disaster response efforts that involve both ground vehicles and multiple drones. The research unfolded in two primary phases. Initially, a novel iterative exact solution algorithm was developed to tackle the intricate problem of simultaneous ground vehicle and multi-drone interventions. This method provides the capacity to resolve larger-scale issues effectively. In the second phase, a hybrid genetic algorithm, leveraging machine learning techniques, was devised, offering a highly efficient alternative for solving medium and large-scale problems. Numerical experiments demonstrate that the hybrid algorithm outperforms other approaches when it comes to coordinating ground vehicles and drones during simultaneous interventions. Particularly in smaller and medium-sized problems, the hybrid genetic algorithm demonstrates quicker solutions than other methods. Nonetheless, its performance slightly falters in the context of larger-scale problems. This study also explored the influence of the number of drones and battery life on response times. While an increased number of drones generally extended the resolution times, a longer battery life enabled more efficient drone utilization. Further investigations are required to gain a more comprehensive understanding of how the number of drones and battery life impact response times.

References

1. Bakirci, M., Ozer, M.M.: Adapting swarm intelligence to a fixed wing unmanned combat aerial vehicle platform. In: Rivera, G., Cruz-Reyes, L., Dorronsoro, B., Rosete, A. (eds.) Data Analytics and Computational Intelligence: Novel Models, Algorithms and Applications. Studies in Big Data, vol. 132, pp. 433–479. Springer, Cham (2023)
2. Liu, H., Yu, Y., Liu, S., Wang, W.: A military object detection model of UAV reconnaissance image and feature visualization. Appl. Sci. **12**(23), 1–17 (2022)
3. Lei, T., Yang, Z., Lin, Z., Zhang, X.: State of art on energy management strategy for hybrid-powered unmanned aerial vehicle. Chin. J. Aeronaut. **32**(6), 1488–1503 (2019)
4. Sharma, S., Solanki, S., Aswal, K., Thakur, E., Malhotra, I.: Review on application of drone systems in agriculture. In: 6th International Conference on Signal Processing, Computing and Control (ISPCC), pp. 40–45. IEEE, India (2021)
5. Ahmed, H., Bakr, M., Talib, M.A., Abbas, S., Nasir, Q.: Unmanned aerial vehicles (uavs) and artificial intelligence (ai) in fire related disaster recovery: analytical survey study. In: International Conference on Business Analytics for Technology and Security (ICBATS), pp. 1–6. IEEE, United Arab Emirates (2022)
6. Emimi, M., Khaleel, M., Alkrash, A.: The current opportunities and challenges in drone technology. Int. J. Electr. Eng. Sustain. **1**(3), 74–89 (2023)
7. Liu, Z., Cai, K., Zhu, Y.: Civil unmanned aircraft system operation in national airspace: a survey from air navigation service provider perspective. Chin. J. Aeronaut. **34**(3), 200–224 (2021)
8. Bakirci, M.: Data-driven system identification of a modified differential drive mobile robot through on-plane motion tests. Electrica. **23**, 619–633 (2023)
9. Otto, A., Agatz, N., Campbell, J., Golden, B., Pesch, E.: Optimization approaches for civil applications of unmanned aerial vehicles (uavs) or aerial drones: a survey. Networks. **72**(4), 11–58 (2018)
10. Bakirci, M., Ozer, M.M.: An avionics system for light-weight multi-rotor unmanned aerial vehicles. In: 1st International Conference of Intelligent Methods, Systems, and Applications, pp. 363–368, Giza, Egypt (2023)
11. Lambey, V., Prasad, A.D.: A review on air quality measurement using an unmanned aerial vehicle. Water Air Soil Pollut. **232**(3), 1–32 (2021)

12. Bouguettaya, A., Zarzour, H., Kechida, A., Taberkit, A.M.: Deep learning techniques to classify agricultural crops through uav imagery: a review. Neural Comput. & Applic. **34**(12), 9511–9536 (2022)
13. Nwilag, B.D., Eyoh, A.E., Ndehedehe, C.E.: Digital topographic mapping and modelling using low altitude unmanned aerial vehicle. Model. Earth Syst. Environ. **9**(2), 1463–1476 (2023)
14. Carrivick, J.L., Smith, M.W.: Fluvial and aquatic applications of structure from motion photogrammetry and unmanned aerial vehicle/drone technology. WIREs Water. **6**(1), 1–17 (2019)
15. Berie, H.T., Burud, I.: Application of unmanned aerial vehicles in earth resources monitoring: focus on evaluating potentials for forest monitoring in Ethiopia. Eur. J. Remote Sens. **51**(1), 326–335 (2018)
16. Dilshad, N., Hwang, J., Song, J., Sung, N.: Applications and challenges in video surveillance via drone: a brief survey. In: International Conference on Information and Communication Technology Convergence (ICTC), pp. 728–732, Korea (South) (2020)
17. Verma, U.: Recent trends and challenges in analysis of uav aerial images for post-disaster scene understanding. In: International Geoscience and Remote Sensing Symposium (IGARSS), pp. 4647–4649. IEEE, Malaysia (2022)
18. Daud, S.M.S.M., Yusof, M.Y.P.M., Heo, C.C., Khoo, L.S., Singh, M.K.C., Mahmood, M.S., Nawawi, H.: Applications of drone in disaster management: a scoping review. Sci. Justice. **62**(1), 30–42 (2022)
19. Rojas Viloria, D., Solano-Charris, E.L., Muñoz-Villamizar, A., Montoya-Torres, J.R.: Unmanned aerial vehicles/drones in vehicle routing problems: a literature review. Int. Trans. Oper. Res. **28**(4), 1626–1657 (2021)
20. Adewumi, A.O., Adeleke, O.J.: A survey of recent advances in vehicle routing problems. Int. J. Syst. Assur. Eng. Manag. **9**(1), 155–172 (2018)
21. Zhu, Y., Liu, M., Jin, D.: A short review of truck and drone collaborative delivery problem. In: International Conference on Cyber Security and Cloud Computing (CSCloud)/ International Conference on Edge Computing and Scalable Cloud (EdgeCom), pp. 18–23. IEEE China (2023)
22. Li, X., Tupayachi, J., Sharmin, A., Martinez Ferguson, M.: Drone-aided delivery methods, challenge, and the future: a methodological review. Drones. **7**(191), 1–26 (2023)
23. Macrina, G., Di Puglia Pugliese, L., Guerriero, F., Laporte, G.: Drone-aided routing: a literature review. Transp. Res. C Emerg. Technol. **120**(102762), 1–25 (2020)
24. Wang, Z., Sheu, J.-B.: Vehicle routing problem with drones. Transp. Res. B Methodol. **122**, 350–364 (2019)
25. Tang, Z., Hoeve, W.J., Shaw, P.: A study on the traveling salesman problem with a drone. In: Rousseau, L.M., Stergiou, K. (eds.) Integration of Constraint Programming, Artificial Intelligence, and Operations Research. CPAIOR 2019. Lecture Notes in Computer Science, vol. 11494, pp. 557–564. Springer, Cham (2019)
26. Kim, S., Moon, I.: Traveling salesman problem with a drone station. IEEE Trans. Syst. Man Cybernet. Syst. **49**(1), 42–52 (2019)
27. Aniceto dos Santos, M.A., Teixeira Vivaldini, K.C.: A review of the informative path planning, autonomous exploration and route planning using uav in environment monitoring. In: International Conference on Computational Science and Computational Intelligence (CSCI), pp. 445–450, Las Vegas, NV, USA (2022)
28. Saleu, R.G.M., Deroussi, L., Grangeon, N., Quilliot, A., Feillet, D.: An iterative two-step heuristic for the parallel drone scheduling traveling salesman problem. Networks. **72**(4), 459–474 (2018)
29. Tu, P.A., Dat, N.T., Dung, P.Q.: Traveling salesman problem with multiple drones. In: 9th International Symposium on Information and Communication Technology (SoICT '18), pp. 46–53. Association for Computing Machinery, New York, NY, USA (2018)
30. Dienstknecht, M., Boysen, N., Briskorn, D.: The traveling salesman problem with drone resupply. OR Spectr. **44**(4), 1045–1086 (2022)

Breast Cancer Diagnosis from Histopathological Images of Benign and Malignant Tumors Using Deep Convolutional Neural Networks

Alime Beyza Arslan and Gökalp Çınarer

1 Introduction

Artificial intelligence (AI) is a set of machines that can perceive and interpret data using a combination of deep learning (DL) and machine learning (ML). DL is a field of AI that has been used in various fields such as data processing, image analysis, and financial transactions, as well as in the medical field in recent years. Especially in cancer diagnosis, there are applications developed with deep learning models. Cancer is caused by the abnormal proliferation of cells with structural differences in development and their accumulation in our body. All types of cancer are caused by several disruptions in the helical structure of DNA. It is seen that 10 percent of cancer types are transmitted through gene transmission, that is, through gene structures passed from parents to children, and the remaining 90 percent of the remaining ninety percent is caused by the disruption of the genetic structure of DNA in various ways, such as disruptions in genes due to mutations [1].

According to World Health Organization (WHO) data, this fatal disease is among the top three causes of child mortality in developed and developing countries. Cancer, which ranks among the top causes of death in the world, is one of the most important health problems in Turkey as it has been all over the world in recent years. It is predicted that cancer-related deaths will increase over time and become the leading cause of death [2]. It is less likely to be caused by ancestral genes than by environmental factors. In some cancer types, such as breast, ovarian, and colon. It has been determined that some diseases are transmitted through gene groups [3].

A. B. Arslan · G. Çınarer (✉)
Yozgat Bozok University, Computer Engineering Department, Yozgat, Turkey
e-mail: gokalp.cinarer@bozok.edu.tr

M. N. Seyman (ed.), *2nd International Congress of Electrical and Computer Engineering*, EAI/Springer Innovations in Communication and Computing,
https://doi.org/10.1007/978-3-031-52760-9_9

In 2018, 2.09 million new cases were detected worldwide and 627,000 deaths occurred due to breast cancer [4]. Compared to the past, the current cancer data have increased considerably. Difficulties have also arisen in controlling these data.

Most people with this disease are detected in the last stages of the disease and these people cannot respond to treatments at this stage. For this reason, the person's life is at risk or the person may die.

Therefore, early detection of tumors, like other types of cancer, is of great importance. Some medical imaging techniques are used to visualize the internal structure of the body and bones. These medical imaging techniques started with x-rays and with the development of technology; magnetic resonance imaging (MRI) and computed tomography (CT). Many medical imaging techniques such as positron emission tomography (PET) and single photon emission computed tomography (TFEBT) have been discovered [5]. In Turkey, studies were conducted to determine the risks in women with cancer, but there are very few studies to determine the risk of cancer in women without cancer. In these few studies, it is aimed to determine the risk level in women who are screened for breast cancer, evaluate the risks that affect cancer with the help of research, determine the risk levels of those who have cancer as a result of this evaluation, and increase their contribution to screening and examinations for early diagnosis [6].

Early diagnosis of cancer is as important as not getting cancer [7]. Primary and secondary prevention are also examples of cancer prevention. Primary prevention is to identify the genetic factors and environmental factors that cause cancer and find a cure for them. For example, a smoker quitting smoking is primary prevention. Secondary prevention aims to detect the tumor in its early stages, i.e., early. Methods for cancer detection fall under secondary prevention. The primary goal in secondary prevention is not to prevent the disease but to diagnose and treat cancer in the early stages [8]. Breast structure is a structure consisting of the areola, areolar gland, cooper ligaments, Spence tail, mammary gland, milk duct, nipple, breast alveoli, breast cleft, and breast cavity [9]. Cancer may be hereditary or may be caused by environmental factors.

The ways to prevent cancer are to be regularly examined and to engage in behaviors that will minimize the risk of cancer. In this way, interventions should be made to try to prevent cancer, detect cancer in early stages, and support the optimization of a healthy and quality life [10]. Therefore, diagnosing cancer in the early stages and carrying out the necessary interventions is of great importance for people to live longer.

In this study, the successes of different models using deep learning architectures for more accurate and faster disease detection are examined. The algorithm performances of deep learning models developed with artificial intelligence technology were analyzed.

2 Material and Method

BreakHis [11] (https://www.kaggle.com/datasets/ambarish/breakhis) data in the study set will be used. Different magnification factors (40×, 100×, 200×, and Histopathologic images collected from 82 patients using 400×) are available. Dataset 3891 microscopic images: 1017 of these images are benign and 2874 of them are images of malignant tumors. The data set is divided into 20% test set and 80% training set. The dataset consists of two classes: benign and malignant. For the images in the dataset, the original size was 700 × 460 and rescaled to 224 × 224. Sample images of benign and malignant images in the Breakhis dataset are given when zoomed at 40×, 100×, 200× and 400× (Figs. 1 and 2).

The dataset's main groups are benign and malign. Benign is a term that refers to a lesion that does not pose any danger. Malignant are lesions that are considered synonymous with cancer. The development of these lesions can spread and destroy neighboring cells and can also spread to distant sites, causing cancer to progress. This has many important consequences, including death.

In some studies conducted with the BreakHis dataset, the following results were obtained. Agarwal et al. performed breast cancer classification using this dataset VGG16 model and achieved 89.67% accuracy [12]. Bayramoğlu et al. used the BreakHis dataset in their study. They achieved an accuracy rate of 83.25% [13]. Matos et al. in experiments on the BreakHis dataset, Inception12 V3 achieved an average accuracy of 0.874 [14]. Albashis et al. used ResNet50 from the CNN architecture for the BreakHis dataset and achieved 88% accuracy [15]. Wang et al. obtained the highest accuracy rate of 91.05% from the MobileNetV3 algorithm with the BreakHis dataset [16]. Voon et al. used the CNN algorithm with the BreakHis dataset. They obtained the highest result of 92.36% with EfficientNetV2B0 [17].

In this study, convolutional neural networks (CNN) are used to compare deep learning methods for breast cancer diagnosis. CNN is widely used in medical imaging, including radiology. In the classification of histopathological images with the CNN model, multiple structures such as convolutional layers, pooling layers, and

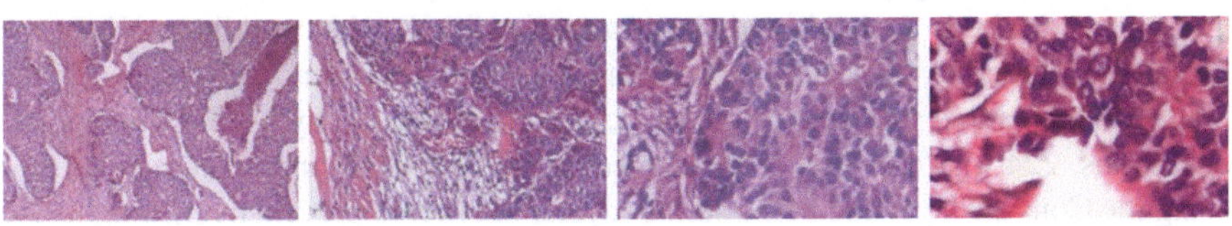

Fig. 1 Dataset malignant images (40×, 100×, 200×, 400×)

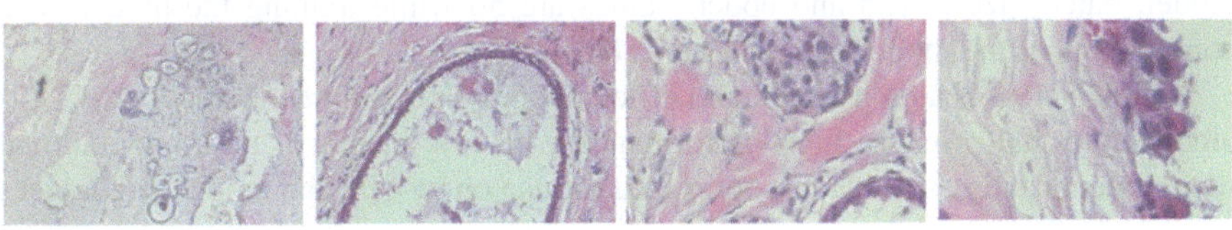

Fig. 2 Dataset benignant images (40×, 100×, 200×, 400×)

Table 1 CNN model network architecture

Layer (type)	Output shape	Param
random_flip (RandomFlip)	(None, 224, 224 3)	0
random_rotation (RandomRotation)	(None, 224, 224, 3)	0
rescaling (Rescalimg)	(None, 224, 224, 3)	0
batch_normalization (BatchNormalization)	(None, 224, 224, 3)	12
conv2d (Conv2D)	(None, 222, 222, 32)	896
max_pooling2d (MaxPooling2D)	(None, 111, 111, 32)	0
conv2d_1 (Conv2D)	(None, 109, 109, 64)	18496
max_pooling2d_1 (MaxPooling2D)	(None, 54, 54, 64)	0
conv2d_2 (Conv2D)	(None, 52, 52, 64)	36928
max_pooling2d_2 (Maxpooling2D)	(None, 26, 26, 64)	0
global_average_pooling2d (GlobalAveragePooling2D)	(None, 64	0
dropout (Dropout)	(None, 64)	0
dense (Dense)	(None, 256)	16640
dropout_1 (Dropout)	(None, 256)	0
dense_1 (Dense)	(None, 64)	16448
dense_2 (Dense)	(None, 1)	65

fully connected layers are effectively used in the learning phase with image processing methods. Among the various deep learning models, the most established algorithm is the convolutional neural network (CNN), a class of artificial neural networks that has been a common method in computerized image processing tasks since the sharing of surprising results in the object recognition competition known as the ImageNet Large Scale [18]. In this study, the dataset was divided into 20% test and 80% train to analyze the model performance while performing operations in the detection of benign and malignant histopathological images. The original size of the images in the dataset is 700 × 460. With rescaling, the images were reduced to 224 × 224, the size that the network can learn. Image processing is a field of computer science that deals with the acquisition, analysis, and processing of digital images. The padding process was used in the study. In the "padded" variable, the image is both resized and its frame is filled with padding. With the flow_from_directory operation used in the study, the numbers and classes of the images in the dataset were determined according to the hierarchy. At the end of these operations, 224 × 224 images were created again. The CNN architecture is defined using the Conv2D layer and Maxpooling layers. In order to prevent overfitting, some neural network randomization layers were added to the training.

The layers of the model used in the study are shown in Table 1. In the CNN model, batch_size is 128 and epoch values are 50 while training the model. The components of the model are given in Table 1.

MobileNetV3 is a lightweight and efficient convolutional neural network (CNN) architecture that provides high performance with low memory and computational requirements and is specifically optimized for use on mobile and embedded devices

Table 2 MobileNetV3 model network architecture

Layer (type)	Output shape	Param
random_flip_1 (RandomFlip)	(None, 224, 224 3)	0
random_rotation (RandomRotation)	(None, 224, 224, 3)	0
lambda (Lambda)	(None, 224, 224, 3)	0
MobilenetV3small (Functional)	(None, 576)	939120
dropout_2 (Dropout)	(None, 576)	0
dense_3 (Dense)	(None, 256)	147712
dropout_3 (Dropout)	(None, 256)	0
dense_4 (Dense)	(None, 64)	16448
dense_5 (Dense)	(None, 1)	65

Table 3 EfficientNetB0 model network architecture

Layer (type)	Output shape	Param
random_flip_2 (RandomFlip)	(None, 224, 224 3)	0
random_rotation_2 (RandomRotation)	(None, 224, 224, 3)	0
efficientnetb0 (Functional)	(None, 1280)	4049571
dropout_4 (Dropout)	(None, 1280)	0
dense_6 (Dense)	(None, 256)	327936
dropout_5 (Dropout)	(None, 256)	0
dense_7 (Dense)	(None, 32)	9224
dense_8 (Dense)	(None, 1)	33

[19]. The hyperparameters of the MobileNetV3 model used in this study are given in Table 2.

EfficientNetB0 is a deep learning model used in computerized image processing and deep learning. The network architecture of the model created within the scope of the study is given in Table 3. The model has a high performance and computationally efficient structure and can be used to achieve effective results in various image classification and feature extraction tasks [20].

VGG16 is a popular convolutional neural network (CNN) model in the field of deep learning. It consists of 16 layers in total, with layers of various depths. VGG16 can be used as a pre-trained model and often gives successful results in image classification tasks [21]. The layers in this model consist of different types, such as convolution layers and fully connected layers, and are known for the repeated use of 3×3 filters, especially in the convolution layers. Table 4 shows the configuration of the layers. The VGG16 model is a widely used deep learning model, especially for its relatively simple and efficient structure.

ResNet50V2 is a convolutional neural network (CNN) model widely used in deep learning. Table 5 shows the network architecture. This model, a member of the ResNet series, consists of 50 layers and is extremely deep and complex. ResNet50V2 uses a connection structure consisting of "identity shortcuts," a unique block structure, to ensure better learning performance and a lower risk of overfitting [22].

Table 4 Vgg16 model network architecture

Layer (type)	Output shape	Param
random_flip_3 (RandomFlip)	(None, 224, 224, 3)	0
random_rotation_3 (RandomRotation)	(None, 224, 224, 3)	0
lambda_1 (Lambda)	(None, 224, 224, 3)	0
Vgg16 (Functional)	(None, 512)	14714688
dropout_6 (Dropout)	(None, 512)	0
dense_9 (Dense)	(None, 256)	131328
dropout_7 (Dropout)	(None, 256)	0
dense_10 (Dense)	(None, 32)	8224
dense_11 (Dense)	(None, 1)	33

Table 5 Resnet50V2 model network architecture

Layer (type)	Output shape	Param
random_flip_6 (RandomFlip)	(None, 224, 224, 3)	0
random_rotation_6 (RandomRotation)	(None, 224, 224, 3)	0
lambda_2 (Lambda)	(None, 224, 224, 3)	0
resnet50V2 (Functional)	(None, 2048)	23564800
dropout_8 (Dropout)	(None, 2048)	0
dense_12 (Dense)	(None, 256)	524544
dropout_9 (Dropout)	(None, 256)	0
dense_13 (Dense)	(None, 32)	8224
dense_14 (Dense)	(None, 1)	33

This improves the performance of the model as depth increases. ResNet50V2 is often used in tasks such as image classification, object detection, and image segmentation, and is a deep-learning model that generally provides high accuracy.

Performance evaluation of algorithms is a process for measuring the functionality, efficiency, and correctness of an algorithm. This evaluation usually involves comparing the performance of an algorithm designed for a specific problem or task with other alternatives.

The evaluation process involves analyzing the algorithm's performance metrics such as processing time, memory usage, and the need for computational resources. It also evaluates the algorithm's success metrics such as accuracy, precision, and recall. Performance evaluation also allows algorithms to be tested on different datasets to measure their generalizability.

This process is used as an important source of feedback for the development, improvement, and optimization of algorithms. As a result, the performance evaluation of algorithms is an important step toward better decision-making, efficient resource utilization, and better results. The Confusion Matrix used to calculate these steps is given in Table 6.

$$\text{Accuracy} = \frac{\text{TP} + \text{TN}}{\text{TP} + \text{TN} + \text{FP} + \text{FN}} \tag{1}$$

Table 6 The Confusion Matrix

	Actual values	
Estimated values	Positive	Negative
Positive	True Positive (TP)	False Positive (FP)
Negative	False Negative (FN)	True Negative (TN)

Table 7 Comparison of test dataset result of the models

Model	Accuracy	Recall	Precision	F1 score
CNN	%91.32	%91.23	%91.27	%91.30
MobileNetV3	%82.91	%82.19	%82.24	%81.84
EfficientNetB0	%91.59	%91.95	%91.40	%91.40
VGG16	%86.20	%86.20	%85.91	%85.49
ResNet50V2	%83.70	%83.07	%83.08	%83.07

$$\text{Sensitivity (Recall)} = \frac{\text{TP}}{\text{TP} + \text{FN}} \tag{2}$$

$$\text{Specificity} = \frac{\text{TN}}{\text{FP} + \text{TN}} \tag{3}$$

$$\text{Precision} = \frac{\text{TP}}{\text{TP} + \text{FP}} \tag{4}$$

$$\text{F1 score} = \frac{2 \times \text{Precision} \times \text{Recall}}{\text{Precision} + \text{Recall}} \tag{5}$$

The ROC (receiver operating characteristic) curve is a graphical tool used to evaluate the performance of classification models. The ROC curve shows the relationship between the sensitivity (recall) and specificity (specificity) of a classification model.

3 Result

The 779 test data, which constitute 20% of the total 3891 samples in the dataset to be used in breast cancer detection, were classified separately with CNN, MobileNetV3, EfficientNetB0, VGG16, and Resnet50V2 models. Confusion matrix results of each model were obtained and accordingly, accuracy, sensitivity, precision, and F1 score values were calculated. These values are shown in Table 7 (Table 8).

According to the table, the EfficientNetB0 model gave the highest accuracy results in the test group. The ROC curve and AP plot of the EfficientNetB0 model are shown in Figs. 3 and 4.

When the ROC curve approaches a value of 1, it indicates a true positive, that is, the algorithm correctly classifies an example and the cases where the example is actually positive are high. Figure 1 shows that the EfficientNetB0 model discriminates more successfully than the other models with an AUC of 0.95.

Table 8 Comparison of train dataset result of the models

Model	Accuracy	Recall	Precision	F1 score
CNN	%93.43	%93.43	%93.41	%93.42
MobileNetV3	%91.69	%86.10	%85.57	%85.41
EfficientNetB0	%92.33	%93.61	%93.54	%93.52
VGG16	%87.30	%87.30	%87.04	%86.69
Resnet50V2	%89.05	%89.05	%88.82	%88.76

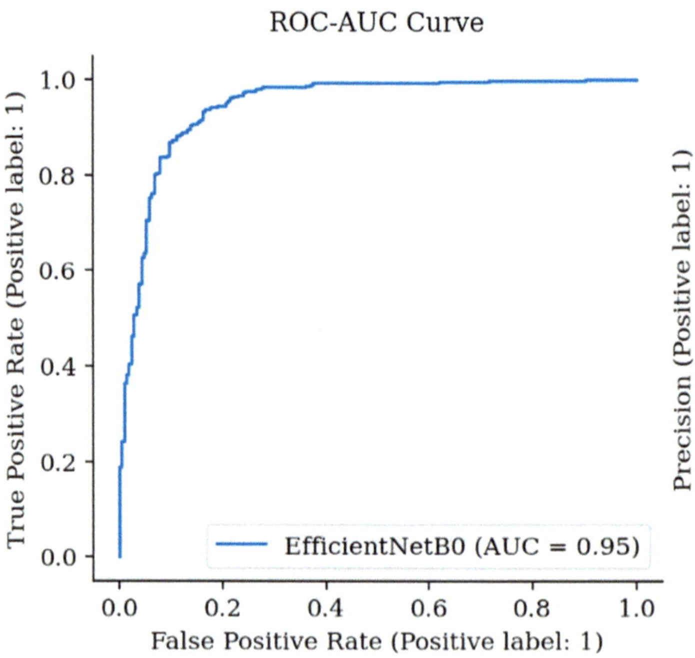

Fig. 3 EfficientNetB0 model ROC AUC

4 Conclusion and Discussion

We focus on the analysis and classification of breast tumors divided into two classes using the BreakHis dataset. The problem of detecting benign and malignant tumors using five different algorithms from current deep learning networks is addressed. The results obtained allowed the performance of each algorithm to be evaluated. The results show that a particular algorithm provides a higher accuracy rate than the others. Furthermore, the use of sampling methods and data augmentation techniques to address the imbalance problem of the dataset can be evaluated in different studies. Compared to the literature studies, the accuracy values obtained by the proposed architectures are quite high. When the results were evaluated, 91.32% accuracy and 91.30% F1 score were obtained with CNN. MobileNetV3 achieved 82.91% accuracy and 81.84% F1 score. With VGG16, 86.20% accuracy and 85.49% F1 score

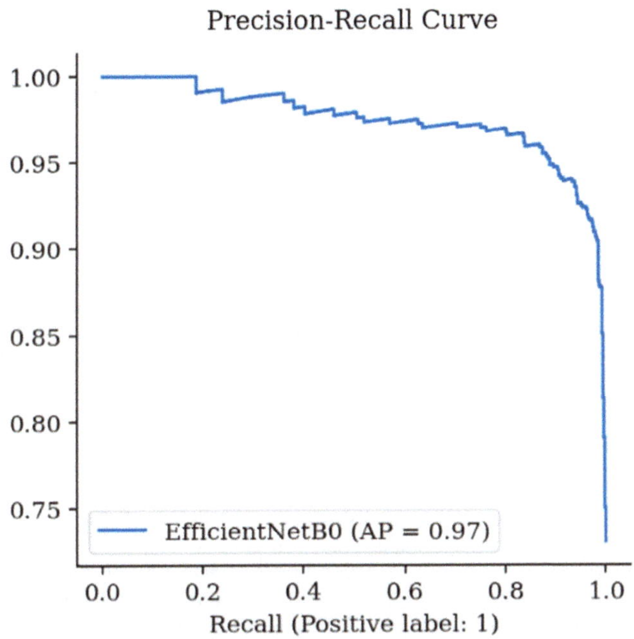

Fig. 4 EfficientNetB0 model and average precision results

were obtained. EfficientNetB0 achieved 91.59% accuracy and 90.40% F1 score. Finally, ResNet50V2 achieved 83.70% accuracy and 83.07% F1 score.

According to these results, the highest performance was obtained with the model proposed with the EfficientNetB0 architecture. The EfficientNetB0 model gave better results compared to other algorithms with 91.59% accuracy and 90.40% F1 score. These results demonstrate the effectiveness and potential of deep learning algorithms in breast cancer diagnosis. In particular, the EfficientNetB0 model stands out with its high accuracy and F1 score.

These results emphasize the importance of using deep learning techniques to achieve more accurate and reliable results in breast cancer diagnosis. It is recommended to conduct a comprehensive study by expanding the datasets and performing a comparative analysis in which different feature extraction methods and classification algorithms are tested. The study provides advances in the field of breast cancer diagnosis and demonstrates the usability of deep learning techniques in clinical applications. Future work could progress by testing these algorithms on larger datasets and investigating their usability in clinical applications.

Conflict of Interest The authors have no relevant financial or non-financial interests to disclose.

Data Availability Training and testing processes have been carried out using the BreakHis Dataset.

BreakHis Dataset is publicly available at https://www.kaggle.com/datasets/ambarish/breakhis.

References

1. Barbaros, M.B., Dikmen, M.: Cancer immunotherapy. Erciyes Univ. Inst. Sci. Technol. J. Sci. **31**(4), 177–182 (2015)
2. Parlar, S., Kaydul, N., Ovayolu, N.: Breast cancer and the importance of breast self-examination. Anatolian J. Nurs. Health Sci. **8**(1), 72–83 (2005)
3. Yokuş, B., Çakır, D.Ü.: Cancer biochemistry. J. Dicle Univ. Faculty Veterin. Med. **1**, 7–18 (2012)
4. Kumar, M., Khatri, S.K., Mohammadian, M.: Breast cancer identification and prognosis with machine learning techniques-an elucidative review. J. Interdiscip. Math. **23**(2), 503–521 (2020)
5. Günay, O., Öztürk, H., Yarar, O.: Project-based learning of the structure of medical imaging devices working with ionizing radiation. J. Health Serv. Educa. **3**(1), 20–27 (2019)
6. Eroglu, C., Eryilmaz, M.A., Civcik, S., Gurbuz, Z.: Breast cancer risk assessment: 5000 cases. Int. J. Hematol. Oncol/UHOD. **20**(1), 27 (2010)
7. Milosevic, M., Jankovic, D., Milenkovic, A., Stojanov, D.: Early diagnosis and detection of breast cancer. Technol. Health Care. **26**(4), 729–759 (2018)
8. Qasim, M., Lim, D.J., Park, H., Na, D.: Nanotechnology for diagnosis and treatment of infectious diseases. J. Nanosci. Nanotechnol. **14**(10), 7374–7387 (2014)
9. Giudice, G., Maruccia, M., Vestita, M., Nacchiero, E., Annoscia, P., Bucaria, V., Elia, R.: The medial-central septum based mammaplasty: a reliable technique to preserve nipple-areola complex sensitivity in post bariatric patients. Breast J. **25**(4), 590–596 (2019)
10. Gençtürk, N.: Protection in breast cancer. Anatolian J. Nurs. Health Sci. **10**(4), 72–82 (2007)
11. Spanhol, F.A., Oliveira, L.S., Petitjean, C., Heutte, L.: A dataset for breast cancer histopathological image classification. I.E.E.E. Trans. Biomed. Eng. **63**(7), 1455–1462 (2016)
12. Agarwal, P., Yadav, A., Mathur, P.: Breast cancer prediction on breakhis dataset using deep cnn and transfer learning model. In: Data Engineering for Smart Systems: Proceedings of SSIC 2021, pp. 77–88 (2022)
13. Bayramoglu, N., Kannala, J., Heikkilä, J.: Deep learning for magnification independent breast cancer histopathology image classification. In: 2016 23rd International Conference on Pattern Recognition (ICPR), pp. 2440–2445. IEEE (2016)
14. De Matos, J., de Souza Britto, A., de Oliveira, L.E., Koerich, A.L.: Texture CNN for histopathological image classification. In: 2019 IEEE 32nd International Symposium on Computer-Based Medical Systems, CBMS, pp. 580–583. IEEE (2019)
15. Albashish, D., Al-Sayyed, R., Abdullah, A., Ryalat, M.H., Almansour, N.A.: Deep CNN model based on VGG16 for breast cancer classification. In: 2021 International Conference on Information Technology, ICIT, pp. 805–810. IEEE (2021)
16. Wang, C., Gong, W., Cheng, J., Qian, Y.: DBLCNN: dependency-based lightweight convolutional neural network for multi-classification of breast histopathology images. Biomed. Sig. Proc. Control. **73**, 103451 (2022)
17. Voon, W., Hum, Y.C., Tee, Y.K., Yap, W.S., Salim, M.I.M., Tan, T.S., et al.: Performance analysis of seven Convolutional Neural Networks (CNNs) with transfer learning for Invasive Ductal Carcinoma (IDC) grading in breast histopathological images. Sci. Rep. **12**(1), 19200 (2022)
18. Russakovsky, O., Deng, J., Su, H., Krause, J., Satheesh, S., Ma, S., Fei-Fei, L.: Imagenet large scale visual recognition challenge. Int. J. Comput. Vis. **115**, 211–252 (2015)
19. Koonce, B., Koonce, B.: MobileNetV3. In: Convolutional Neural Networks with Swift for Tensorflow. Image Recognition and Dataset Categorization, pp. 125–144. Apress, New York (2021)
20. Zuber Khan, T.S., Arya, R.K.: Skin cancer detection using computer vision. In: Topical Drifts in Intelligent Computing: Proceedings of International Conference on Computational Techniques and Applications, ICCTA 2021, vol. 426, pp. 3–11. Springer Nature (2022)

21. Jahangeer, G.S.B., Rajkumar, T.D.: Early detection of breast cancer using hybrid of series network and VGG-16. Multimed. Tools Appl. **80**, 7853–7886 (2021)
22. Karlsson, J., Ramkull, J., Arvidsson, I., Heyden, A., Åström, K., Overgaard, N.C., Lång, K.: Machine learning algorithm for classification of breast ultrasound images. In: Medical Imaging 2022 Computer-Aided Diagnosis, vol. 12033, pp. 473–483. SPIE, California (2022)

Enhancing Skin Lesion Classification with Ensemble Data Augmentation and Convolutional Neural Networks

Aytug Onan, Vahide Bulut, and Ahmet Ezgi

1 Introduction

In recent years, convolutional neural networks (CNNs) and their derived architectures have garnered significant attention in the realm of image classification [1, 2]. Their notable accomplishments across various domains, from computer vision to medical imaging, have elevated them as a prominent subject of research [3]. CNNs' capacity to autonomously acquire hierarchical features from raw image data, combined with their ability to handle extensive datasets, has transformed the landscape of image classification tasks [4]. Furthermore, CNN derivatives, such as ResNet50, MobileNetV2, EfficientNet, and DenseNet, have emerged as potent tools, offering a diverse range of capabilities tailored to specific classification challenges [5]. The ongoing refinement and adaptation of CNNs for image classification underscore their sustained prominence as a dynamic and evolving area of study.

CNNs undergo a training process, reliant on extensive data for effective outcomes [3, 5]. In the training phase, CNNs commence with randomly initialized weights and biases. They learn by iteratively fine-tuning these parameters while exposed to a large, labeled dataset [4]. Optimization techniques, such as gradient descent, adjust the weights to minimize prediction errors when comparing their output to the actual

A. Onan (✉)
İzmir Katip Celebi University, Department of Computer Engineering, Izmir, Turkey

İzmir Katip Celebi University, Division of Software Engineering, Izmir, Turkey
e-mail: aytug.onan@ikcu.edu.tr

V. Bulut
İzmir Katip Celebi University, Division of Software Engineering, Izmir, Turkey

İzmir Katip Celebi University, Department of Engineering Sciences, Izmir, Turkey

A. Ezgi
İzmir Katip Celebi University, Division of Software Engineering, Izmir, Turkey

M. N. Seyman (ed.), *2nd International Congress of Electrical and Computer Engineering*, EAI/Springer Innovations in Communication and Computing,
https://doi.org/10.1007/978-3-031-52760-9_10

labels [4, 5]. This learning process unfolds through various layers within the network, encompassing convolutional and fully connected layers, which enable CNNs to autonomously identify intricate features in images [5]. The requirement for extensive data arises from the inherent complexity of CNNs, constituted by millions of parameters distributed across layers. A substantial dataset equips CNNs with robust generalization capabilities, enabling them to capture diverse image patterns and mitigate the risks of overfitting [6].

The manual annotation of image datasets, particularly in the medical domain, presents notable challenges [7, 8]. Medical images, including those depicting skin lesions, necessitate expert knowledge for precise labeling [9]. Trained medical professionals must meticulously scrutinize each image, identify crucial features, and provide precise annotations, often detailing lesion type, size, and other critical attributes. This process consumes time and effort, demanding a high level of expertise [10]. Additionally, ensuring consistency and reliability in annotations across different annotators proves to be a formidable task. Variations in interpretations and inter-annotator disagreements can further complicate the annotation process [11]. Moreover, privacy and ethical considerations mandate careful handling of sensitive medical data [12]. These challenges underscore the significance of data augmentation techniques, as they serve to maximize the utility of limited annotated data, thereby addressing the shortage of properly labeled medical image datasets [13].

Data augmentation emerges as a pivotal technique in the realm of image classification, offering a strategic means to enrich training datasets [14]. It encompasses the application of diverse transformations and alterations to the original images, effectively generating novel data instances with minor variations. These transformations encompass a wide array of operations, spanning geometric modifications like rotation, scaling, and mirroring, as well as alterations in color space, contrast, and brightness [13, 14]. In the context of medical image analysis, data augmentation proves particularly valuable when confronted with limited annotated data [15]. By augmenting the dataset with diverse variations of existing images, the model gains enhanced robustness and improved generalization capabilities to handle unseen examples. Data augmentation effectively mitigates overfitting, boosts model performance, and facilitates the training of deep learning models such as convolutional neural networks (CNNs) in scenarios where obtaining extensive annotated data may prove challenging or costly [13].

The remainder of this chapter follows a structured approach. Section 2 presents the related work. In Sect. 3, an exhaustive exploration of data augmentation techniques is provided. Section 4 outlines the neural network architectures deployed in our study, encompassing ResNet50, MobileNetV2, EfficientNet, and DenseNet, which form the basis for our empirical analysis. Section 5 presents the results of our empirical analysis. Concluding remarks are presented in Sect. 6.

2 Related Work

In the quest for accurate skin lesion identification, researchers have explored various strategies to improve the performance of deep learning models. One crucial approach has been the careful application of data augmentation techniques, which serve as a vital tool in addressing the difficulties posed by limited annotated datasets [13]. In this section, we provide a brief overview of related work in the field of skin lesion classification, with a particular focus on the integration of data augmentation methods.

In a study conducted by Khan et al. [16], they introduced an ensemble-learning framework that incorporates the EfficientNetB3 deep learning model for analyzing skin lesions. To tackle the issue of data imbalance within the PAD-UFES-20 dataset, which covers six skin cancer categories, they utilized data augmentation methods. Their findings indicate that combining clinical data with information about skin lesions enhances the accuracy of automated diagnosis.

In the research by Zhang et al. [17], the identification of skin cancer was explored through an optimized convolutional neural network (CNN). Researchers integrated an optimization algorithm to effectively select biases and weights, aiming to minimize errors in network output and align the desired output within the networks.

Goceri [18] made significant contributions to skin cancer classification by incorporating adjustable and fully convolutional capsule layers. This approach aimed to enhance classification performance by leveraging capsule networks, known for their ability to capture hierarchical feature relationships. The publicly available HAM10000 dataset was used for evaluation, with results demonstrating the superiority of the capsule network over baseline models.

Adla et al. [19] concentrated on skin cancer classification and detection, introducing an innovative approach that combined a full-resolution convolutional network with a dynamic graph cut algorithm. Their primary goal was to achieve precise detection and classification of skin cancer. Rigorous testing on a publicly accessible dataset highlighted the approach's superiority over existing methods, showing increased accuracy and sensitivity.

Liu et al. [20] introduced the Anti Coronavirus Optimized Kernel-based Softplus Extreme Learning Machine (ACO-KSELM) for accurately predicting various types of skin cancer from high-dimensional datasets. The method employed feature extraction techniques to address redundancy and irrelevant features within biomedical datasets while uncovering underlying data patterns.

Yang et al. [21] presented the Multi-scale Fully-shared Fusion Network (MFF-Net) for classifying skin lesions, which amalgamated features from both dermoscopic and clinical images. The MFF-Net incorporated a multi-scale fusion architecture to unite deep and shallow features within each modality, mitigating the loss of spatial information in high-level feature maps. Furthermore, the introduction of the Dermo-Clinical Block (DCB) facilitated the integration of feature maps from dermoscopic and clinical images, enhancing information exploration across various stages.

Onan et al. [22] examined the predictive performance of combining conventional data augmentation methods with three different deep learning architectures, namely, DenseNet-201, ResNet-152, and InceptionV3.

These studies collectively contribute to the ongoing advancement of skin lesion identification through deep learning approaches, with data augmentation playing a critical role in addressing the challenges associated with limited annotated datasets.

3 Data Augmentation Methods

In this section, we have explored two categories of data augmentation techniques: conventional and ensemble. Conventional methods entail basic transformations applied to individual images to enrich dataset diversity. In contrast, ensemble methods combine multiple augmentation strategies, introducing intricacy and richness to the training process.

3.1 Conventional Data Augmentation Methods

Traditional methods of data augmentation have their roots in prior research. In the following section, we offer a succinct overview of standard data augmentation techniques:

AUG1: This approach generates three additional images through geometric transformations [13, 23]. It begins with an initial image and applies the first transformation by randomly flipping it vertically, followed by a horizontal flip as the second transformation. The third transformation involves linear scaling of the original image along both its horizontal and vertical axes.

AUG2: Another technique involving geometric transformations, AUG2 produces six extra images by replicating the operations from AUG1 and introducing three more: rotation, translation, and shearing [13, 23].

AUG3: This method, also rooted in geometric transformations, creates four supplementary images by following the process of AUG2 but excluding the shearing operation [13, 23].

AUG4: It utilizes kernel filter-based techniques to generate three additional images by implementing transformations based on principal component analysis (PCA) [24].

AUG5: Similar to AUG4, AUG5 is a kernel filter-based method that generates three extra images by applying transformations based on the discrete cosine transform (DCT) [13, 24].

AUG6: This is a color-space transformation technique that generates three fresh images by manipulating the color space. These images are crafted by adjusting contrast, sharpness, and introducing color shifts [13, 24].

AUG7: By combining color-space transformation and kernel filters, AUG7 creates seven additional images. The initial four augmented images are crafted by modifying the pixel colors in the original image. Additionally, two images are generated through a combination of sharpening and the application of a Gaussian filter, while another image is created by introducing color shifts [23].

AUG8: Another technique rooted in color-space transformation and kernel filters, AUG8 generates two more images by manipulating the color space and applying two nonlinear mappings [22, 24].

AUG9: This method is a transformation technique based on geometric transformation and kernel filters, generating six extra images by employing elastic deformations in conjunction with low-pass filters [23].

AUG10: It is a kernel filter-based transformation method that produces three fresh images by introducing perturbations to the matrices derived from the discrete wavelet transform (DWT) [24].

AUG11: Another kernel filter-based method, AUG11, generates three additional images through the constant-Q transform (CQT) [25].

AUG12: This augmentation method is based on kernel-based filters and image mixing, yielding five new images by merging the discrete cosine transform (DCT) with the random selection of other images [23].

AUG13: It is a kernel filter-based method that generates three supplementary images by applying the radon transform (RT) in a distinct manner [23].

AUG14: This technique produces two extra images by employing the fast Fourier transform (FFT) and the discrete Cosine transform (DCT) [23].

3.2 *Ensemble Data Augmentation Methods*

In this section, we delve into ensemble data augmentation techniques designed to enhance the diversity and complexity of training data for skin lesion identification:

AUG15: This ensemble method creates 11 new images with distinct modifications [23]. It starts with the initial image and applies two consecutive discrete cosine transform (DCT) operations to each matrix within the color planes. Algorithms are then used to reduce haze and equalize histograms, improving image clarity. FFT and other transformations are also applied to create a diverse set of images.

AUG16: Building on the DCT-based approach from AUG15, AUG16 incorporates the first distorted image from AUG9.

AUG17: This comprehensive method combines various augmentation techniques [23], including FFT, Hilbert, and Hampel augmentations, as well as a combination of AUG1 through AUG3. These methods encompass flips, rotations, noise addition, cropping, and adjustments to hue, saturation, brightness, and contrast, adding diversity to the dataset.

These ensemble augmentation techniques enhance dataset diversity and complexity, benefiting the training of skin lesion identification models.

4 Convolutional Neural Network Models

In the domain of deep learning and convolutional neural networks (CNNs), the selection of architecture significantly influences the performance and efficacy of image classification tasks. This section offers an overview of four notable CNN architectures: ResNet50, MobileNetV2, EfficientNet, and DenseNet. The following part of this section provides a brief introduction to these convolutional neural network models.

4.1 ResNet50

Residual Networks, often referred to as ResNets, have had a significant impact on image classification, particularly by tackling the problem of vanishing gradients during training [26]. Among the ResNet variants, ResNet50 stands out due to its remarkable depth, consisting of 50 layers, and its incorporation of skip connections that facilitate gradient flow during back-propagation. This architectural innovation has made it possible to train exceptionally deep networks, establishing ResNet50 as a powerful option for tasks demanding intricate feature extraction.

4.2 MobileNetV2

In resource-constrained settings like mobile devices, MobileNetV2 presents itself as an attractive choice [27]. Notable for its efficiency and lightweight architecture, MobileNetV2 places a strong emphasis on optimizing inference speed while maintaining high classification accuracy. It achieves this by utilizing depth-wise separable convolutions and inverted residual blocks, making MobileNetV2 highly valuable for real-time and on-device image analysis tasks.

4.3 EfficientNet

EfficientNet models excel at delivering state-of-the-art results while requiring fewer parameters, making them well-suited for situations with constrained computational resources [28]. The architecture employs a compound scaling method that meticulously manages network depth, width, and resolution. This systematic approach yields models that achieve an optimal balance between accuracy and computational efficiency.

4.4 DenseNet

Dense convolutional networks, or DenseNets, introduce a novel connectivity pattern wherein each layer receives direct input from all preceding layers [29]. This densely interconnected structure fosters feature reuse and gradient flow, mitigating vanishing gradients and promoting robust learning. DenseNet architectures are known for their parameter efficiency and remarkable performance on various image classification tasks.

5 Experimental Results and Discussion

The evaluation of the methods under consideration was conducted using three benchmark datasets, namely, ISIC 2016, ISIC 2017, and HAM10000:

1. *ISIC 2016 Dataset*: This dataset, short for the International Skin Imaging Collaboration 2016 dataset, consists of high-resolution images of skin lesions, designed for dermatology and medical image analysis research [30]. It comprises a total of 900 images, with 727 for training (173 melanoma, 727 non-melanoma) and 379 for testing (75 melanoma, 304 non-melanoma). Image dimensions vary from 556 × 679 pixels to 2848 × 4828 pixels.
2. *ISIC 2017 Dataset*: Released in 2017, the ISIC 2017 dataset is an extension of the ISIC project and offers a valuable resource for dermatology and medical image analysis. It includes 2000 training images and 600 testing images. The dataset contains 374 melanoma cases, 1372 benign nevi cases, and 254 seborrheic keratosis cases [31].
3. *HAM10000 Dataset*: The HAM10000 dataset is a collection of images representing seven distinct categories of skin lesions, each associated with specific medical conditions [32]. It includes 10,015 images and serves as a substantial resource for dermatology and medical image classification.

These datasets were split into training and testing sets with a 70:30 ratio. To evaluate melanoma classification performance, accuracy (ACC) and F1-score were employed as metrics. Data augmentation techniques were implemented using Python-based Keras and the Sklearn library. The selection of neural network parameters is pivotal to the overall model performance and generalization capabilities. To ensure an optimal configuration for each of the architectures discussed, a systematic approach was adopted. Initially, we began with commonly recommended parameter values from the literature for each respective network. These values served as a starting point. Subsequently, a combination of grid search and random search was employed to fine tune hyperparameters such as learning rate, batch size, and weight decay. Regularization techniques, like dropout rates, were also optimized through a series of cross-validation runs. Additionally, for networks that offered architectural flexibility, like EfficientNet's depth, width, and resolution scaling, a heuristic

Table 1 ACC values obtained by the models on the ISIC 2016 dataset

Data augmentation model	ResNet50	MobileNetV2	EfficientNet	DenseNet
No augmentation	79,123	78,456	80,234	81,789
AUG1	78,234	78,567	79,890	81,345
AUG2	78,678	78,123	79,456	81,901
AUG3	78,789	78,901	79,678	81,234
AUG4	78,567	78,456	79,234	81,678
AUG5	78,456	78,678	79,567	81,789
AUG6	78,901	78,234	79,901	81,567
AUG7	78,345	78,789	79,345	81,123
AUG8	79,567	79,678	82,012	83,678
AUG9	79,234	79,345	82,456	83,456
AUG10	79,901	79,567	82,123	83,789
AUG11	79,345	79,789	82,567	83,234
AUG12	79,789	79,234	82,789	83,901
AUG13	79,012	79,901	82,345	83,345
AUG14	80,123	80,345	83,123	84,012
AUG15	80,789	80,901	83,567	84,234
AUG16	80,456	80,012	83,789	84,789
AUG17	84,234	83,901	83,234	84,567

approach based on preliminary experiment results was taken. All models were trained on a validation set, and the configurations yielding the highest validation accuracy were chosen for our experiments. This rigorous and iterative methodology ensured that each network operated near its peak performance capability for the given task. In Table 1, the empirical results of four different models (ResNet50, MobileNetV2, EfficientNet, and DenseNet) are presented, showcasing their performance on the ISIC 2016 dataset across various data augmentation settings (AUG1 to AUG17). The evaluation metric used for comparison is accuracy (ACC), reported as percentages. DenseNet consistently outperforms the other models across all data augmentation levels, achieving accuracy values in the range of 81.345–84.567%.

EfficientNet also demonstrates competitive performance, consistently ranking second across all augmentation models. The impact of data augmentation on model performance is evident, with higher levels of augmentation generally leading to improved accuracy. This suggests that augmenting the training data enhances model generalization.

For DenseNet, the highest accuracy of 84.567% is achieved at AUG17, indicating that ensemble data augmentation can lead to superior model performance. ResNet50 and MobileNetV2 exhibit relatively stable performance across different augmentation levels but tend to lag behind DenseNet and EfficientNet in terms of accuracy. This suggests that the latter models may be better suited for this specific dataset.

Table 2 provides a comprehensive assessment of the performance of four distinct models—ResNet50, MobileNetV2, EfficientNet, and DenseNet—on the ISIC 2017 dataset. The table emphasizes a noteworthy enhancement in model accuracy when using ensemble data augmentation models. This underscores the significance of data

Table 2 ACC values obtained by the models on the ISIC 2017 dataset

Data augmentation model	ResNet50	MobileNetV2	EfficientNet	DenseNet
No augmentation	73,848	74,540	75,861	76,242
AUG1	75,509	77,787	75,773	80,849
AUG2	76,627	76,734	77,295	77,469
AUG3	76,687	77,268	77,414	78,081
AUG4	78,229	78,382	78,250	78,367
AUG5	78,429	78,682	78,445	78,453
AUG6	78,718	78,980	78,771	78,844
AUG7	79,023	79,228	79,105	79,108
AUG8	79,293	79,605	79,345	79,419
AUG9	79,651	79,857	79,748	79,787
AUG10	80,034	80,442	80,191	80,272
AUG11	80,447	80,750	80,481	80,663
AUG12	80,931	81,432	81,080	81,161
AUG13	81,484	81,773	81,609	81,624
AUG14	81,904	82,773	82,440	82,445
AUG15	82,831	83,655	83,368	83,505
AUG16	83,691	84,362	83,829	84,335
AUG17	84,511	87,268	84,701	85,393

augmentation techniques in improving the models' ability to accurately classify skin lesions. DenseNet consistently outperforms the other models across all augmentation levels, achieving the highest accuracy of 85.393% at AUG17. MobileNetV2, EfficientNet, and ResNet50 also demonstrate competitive performance, with accuracy values ranging from 84.511% to 87.268% at AUG17. This highlights the effectiveness of ensemble data augmentation techniques in enhancing the classification accuracy of skin lesion models on the ISIC 2017 dataset.

Table 3 presents the accuracy values obtained on the HAM10000 dataset. Across all data augmentation levels, the DenseNet model consistently achieves the highest accuracy, reaching an impressive 92.098% in AUG17. This suggests that DenseNet is well suited for the complex task of skin lesion classification in the HAM10000 dataset. EfficientNet and ResNet50 also deliver competitive accuracy values, particularly at the highest augmentation levels, surpassing 90% accuracy. In contrast, MobileNetV2 seems to perform less effectively than the other models on this dataset, with an accuracy of around 88–89% even with ensemble data augmentation. In Tables 4, 5, and 6, we present the F1-score values obtained by ISIC 2016, ISIC 2017, and HAM10000 datasets, respectively. The same patterns obtained by the models in terms of classification accuracy are also valid for the predictive performance values presented in Tables 1, 2, and 3.

In Figs. 1, 2, and 3, the main effect plot of accuracy for data augmentation models, CNN architectures, and different augmentation approaches, respectively, are shown. As illustrated in Fig. 1, image data augmentation can enhance the predictive performance of deep neural network architectures for skin lesion classification. Regarding the results summarized in Fig. 2, the highest average predictive

Table 3 ACC values obtained by the models on the HAM10000 dataset

Data augmentation model	ResNet50	MobileNetV2	EfficientNet	DenseNet
No augmentation	87,812	78,543	77,123	77,123
AUG1	77,123	77,123	88,045	88,769
AUG2	78,543	78,543	78,543	78,543
AUG3	78,543	78,543	78,543	78,543
AUG4	78,543	78,543	78,543	78,543
AUG5	78,543	78,543	78,543	78,543
AUG6	78,543	78,543	78,543	78,543
AUG7	78,543	78,543	78,543	78,543
AUG8	89,456	88,789	89,123	90,122
AUG9	89,567	88,890	89,234	90,188
AUG10	89,678	88,901	89,345	90,233
AUG11	89,789	88,912	89,456	90,329
AUG12	89,900	88,923	89,567	90,624
AUG13	90,000	88,934	89,678	90,412
AUG14	90,056	88,945	89,789	90,590
AUG15	90,067	88,956	89,900	90,076
AUG16	90,078	88,967	90,213	91,087
AUG17	90,089	88,978	90,011	92,098

Table 4 F1-score values obtained by the models on the ISIC 2016 dataset

Data augmentation model	ResNet50	MobileNetV2	EfficientNet	DenseNet
No augmentation	0.633	0.628	0.642	0.654
AUG1	0.626	0.629	0.639	0.651
AUG2	0.629	0.625	0.636	0.655
AUG3	0.630	0.631	0.637	0.650
AUG4	0.629	0.628	0.634	0.653
AUG5	0.628	0.629	0.637	0.654
AUG6	0.631	0.626	0.639	0.653
AUG7	0.627	0.630	0.635	0.649
AUG8	0.637	0.637	0.656	0.669
AUG9	0.634	0.635	0.660	0.668
AUG10	0.639	0.637	0.657	0.670
AUG11	0.635	0.638	0.661	0.666
AUG12	0.638	0.634	0.662	0.671
AUG13	0.632	0.639	0.659	0.667
AUG14	0.641	0.643	0.665	0.672
AUG15	0.646	0.647	0.669	0.674
AUG16	0.644	0.640	0.670	0.678
AUG17	0.674	0.671	0.666	0.677

Table 5 F1-score values obtained by the models on the ISIC 2017 dataset

Data augmentation model	ResNet50	MobileNetV2	EfficientNet	DenseNet
No augmentation	0.591	0.596	0.607	0.610
AUG1	0.604	0.622	0.606	0.647
AUG2	0.613	0.614	0.618	0.620
AUG3	0.613	0.618	0.619	0.625
AUG4	0.626	0.627	0.626	0.627
AUG5	0.627	0.629	0.628	0.628
AUG6	0.630	0.632	0.630	0.631
AUG7	0.632	0.634	0.633	0.633
AUG8	0.634	0.637	0.635	0.635
AUG9	0.637	0.639	0.638	0.638
AUG10	0.640	0.644	0.642	0.642
AUG11	0.644	0.646	0.644	0.645
AUG12	0.647	0.651	0.649	0.649
AUG13	0.652	0.654	0.653	0.653
AUG14	0.655	0.662	0.660	0.660
AUG15	0.663	0.669	0.667	0.668
AUG16	0.670	0.675	0.671	0.675
AUG17	0.676	0.698	0.678	0.683

Table 6 F1-score values obtained by the models on the HAM10000 dataset

Data augmentation model	ResNet50	MobileNetV2	EfficientNet	DenseNet
No augmentation	0.702	0.628	0.617	0.617
AUG1	0.617	0.617	0.704	0.710
AUG2	0.628	0.628	0.628	0.628
AUG3	0.628	0.628	0.628	0.628
AUG4	0.628	0.628	0.628	0.628
AUG5	0.628	0.628	0.628	0.628
AUG6	0.628	0.628	0.628	0.628
AUG7	0.628	0.628	0.628	0.628
AUG8	0.716	0.710	0.713	0.721
AUG9	0.717	0.711	0.714	0.722
AUG10	0.717	0.711	0.715	0.722
AUG11	0.718	0.711	0.716	0.723
AUG12	0.719	0.711	0.717	0.725
AUG13	0.720	0.711	0.717	0.723
AUG14	0.720	0.712	0.718	0.725
AUG15	0.721	0.712	0.719	0.721
AUG16	0.721	0.712	0.722	0.729
AUG17	0.721	0.712	0.720	0.737

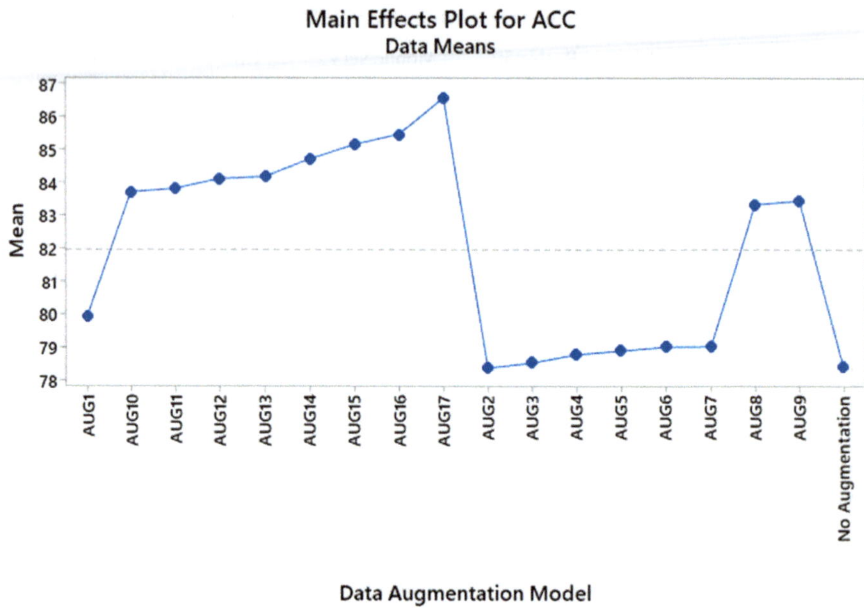

Fig. 1 The main effect plot of accuracy for data augmentation models

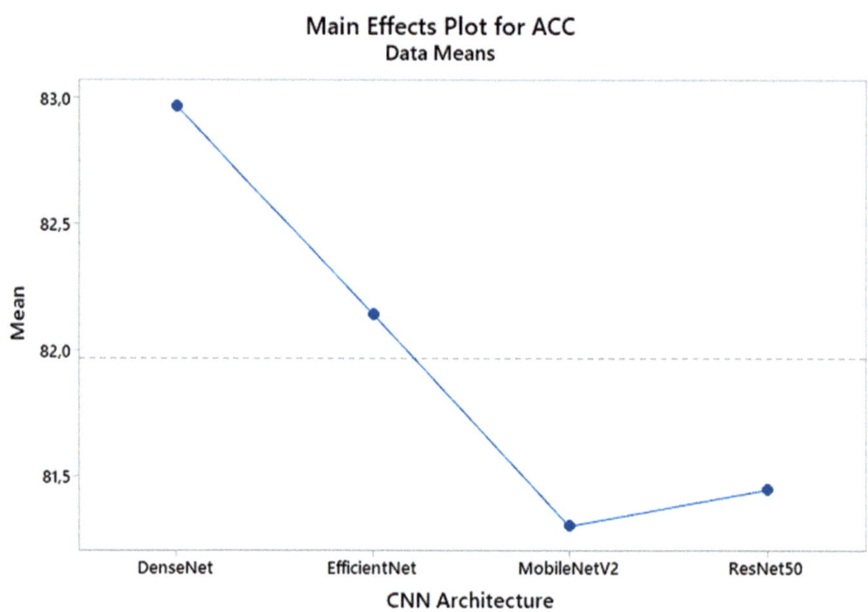

Fig. 2 The main effect plot of accuracy for CNN architectures

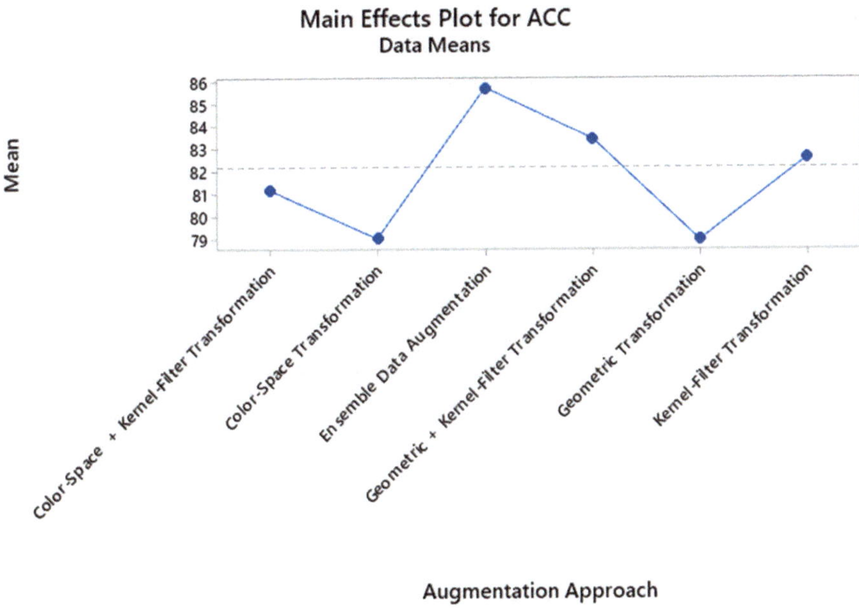

Fig. 3 The main effect plot of different augmentation approaches

performances have been achieved by the DenseNet model and the lowest average predictive performances have been achieved by the MobileNetV2 model. Regarding the predictive performances of augmentation approaches as illustrated in Fig. 3, geometric transformation approaches (such as AUG1, and AUG3) yield relatively low predictive performance. Similarly, color-space transformation approaches (such as AUG6) yield relatively low predictive performance. The ensemble data augmentation approaches and augmentation models combining several approaches yield better predictive performances.

6 Conclusion

The increasing need for precise and effective techniques in medical imaging, especially for skin lesion identification, highlights the importance of improving the reliability and precision of convolutional neural networks (CNNs). This study delved deeply into the realm of data augmentation, addressing issues such as the lack of extensive annotated datasets and the complexity of annotations. From our empirical examination, several crucial insights were gleaned: Data augmentation is essential to compensate for the challenges posed by limited datasets. These augmentation methods serve to expand the training data, subsequently enhancing the adaptability of models, a factor particularly vital for niche areas like skin lesion detection. There's a variance in the effectiveness of augmentation techniques. While traditional geometric and color-based transformations provide certain benefits, they often lag in performance compared to more advanced methods. In terms of model performance, DenseNet consistently outperformed others across different augmentation levels, achieving a notable accuracy of 92.098% at AUG17. Ensemble data augmentation techniques, which utilize a combination of different methods, were particularly effective. They consistently showcased better outcomes, emphasizing their potential to develop robust models. Such findings suggest that augmented data can pave the way for powerful automated skin lesion detection systems, offering immense benefits in clinical scenarios by reducing manual labeling and speeding up diagnosis. In essence, while data augmentation is a game-changer for medical imaging tasks, the choice of techniques and models, like DenseNet, can significantly shape the results.

Acknowledgments This study has been supported within the scope of TÜBİTAK Project Number 122E601.

Conflict of Interest The authors have no relevant financial or non-financial interests to disclose.

Data Availability Training and testing processes have been carried out using the 1. ISIC 2016 Dataset, 2. ISIC 2017 Dataset, and 3. HAM10000 Dataset.

References

1. Elngar, A.A., Arafa, M., Fathy, A., Moustafa, B., Mahmoud, O., Shaban, M., Fawzy, N.: Image classification based on CNN: a survey. J. Cybersecur. Inf. Manag. **6**(1), 18–50 (2021)
2. Rawat, W., Wang, Z.: Deep convolutional neural networks for image classification: a comprehensive review. Neural Comput. **29**(9), 2352–2449 (2017)
3. Li, Z., Liu, F., Yang, W., Peng, S., Zhou, J.: A survey of convolutional neural networks: analysis, applications, and prospects. IEEE Trans. Neural Netw. Learn. Syst. **33**, 6999–7019 (2021)
4. Chen, L., Li, S., Bai, Q., Yang, J., Jiang, S., Miao, Y.: Review of image classification algorithms based on convolutional neural networks. Remote Sens. **13**(22), 4712 (2021)
5. Sharma, N., Jain, V., Mishra, A.: An analysis of convolutional neural networks for image classification. Procedia Comput. Sci. **132**, 377–384 (2018)
6. Hernández-García, A., König, P.: Further advantages of data augmentation on convolutional neural networks. In: Artificial Neural Networks and Machine Learning–ICANN 2018: 27th International Conference on Artificial Neural Networks, pp. 95–103. Springer International Publishing (2018)
7. Grünberg, K., Jimenez-del-Toro, O., Jakab, A., Langs, G., Salas Fernandez, T., Winterstein, M., Krenn, M.: Annotating medical image data. In: Cloud-Based Benchmarking of Medical Image Analysis, pp. 45–67. Springer (2017)
8. Dgani, Y., Greenspan, H., Goldberger, J.: Training a neural network based on unreliable human annotation of medical images. In: 2018 IEEE 15th International Symposium on Biomedical Imaging (ISBI 2018), pp. 39–42. IEEE (2018)
9. Mikołajczyk, A., Majchrowska, S., Carrasco Limeros, S.: The (de)biasing effect of gan-based augmentation methods on skin lesion images. In: International Conference on Medical Image Computing and Computer-Assisted Intervention, pp. 437–447. Springer Nature Switzerland (2022)
10. Adegun, A., Viriri, S.: Deep learning techniques for skin lesion analysis and melanoma cancer detection: a survey of state-of-the-art. Artif. Intell. Rev. **54**, 811–841 (2021)

11. Zhang, D., Islam, M.M., Lu, G.: A review on automatic image annotation techniques. Pattern Recogn. **45**(1), 346–362 (2012)
12. Hanbury, A.: A survey of methods for image annotation. J. Vis. Lang. Comput. **19**(5), 617–627 (2008)
13. Bravin, R., Nanni, L., Loreggia, A., Brahnam, S., Paci, M.: Varied image data augmentation methods for building ensemble. IEEE Access. **11**, 8810–8823 (2023)
14. Shorten, C., Khoshgoftaar, T.M.: A survey on image data augmentation for deep learning. J. Big Data. **6**(1), 1–48 (2019)
15. Garcea, F., Serra, A., Lamberti, F., Morra, L.: Data augmentation for medical imaging: a systematic literature review. Comput. Biol. Med. **152**, 106391 (2022)
16. Khan, I.U., Aslam, N., Anwar, T., Aljameel, S.S., Ullah, M., Khan, R., Akhtar, N.: Remote diagnosis and triaging model for skin cancer using EfficientNet and extreme gradient boosting. Complexity. **2021**, 1–13 (2021)
17. Zhang, N., Cai, Y.X., Wang, Y.Y., Tian, Y.T., Wang, X.L., Badami, B.: Skin cancer diagnosis based on optimized convolutional neural network. Artif. Intell. Med. **102**, 101756 (2020)
18. Goceri, E.: Classification of skin cancer using adjustable and fully convolutional capsule layers. Biomed. Signal Process. Control. **85**, 104949 (2023)
19. Adla, D., Reddy, G.V.R., Nayak, P., Karuna, G.: A full-resolution convolutional network with a dynamic graph cut algorithm for skin cancer classification and detection. Healthc. Anal. **3**, 100154 (2023)
20. Liu, N., Rejeesh, M.R., Sundararaj, V., Gunasundari, B.: ACO-KELM: Anti Coronavirus Optimized Kernel-based Softplus Extreme Learning Machine for classification of skin cancer. Expert Syst. Appl. **232**, 120719 (2023)
21. Yang, Y., Xie, F., Zhang, H., Wang, J., Liu, J., Zhang, Y., Ding, H.: Skin lesion classification based on two-modal images using a multi-scale fully-shared fusion network. Comput. Methods Prog. Biomed. **229**, 107315 (2023)
22. Onan, A., Bulut, V., Ezgi, A.: Dermoskopik Görüntü Sınıflandırmada Temel Veri Artırım Yöntemlerinin Değerlendirilmesi. In: International Conference on Recent and Innovative Results in Engineering and Technology, pp. 78–86. All Sciences Proceedings (2023)
23. Nanni, L., Paci, M., Brahnam, S., Lumini, A.: Comparison of different image data augmentation approaches. J. Imag. **7**(12), 254 (2021)
24. Nanni, L., Paci, M., Brahnam, S., Lumini, A.: Feature transforms for image data augmentation. Neural Comput. & Applic. **34**(24), 22345–22356 (2022)
25. Velasco, G.A., Holighaus, N., Dörfler, M., Grill, T.: Constructing an invertible constant-Q transform with non-stationary Gabor frames. Proc. DAFx-11. **33**, 81 (2011)
26. He, K., Zhang, X., Ren, S., Sun, J.: Deep residual learning for image recognition. In: Proceedings of the IEEE Conference on Computer Vision and Pattern Recognition, pp. 770–778. IEEE (2016)
27. Sandler, M., Howard, A., Zhu, M., Zhmoginov, A., Chen, L.C.: Mobilenetv2: inverted residuals and linear bottlenecks. In: Proceedings of the IEEE Conference on Computer Vision and Pattern Recognition, pp. 4510–4520. IEEE (2018)
28. Tan, M., Le, Q.: Efficientnet: rethinking model scaling for convolutional neural networks. In: International Conference on Machine Learning, pp. 6105–6114. PMLR (2019)
29. Huang, G., Liu, Z., Pleiss, G., Van Der Maaten, L., Weinberger, K.Q.: Convolutional networks with dense connectivity. IEEE Trans. Pattern Anal. Mach. Intell. **44**(12), 8704–8716 (2019)
30. Codella, N.C., Gutman, D., Celebi, M.E., Helba, B., Marchetti, M.A., Dusza, S.W., Halpern, A.: Skin lesion analysis toward melanoma detection: a challenge at the 2017 international symposium on biomedical imaging (ISBI), hosted by the international skin imaging collaboration (ISIC). In: 2018 IEEE 15th International Symposium on Biomedical Imaging (ISBI 2018), pp. 168–172. IEEE (2018)
31. Berseth, M.: ISIC 2017-skin lesion analysis towards melanoma detection. arXiv preprint arXiv:1703.00523. (2017)
32. Tschandl, P., Rosendahl, C., Kittler, H.: The HAM10000 dataset, a large collection of multi-source dermatoscopic images of common pigmented skin lesions. Sci. Data. **5**(1), 1–9 (2018)

Open-Source Visual Target-Tracking System Both on Simulation Environment and Real Unmanned Aerial Vehicles

Celil Yılmaz, Abdulkadir Ozgun, Berat Alper Erol, and Abdurrahman Gumus

1 Introduction

Unmanned aerial vehicles (UAV) with advanced vision techniques offer substantial advantages, particularly for tasks requiring clear visualization and reliable perception, such as aerial surveillance. Surveillance is crucial for ensuring safety and security by identifying and preventing unusual events [1]. Tasks like information gathering, military reconnaissance, target tracking, and traffic management are closely tied to surveillance technologies [2]. However, traditional surveillance methods involve the manual identification of targets, which is time-consuming, labor-intensive, costly, and risky in inaccessible areas. To address these challenges, UAVs equipped with vision capabilities are gaining popularity as surveillance tools [3]. These UAVs are agile, allowing them to access confined spaces, and possess real-time visual capabilities to capture remote scenes. Autonomous UAVs, operating without manual intervention, offer cost-effective solutions for routine surveillance across industries. They can continuously monitor distant moving objects, significantly reducing human effort [4].

Certain tracking approaches involve computational overhead, limiting real-time performance due to constrained onboard resources. Incorporating multiple sensor

C. Yılmaz · A. Gumus (✉)
Electrical and Electronics Engineering, Izmir Institute of Technology, Izmir, Turkey
e-mail: abdurrahmangumus@iyte.edu.tr

B. A. Erol
Computer Engineering, Izmir Institute of Technology, Izmir, Turkey
e-mail: beraterol@iyte.edu.tr

A. Ozgun
Meshine Swarm Technologies, Izmir, Turkey
e-mail: kadir@meshine.tech

M. N. Seyman (ed.), *2nd International Congress of Electrical and Computer Engineering*, EAI/Springer Innovations in Communication and Computing,
https://doi.org/10.1007/978-3-031-52760-9_11

data fusion can increase payload and power consumption, contradicting low-cost UAV solutions. Considering these limitations, traditional image post-processing methods for UAVs are deemed inadequate for real-time surveillance. With size, weight, and power limitations, single cameras are commonly favored sensors in state-of-the-art UAV technology. That said, deep learning-based systems have made significant strides in UAV research, particularly in the realm of computer vision [5]. Given the restricted field of view of cameras, effective UAV surveillance requires dynamic maneuvering to track and maintain a target within the camera's sight. This drives the need for a deep learning-based UAV system capable of real-time, dynamic object tracking. Such a system can autonomously monitor target activity during flights, using deep learning-based perception and filter-based 3D object pose tracking methods [6].

However, the issue lies in the fact that utilizing a microcontroller-based flight control card such as 'Pixhawk' or 'Ardupilot' is inadequate for executing computer vision tasks on unmanned aerial vehicles. This inadequacy stems from the presence of pre-existing flight control software on the microcontroller. Consequently, to enable real-time onboard computation required for vision applications, an advanced embedded hardware system like the NVIDIA Jetson Nano is employed as a companion computer on the UAV. This companion Jetson Nano computer must maintain constant communication with the main flight control computer. Nonetheless, the challenge at hand pertains to establishing the necessary communication infrastructure and data flow. Furthermore, it is essential to identify the tools that offer the most efficient and portable solution encompassing both software and hardware aspects.

Many papers on this topic propose a communication network between the main and companion computers utilizing a closed-loop or off-the-shelf system [7–10]. However, these proposals lack insights into the specifics of how this communication network was established. Some articles run both flight control software and deep learning-based applications on a single Nvidia computer [11], while others merely present target-tracking algorithms, omitting any mention of the underlying network system [12]. A few papers delve into the mechanics of how vision-based applications on the companion computer maintain connectivity with the main computer, guiding the UAV through data exchange facilitated by an open-source network system [6]. Several academic papers have explored a scenario similar to our own, involving the deployment of both onboard and flight computers on drones for target-tracking tasks [13, 14]. While these papers have showcased their vision-related tasks, they have primarily relied on software-in-the-loop (SIL) simulations and have not conducted real-world experiments. In some instances, real-life experiments were carried out, yet using different hardware such as Raspberry Pi and software communication tools such as MavProxy [15]. Despite achieving their objectives, the tools employed in these papers are considered somewhat outdated when compared to the more contemporary alternatives offered by Robot Operating Systems (ROS) and Jetson products. Furthermore, a subset of these papers introduced comprehensive and successful algorithms for managing target-tracking tasks [16, 17]. However, they did not provide open-source implementations of these algorithms. Additionally, although these papers share a conceptual similarity with

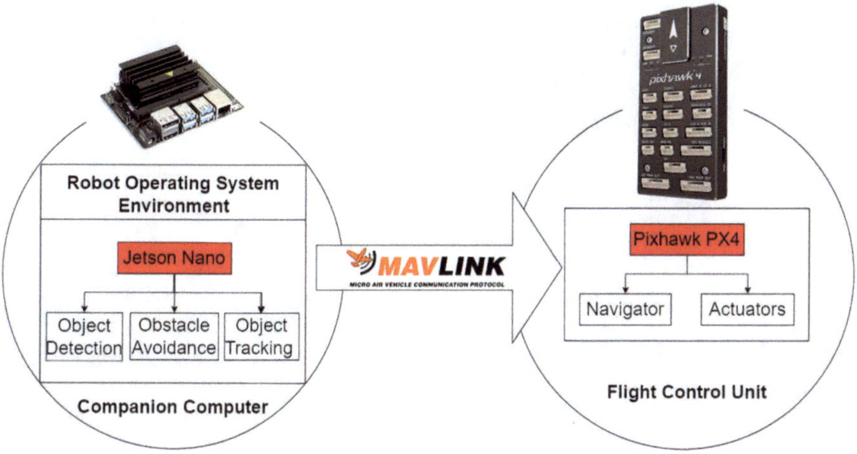

Fig. 1 Overview of the proposed system

our work and have conducted numerous successful online experiments, their test scenarios have largely been based on well-known aerial data sets [18]. Consequently, our paper explores an almost identical onboard target-tracking scenario as those previously discussed. However, we distinguish our research by utilizing open-source tools and cutting-edge vision algorithms, and importantly, by conducting experiments in both simulated and real-world environments.

Hence, it is imperative to introduce constructive approaches to address the challenge of interconnecting two computers and enabling offboard control via NVIDIA systems as depicted in Fig. 1, employing open-source tools and operational environments like ROS and Gazebo. These tools seamlessly integrate, boast user-friendly interfaces, and come with extensive support from the open-source community. This endeavor holds the potential to unlock more avenues for research and enhance productivity within the realm of computer vision and offboard guidance for unmanned autonomous vehicles.

The following content of the chapter is organized as follows: Sect. 2 introduces the proposed solution while Sect. 3 describes the results and discussions of the UAV system. Section 4 concludes the paper. Video footage of the experiments and implementation codes are attached as supplementary material.

2 Proposed Solution

In this context, we aim to contribute within the domain that involves integrating vision-based applications with UAVs, employing ROS open-source tools. ROS, short for Robot Operating System, is a middleware software development kit tailored for various robotics applications, including motion planning, computer vision,

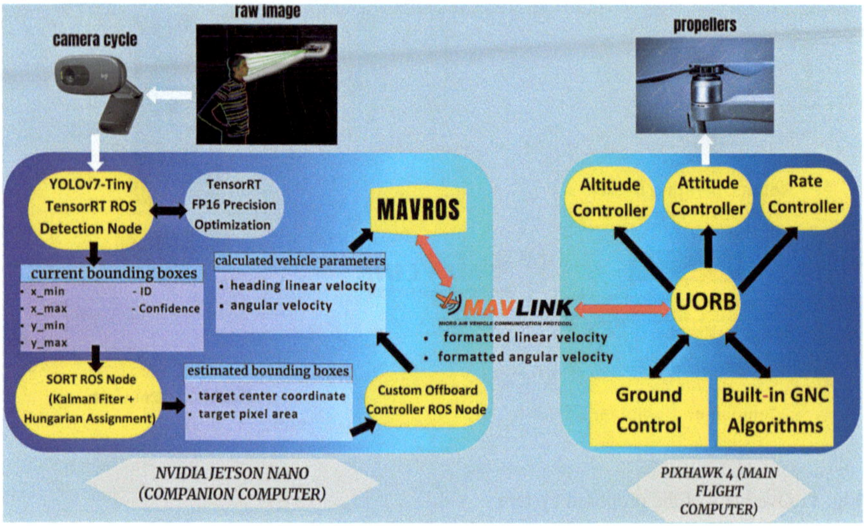

Fig. 2 Starting from the very first raw image data flows through ROS nodes on the companion computer qualifying more and more as moving on to the main controller

simulation, and robot control, among others, that facilitates communication between processes on the companion computer's side and the UAV. This is efficiently achieved through the ROS environment, aided by the MAVROS package which facilitates the conversion of messages into Micro Air Vehicle Link (MAVLink) message packages.

Communication with UAVs and intercommunication between their onboard components is facilitated by a protocol known as MAVLink [19]. This protocol utilizes a publish-subscribe approach, where data streams are disseminated as topics. While the Pixhawk PX4 environment employs a distinct communication protocol called uORB, specifically designed by PX4 developers for Pixhawk hardware, MAVLink serves as a versatile serial protocol commonly employed to transmit data and directives between vehicles and ground stations as shown in Fig. 2.

2.1 Software Part

This project essentially uses many ROS tools to be able to develop our vision applications in a fast way. Communication between the processes on the companion computer side is established easily using a ROS package (MAVROS) which can reformat the messages into a suitable version. With ROS, we will be able to easily separate our code base into packages containing small executable programs, called nodes. These nodes are publishing and subscribing to the data flow via ROS topics. For example, a Universal Serial Bus (USB) camera publishes images via the topic

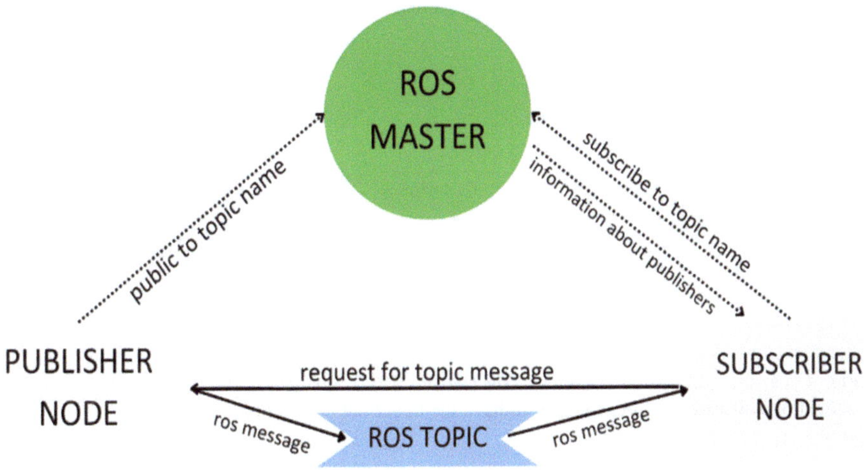

Fig. 3 ROS data distribution mechanism

'image_raw' and YOLOv7-Tiny TensorRT ROS Node subscribes to this topic to get the raw image data. There is a master node called 'roscore' in the ROS environment. In order to provide communication between the slave nodes, there must be communication with the master node first. Master node can be thought of as a kernel and their relationship can be seen in Fig. 3.

Object tracking consists of three different nodes. They can be listed as object detection, estimation, and controller nodes. An object detection node is necessary for target tracking because it detects the pixel coordinates of the target in the raw image provided by a real USB camera. This node uses the YOLOv7-Tiny model accelerated with the TensorRT engine. During the time of this research, YOLOv7 stood as the most recent iteration within the YOLO series. YOLOv7 represents a groundbreaking advancement in the computer vision field due to its remarkable features as a real-time object detection system. YOLOv7 outperforms all known object detectors in terms of both speed and accuracy, spanning the range from 5 FPS to 160 FPS. It boasts the highest accuracy, with a 56.8% average precision (AP), among all real-time object detectors operating at 30 FPS or higher on graphics processing units (GPU) V100. Furthermore, YOLOv7 achieved significant enhancements in real-time object detection accuracy without increasing the inference cost. It managed to reduce parameters by approximately 40% and computation by 50% compared to the state-of-the-art real-time object detector [20]. The first job of the YOLOv7-Tiny object detection node is to convert the darknet model to an *onnx* frozen graph and then convert this onnx to the TensorRT engine so that we can use the optimized version of the YOLOv7-Tiny model. This model publishes the coordinates of the bounding boxes of the target it detects, and also publishes the box identification number and the confidence value as a probability format. The estimation node receives the bounding box coordinates of the target in real time and makes an assignment to the most likely target which is determined from the predictions. This case is only

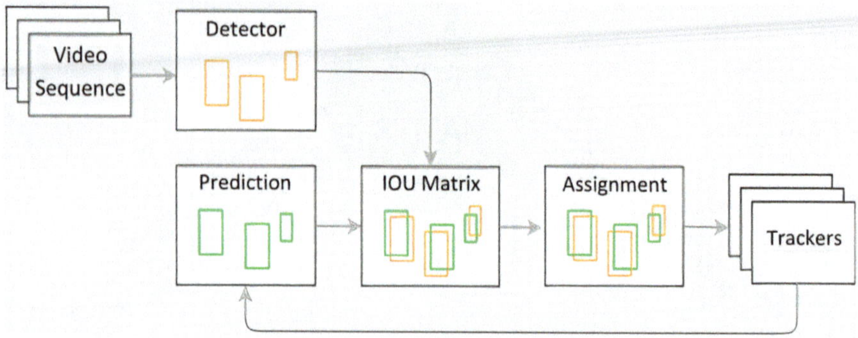

Fig. 4 On the companion computer side (Jetson Nano), image data are combined with SORT estimation software modules in order to have the target out of potentials

valid for single object tracking. The controller node takes the estimated bounding box coordinates of the target in real time and using these coordinates it calculates the area of the box and center of the box to modify the yaw angle and forwarding velocity. This way, the controller node publishes these updated messages to MAVROS topics. The overall data flow between the modules starting from the left to the right can be seen in Fig. 4.

The controller node takes the estimated detection results into the algorithm and generates the angular and linear velocity to keep the clearance distance of the target still. In this way, the target is followed offboard at the center of the frame during its mission. The provided pseudocode illustrates the internal algorithm of the offboard controller node. The resulting outputs are channeled into the MAVROS node, enabling the conversion of ROS messages into the MAVLink format. This makes Pixhawk distribute the incoming data into its built-in PX4 modules (Fig. 5).

2.2 *Hardware Part*

PX4 is the flight control unit of our system, and it is responsible for carrying out the guidance, navigation, and flight tasks. Thanks to its strong processor units and communication ports, the open-source autopilot system Pixhawk PX4 is an essential device for developers [21].

On the other hand, NVIDIA Jetson Nano is the GPU-based onboard companion computer in our system. Jetson Nano contributes our vision algorithms to the system by taking advantage of its CUDA GPU cores which are absent on the Pixhawk side. Therefore, target-tracking algorithms are run on Jetson Nano which is connected to Pixhawk via the Universal Asynchronous Receiver-Transmitter (UART) protocol physically (Fig. 6a, b).

```
Algorithm 1 Offboard Tracker
1: Input (Output of Tracker) : Class for person L = 1
2: Area of the estimated bounding box a = (height x weight)
3: Center of the estimated bounding box c = (height, weight)
4: A = constant (fixed to 5m in real world
5: C = constant (center of the pixel frame)
6:
7: Output (Input of the MAVROS) :
8: ErrorCenter (Angular Velocity) = C - c
9: ErrorArea (Linear Velocity) = A - a
10:
11: for iteration = 1, 2, ... do
12:     Look for target N
13:     if N = 1 then
14:         if L = 1 then
15:             Retrieve A and C from input
16:             Calculate ErrorArea = A - a
17:             Calculate ErrorCenter = C - c
18:             Repeat Step 16 until a ≤ A, and use ErrorArea for setting
    forward velocity
19:             Repeat Step 17 until ErrorCenter = 0, and use ErrorCenter
    for setting angular velocity
20:         else
21:             Do nothing
22:
```

Fig. 5 Pseudo code for offboard controller ROS node

3 Results and Discussions

We devised an independent system for monitoring using an unmanned aerial vehicle (UAV) that employs deep learning, SORT, PX4, and ROS tools and software packages. The YOLOv7-Tiny model, trained on the COCO data set, was utilized for recognizing objects, with an added step of eliminating irrelevant objects. To enhance tracking performance, SORT (employing Kalman filter, IOU Matrix, and Hungarian Assignment) was integrated with YOLOv7-Tiny to estimate the pose of targets.

Specifically for the graphical side, while Nvidia offers a comprehensive Jetpack Software Development Kit along with preconfigured tools and accelerated libraries for deep learning and computer vision [22], it was crucial for all application versions running in this environment to be compatible with each other. At the time we wrote this chapter, we employed Jetpack 4.5.1, TensorRT 7.1.3, and CUDA 10.2 to ensure compatibility. Furthermore, we set up the necessary environment by installing PyTorch v1.7.0 and TorchVision v.0.8.0 deep learning libraries, along with OpenCV version 3.4 and ROS Melodic.

Before the real-world tests, we needed to integrate the software and hardware components on the companion computer and flight control unit. Continuous a

b

Fig. 6 (**a**) Physical connection between the main flight controller unit (Pixhawk 4) and graphical unit (Jetson Nano). (**b**) Footage of the complete UAV waiting for takeoff

monitoring of the UAV's actions and computer activities was essential. To achieve this, we installed the Gazebo simulation environment and the software stack of Pixhawk PX4 on the host computer. To ensure the proper functioning of the Gazebo simulation environment and Pixhawk PX4's hardware, we updated the PX4 autopilot version and MAVLink protocol. For making the simulation part happen, we capitalized on the benefits of employing SITL (Simulation in the Loop) during the initial testing phase, primarily because it operates independently of hardware. Following that, we transitioned to HITL (Hardware in the Loop), which fully engages all the sensor and software components on physical hardware.

Test results in both Gazebo and the real world demonstrated that the settling time of the center coordinates of the target is short enough to keep the target alive on the frame center (Fig. 7a, b).

The results also showed that UAVs could complete their mission by processing about 18–20 FPS along with high accuracy. The object detection + SORT modules started to run with the very first frame which include the walking person in the scene (both simulation and real world). Just as the offboard controller ROS node takes the target information, angular and linear velocities are generated to guide the UAV offboard as shown in Fig. 8.

In other words, when the person tended to move forward-backward or left-right, the first modules precepted this movement and informed the offboard controller node. After that, inside the offboard controller node, the algorithm calculated the necessary pitch rate and yaw rate values to guide the UAV toward the target. The same scenario applies to the car as illustrated in Fig. 9.

These calculated parameters were sent to MAVROS topics to be able to reach the MAVLink side. Whenever MAVLink took these messages from MAVROS, it shared these messages with the low-level controller modules inside the Pixhawk via uORB. The simulation of this application runs on the host computer. The Pixhawk hardware is connected to the host computer for sharing the external commands coming from the companion computer. To control the UAV effectively, the Pixhawk controller needed to be in the offboard mode within the Gazebo simulation environment. During system testing, we encountered an issue where the UAV couldn't maintain offboard mode consistently. This was due to the low frequency of messages continuously transmitted from MAVLink to the companion computer. To address this, we increased the overall publication frequency of the topics. Overall, we incorporated a deep learning-based visual system into a flight controller, leveraging widely adopted open-source tools commonly utilized in both industry and academia. Given that this system encompasses various interconnected concepts, individuals have the flexibility to take advantage of its modularity and customize the system further according to their specific needs.

For our upcoming works, in order not to have problems within the data exchange, we plan to utilize ROS2 tools and enhance the tracking capabilities by incorporating the Deep SORT tracking algorithm alongside the detection algorithm, thereby increasing the robustness of the tracking process.

4 Conclusion

Our approach encompassed several key components. First, we deployed the YOLOv7-Tiny model to perform object detection, drawing on the COCO data set and implementing filters to eliminate unnecessary objects. To further enhance tracking accuracy, we harnessed the power of SORT, integrating elements like Kalman filters, IOU Matrix, and Hungarian Assignment to facilitate 3D target pose estimation alongside YOLOv7-Tiny. Our offboard ROS controller node

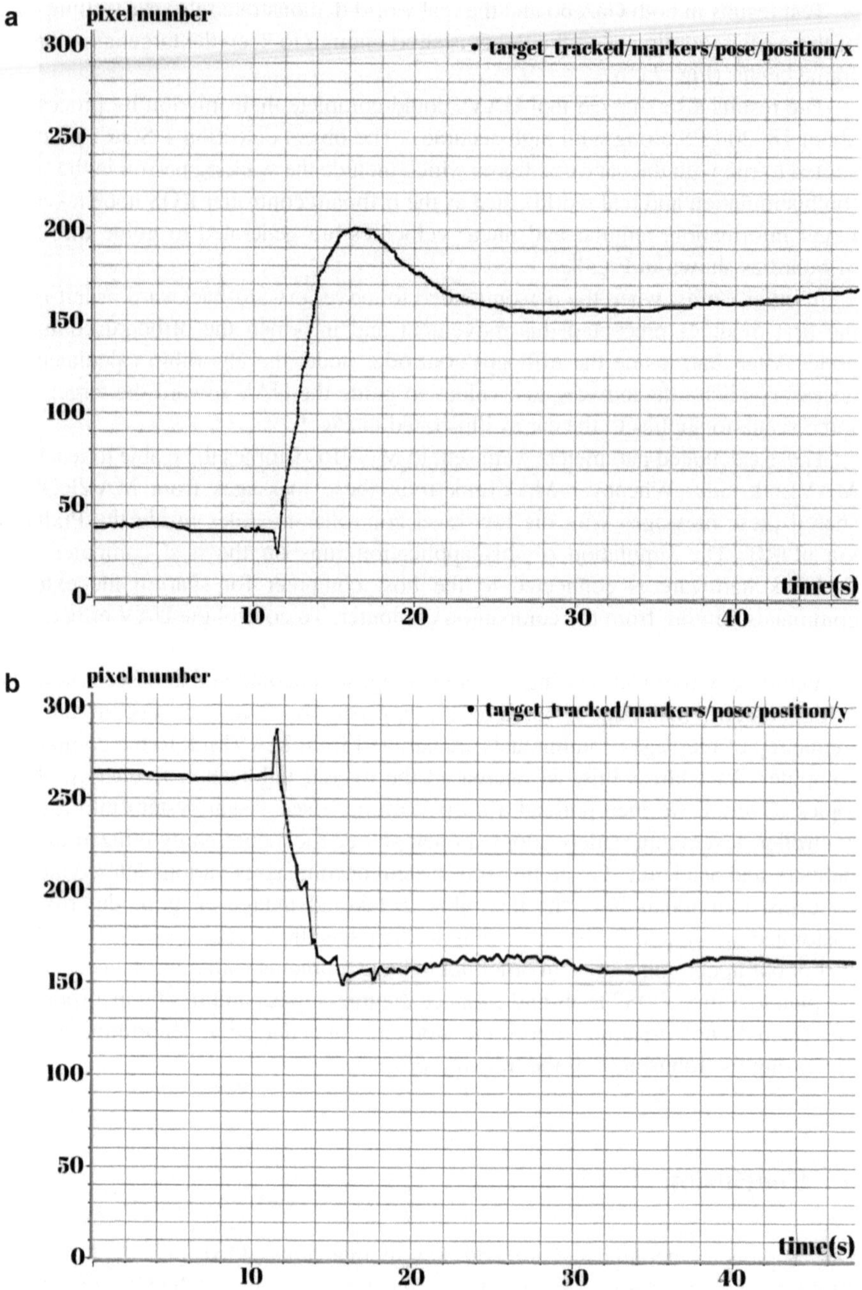

Fig. 7 (**a**) Settling time of the horizontal center coordinate of the target. (**b**) Settling time of the vertical center coordinate of the target

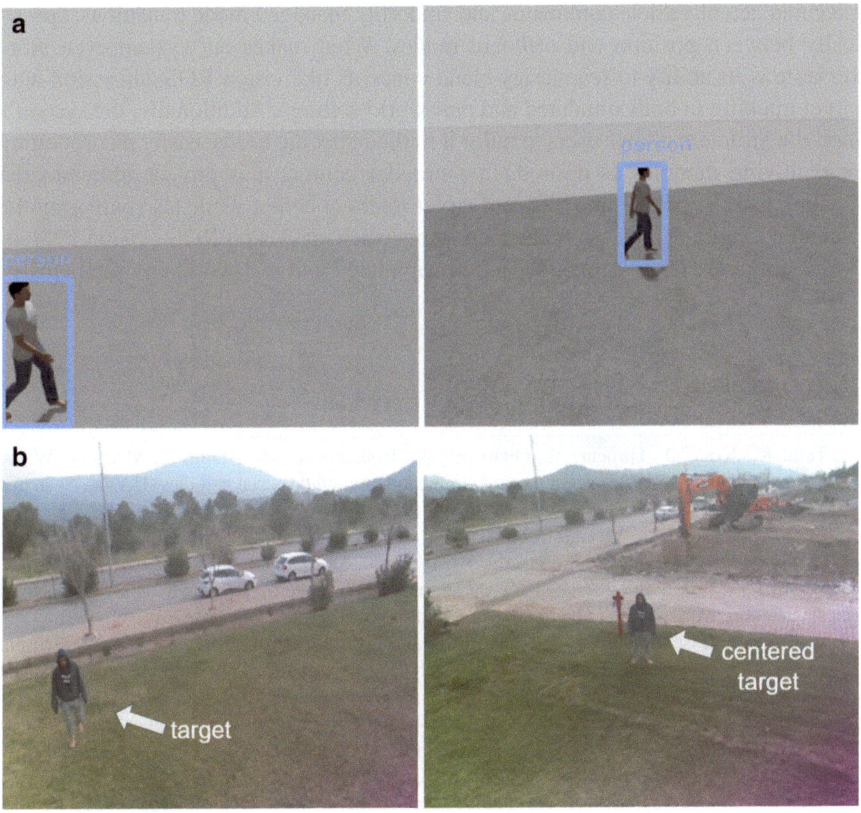

Fig. 8 (**a**) Gazebo simulation demonstrations of the person tracking. (**b**) Real-world tracking demonstration tests both longitudinal and lateral control parameters of the offboard ROS node while following the person

Fig. 9 Real-world tracking demonstration tests the longitudinal control parameters of the offboard ROS node while approaching the car

executed actual vehicle commands and smoothly managed mode transitions, specifically between position and offboard modes. What makes our system even more versatile is its ability to seamlessly blend concepts like vision-ROS and vision-auto target tracking in both simulated and real-world settings. Additionally, the system's modular structure allows users to tailor it to their specific needs, easily incorporating or removing modules as desired. In a broader context, this project addresses the existing gap in open-source offboard target-tracking codes, while also exploring the potential combinations of related concepts such as vision-ROS, vision-PX4 or ArduPilot, and vision-estimation in both simulated and real-world environments.

References

1. Telli, K., Kraa, O., Himeur, Y., Ouamane, A., Boumehraz, M., Atalla, S., Mansoor, W.: A comprehensive review of recent research trends on unmanned aerial vehicles (UAVs). Systems. **11**(8), 400 (2023)
2. Emimi, M., Khaleel, M., Alkrash, A.: The current opportunities and challenges in drone technology. Int. J. Electr. Eng. Sustain. **1**, 74–89 (2023)
3. Petkova, M.: Deploying drones for autonomous detection of pavement distress. Doctoral dissertation, Massachusetts Institute of Technology (2016)
4. Albani, D., IJsselmuiden, J., Haken, R., Trianni, V.: Monitoring and mapping with robot swarms for agricultural applications. In: 14th IEEE International Conference on Advanced Video and Signal Based Surveillance (AVSS) 2017, pp. 1–6. IEEE (2017)
5. Loce, R.P., Bala, R., Trivedi, M.: Computer vision and imaging in intelligent transportation systems. In: Wiley, J. (ed.) . Wiley, Hoboken (2017)
6. Lo, L.Y., Yiu, C.H., Tang, Y., Yang, A.S., Li, B., Wen, C.Y.: Dynamic object tracking on autonomous UAV system for surveillance applications. Sensors. **21**(23), 7888 (2021)
7. Hossain, S., Lee, D.J.: Deep learning-based real-time multiple-object detection and tracking from aerial imagery via a flying robot with GPU-based embedded devices. Sensors. **19**(15), 3371 (2019)
8. Fang, R., Cai, C.: Computer vision-based obstacle detection and target tracking for autonomous vehicles. In: MATEC Web of Conferences 2021, vol. 336, p. 07004. EDP Sciences (2021)
9. Paul, M., Danelljan, M., Mayer, C., Van Gool, L.: Robust visual tracking by segmentation. In: European Conference on Computer Vision 2022, pp. 571–588. Springer Nature Switzerland, Cham (2022)
10. Suchan, J., Bhatt, M., Varadarajan, S.: Commonsense visual sensemaking for autonomous driving–on generalised neurosymbolic online abduction integrating vision and semantics. Artif. Intell. **299**, 103522 (2021)
11. Jiang, Y., Jingliang, G., Yanqing, Z., Min, W., Jianwei, W.: Detection and tracking method of small-sized UAV based on YOLOv5. In: 19th International Computer Conference on Wavelet Active Media Technology and Information Processing (ICCWAMTIP) 2022, pp. 1–5. IEEE (2022)
12. Hao, J., Zhou, Y., Zhang, G., Lv, Q., Wu, Q.: A review of target tracking algorithm based on UAV. In: IEEE International Conference on Cyborg and Bionic Systems (CBS) 2018, pp. 328–333. IEEE (2018)
13. Nguyen, K.D., Nguyen, T.T.: Vision-based software-in-the-loop-simulation for Unmanned Aerial Vehicles using gazebo and PX4 open source. In: International Conference on System Science and Engineering (ICSSE) 2019, pp. 429–432. IEEE (2019)
14. Varatharasan, V., Rao, A.S.S., Toutounji, E., Hong, J.H., Shin, H.S.: Target detection, tracking and avoidance system for low-cost uavs using ai-based approaches. In: Workshop on Research,

Education and Development of Unmanned Aerial Systems (RED UAS) 2019, pp. 142–147. IEEE (2019)
15. Choi, H., Geeves, M., Alsalam, B., Gonzalez, F.: Open-source computer-vision based guidance system for UAVs on-board decision making. In: IEEE aerospace conference 2016, pp. 1–5. IEEE (2016)
16. Cheng, H., Lin, L., Zheng, Z., Guan, Y., Liu, Z.: An autonomous vision-based target tracking system for rotorcraft unmanned aerial vehicles. In: IEEE/RSJ International Conference on Intelligent Robots and Systems (IROS) 2017, pp. 1732–1738. IEEE (2017)
17. Wang, S., Jiang, F., Zhang, B., Ma, R., Hao, Q.: Development of UAV-based target tracking and recognition systems. IEEE Trans. Intell. Transp. Syst. **21**(8), 3409–3422. IEEE (2019)
18. Xiang, T., Jiang, F., Lan, G., Sun, J., Liu, G., Hao, Q., Wang, C.: UAV based target tracking and recognition. In: IEEE International Conference on Multisensor Fusion and Integration for Intelligent Systems (MFI) 2016, pp. 400–405. IEEE (2016)
19. Stateczny, A., Gierlowski, K., Hoeft, M.: Wireless local area network technologies as communication solutions for unmanned surface vehicles. Sensors. **22**(2), 655 (2022)
20. Olorunshola, O.E., Irhebhude, M.E., Evwiekpaefe, A.E.: A comparative study of YOLOv5 and YOLOv7 object detection algorithms. J. Comput. Soc. Inf. **2**(1), 1–12 (2023)
21. Gustafsson, J., Mogensen, D.: Streamlining UAV Communication: investigating and implementing an accessible communication interface between a ground control station and a companion computer (2023)
22. Kang, P., Somtham, A.: An evaluation of modern accelerator-based edge devices for object detection applications. Mathematics. **10**(22), 4299 (2022)

Semi-supervised Deep Learning for Liver Tumor and Vessel Segmentation in Whole-Body CT Scans

Hao-Liang Wen, Maxim Solovchuk, and Po-chin Liang

1 Introduction

The application of deep learning techniques has significantly advanced in the field of medical image analysis in recent years. The segmentation of medical images like CT scans has exhibited encouraging outcomes using these techniques in a variety of applications. Segmentation of diseased liver arteries, which is a key component of preoperative liver surgery simulation, is one area of particular interest [1].

Existing methods based on deep learning have been successful in segmenting general natural images and medical images [2]. However, the specific attributes of the pathological liver vessels make it challenging to apply these methods directly [1]. For example, the distinguishing features of the hepatic and portal veins require a more robust approach to assign labels accurately. To address the challenges associated with pathological segmentation of liver vessels, researchers have proposed the use of deep vision transformers. Deep vision transformers combine the capabilities of deep learning and transformers to improve the accuracy and robustness of segmentation results.

One of the key advantages of deep vision transformers is their ability to capture long-range dependencies in the image [3]. This is particularly important in the case

H.-L. Wen · M. Solovchuk (✉)
Institute of Biomedical Engineering and Nanomedicine, National Health Research Institutes, Zhunan, Taiwan

Department of Engineering Science and Ocean Engineering, National Taiwan University, Taipei, Taiwan
e-mail: solovchuk@nhri.edu.tw; solovchuk@gmail.com

P.-c. Liang
Department of Medical Imaging, National Taiwan University Hospital, Hsin-Chu Branch, Hsinchu, Taiwan

M. N. Seyman (ed.), *2nd International Congress of Electrical and Computer Engineering*, EAI/Springer Innovations in Communication and Computing,
https://doi.org/10.1007/978-3-031-52760-9_12

of pathological liver vessel segmentation, where the veins can extend across the entire liver. By considering the global context of the image, deep vision transformers can better distinguish between hepatic and portal veins and accurately assign labels. Another benefit of deep vision transformers is their capability to handle complex and ambiguous cases of vessel segmentation [4]. Unlike traditional machine learning methods, deep vision transformers can capture intricate patterns and subtle differences, allowing for more accurate classification of vessels.

In addition, deep vision transformers overcome the limitations of traditional deep learning approaches by addressing the issue of noisy misclassifications. By considering local 3D patches and tracing vessels to their source, deep vision transformers can reduce misclassifications and improve the overall accuracy of the segmentation results. In general, deep vision transformers show great potential in the field of pathological segmentation of liver vessels. Their ability to capture long-range dependencies, handle complex cases, and reduce misclassifications makes them a promising approach for accuracy and robustness.

Vessel segmentation has seen substantial advancements in recent years, transitioning from conventional image processing techniques to more advanced machine learning and deep learning methodologies. These progressions are pivotal in medical imaging applications, where precise vessel segmentation is crucial for both diagnostic and therapeutic purposes. Initial approaches to vessel segmentation largely utilized traditional image processing and machine learning techniques. However, these methods often found it challenging to cope with the complex and variable nature of vascular structures in different imaging modalities and individual subjects. The need for reduced manual intervention and enhanced automated segmentation drove the exploration toward deep learning methodologies.

With the advent of deep learning, convolutional neural networks (CNNs) emerged as a cornerstone for vessel segmentation tasks. The study by Yang et al. (2022) is a testament to the application of CNNs in improving image quality, a crucial aspect of precise vessel segmentation [2]. Despite their remarkable success, CNNs sometimes falter in handling long-range dependencies and intricate vascular structures, particularly in pathological scenarios. The limitations inherent in CNNs inspired further innovation, ushering in the era of deep-vision transformers. These architectures, as elaborated in the works of Zhou et al. (2023) and Kuang et al. (2022), show promise in capturing long-range dependencies and managing complex segmentation scenarios [3, 4]. Deep vision transformers exhibit an enhanced ability to differentiate between various vascular structures, such as the hepatic and portal veins, a critical requirement in pathological liver vessel segmentation.

Furthermore, the study by Liu et al. (2022) explores the potential of deep learning algorithms in a wider spectrum of medical imaging, signifying the evolving landscape of methodologies aimed at addressing the unique challenges posed by vessel segmentation tasks [1]. The ability of deep vision transformers to minimize misclassifications and improve overall accuracy, as discussed in the studies, indicates a promising trajectory for the field. In summary, the evolution of methodologies from classical machine learning to deep vision transformers underscores the ongoing quest for more precise and robust vessel segmentation techniques. The comparative advantages of deep vision transformers, highlighted in recent studies, mark significant strides toward addressing the inherent challenges of vessel segmentation, laying a solid foundation for future explorations in this domain.

2 Methodology

2.1 Study Design

The objective of this study is to develop a neural network model from the ground up, utilizing a semi-supervised learning pipeline, to perform the segmentation of liver tumors and hepatic vessels within whole-body CT scans.

2.2 Data Source

The dataset used in this research is a valuable resource for researchers and developers in the field of medical imaging. This dataset includes 443 CT scans (303 with ground truth and 140 without ground truth) of liver tumors and vessels with ground truth, which were obtained with specific criteria and semi-automatically segmented using the Scout application. The dataset was provided by Memorial Sloan Kettering Cancer Center and previously reported. This dataset is one of the ten datasets available for download and reuse in the MONAI dataset collection. MONAI is an open-source framework for deep learning in healthcare imaging that provides a unified API, data handlers for common medical imaging formats, and a large collection of pre-trained models and datasets. The MONAI dataset collection includes a variety of medical image datasets, including brain tumors, heart, liver, prostate, lung, pancreas, spleen, colon, and hepatic vessels. These datasets are designed to aid researchers and developers in the development and evaluation of segmentation algorithms for medical imaging. The dataset, in particular, is a valuable resource for those working on liver tumor and vessel segmentation.

2.3 Data Partition

Our dataset comprises 443 CT scans, divided into two subsets. The first subset consists of 303 CT scans with ground truth annotations, designated for model training (80%) and testing (20%). The second subset includes 140 CT scans without annotations, which are harnessed for our self-training approach. In self-training, these unlabeled scans contribute to dynamic data augmentation by generating pseudo-labels for model refinement. This strategy allows us to leverage unannotated data effectively, enhancing the model's segmentation performance on both annotated and unannotated medical images.

2.4 Model Architecture

UNETR [5], an innovative transformer-based architectural framework designed for semantic segmentation in volumetric medical imaging, fundamentally redefines the segmentation task as a 1D sequence-to-sequence prediction challenge. Leveraging the transformative capabilities of a transformer encoder, UNETR empowers the model to effectively grasp long-range dependencies, facilitating the capture of comprehensive global contextual representations across multiple scales. A notable departure from convention is UNETR's elimination of the reliance on a convolutional neural network (CNN) backbone for input sequence generation, opting instead for direct utilization of tokenized patches derived from volumetric data. This pioneering approach has demonstrated remarkable potential, establishing UNETR as a leader in both the Standard and Free Competitions on the BTCV leaderboard for multi-organ segmentation, achieving state-of-the-art performance.

The UNETR architecture, see Fig. 1, employs a contracting-expanding framework that integrates a sequence of transformers constituting the encoder, connected to a decoder through skip connections. Operating on a 1D sequence of input embeddings, these transformers are generated by partitioning the 3D input volume into non-overlapping, uniformly flattened patches. These patches undergo projection into a fixed-dimensional embedding space via a constant linear layer throughout the transformer layers. The preservation of spatial information from these patches is ensured through the incorporation of a learnable 1D positional embedding into the projected patch embeddings. Subsequently, a decoder facilitates the upscaling of the

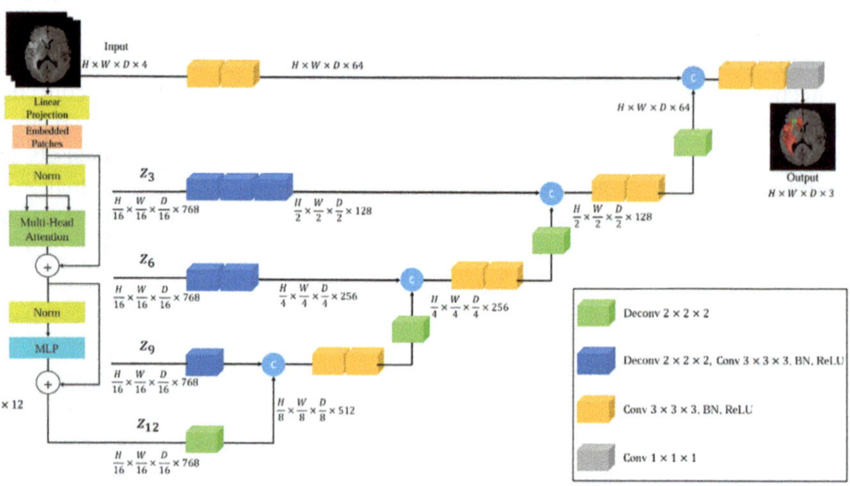

Fig. 1 The model architecture of UNETR. (The figure is sourced from Hatamizadeh et al. [5])

acquired representations to the input resolution, enabling pixel/voxel-wise semantic prediction. The strategic inclusion of skip connections merges encoder and decoder outputs at different resolutions, thereby facilitating the recovery of spatial information that may be compromised during down-sampling. UNETR's innovative architectural design holds promise as a foundation for a new class of transformer-based segmentation models within the domain of medical image analysis.

The architectural strengths of UNETR make it particularly well-suited for 3D image segmentation tasks. Its direct utilization of volumetric data and the pivotal role of transformers in the encoder, connected to the decoder via skip connections [6], enable the model to capture spatial context and long-range dependencies intrinsic to 3D medical images with exceptional efficacy. By incorporating transformers into the encoder, UNETR adeptly models global contextual information across multiple scales, a critical aspect for the precise segmentation of intricate anatomical structures. Furthermore, the utilization of tokenized patches as input sequences enhances efficiency in processing large 3D volumes while preserving vital spatial information. In summary, UNETR exhibits substantial potential for enhancing the precision and efficiency of 3D medical image segmentation tasks.

Notably, UNETR surpasses other CNN-based models, achieving state-of-the-art performance in both Standard and Free Competitions on the BTCV leaderboard. In the Standard Competition, UNETR attains a new pinnacle with an average Dice score of 85.3% across all organs. In the Free Competition, UNETR secures an overall average Dice score of 0.899, surpassing the second, third, and fourth top-ranked methodologies by 1.238%, 1.696%, and 5.269%, respectively. Additionally, UNETR outperforms the second-best baselines by 1.043%, 0.830%, and 2.125%, respectively, in terms of the Dice score for large organs like the spleen, liver, and stomach. For small organ segmentation, UNETR maintains a significant performance advantage over the second-best model.

However, UNETR is not without limitations. Transformers, despite their adeptness in capturing global information, struggle with localized details. To address this, UNETR employs a CNN-based decoder for capturing localized information. While UNETR maintains moderate model complexity and faster processing speeds compared to certain transformer-based models, it may not be the fastest option for 3D medical image segmentation. Furthermore, optimal performance with UNETR often necessitates a substantial volume of training data, which may not always be available in medical imaging applications. Lastly, UNETR's suitability may vary depending on the characteristics of the imaging modality and the specific anatomical structures being segmented.

Despite these limitations, UNETR consistently demonstrates its effectiveness in achieving state-of-the-art performance across a diverse range of 3D medical image segmentation tasks.

2.5 Data Augmentation

Data augmentation plays a pivotal role in medical image segmentation, addressing the perennial challenge of limited training data availability in the medical domain [7, 8]. Medical images, often characterized by their high dimensionality and complexity, frequently suffer from an insufficient number of samples to effectively train deep learning models. This technique's significance lies in its ability to artificially augment the training set by generating synthetic samples closely resembling actual data. Such augmentation not only enhances segmentation results' quality but also mitigates the pervasive problem of overfitting in deep learning applications constrained by limited training data. Beyond this, data augmentation effectively tackles the challenge of missing modalities in multi-modal image segmentation. Synthesizing additional samples becomes instrumental in improving the accuracy and robustness of medical image segmentation models, potentially revolutionizing clinical practice.

Moreover, data augmentation extends its utility beyond bolstering segmentation model accuracy and robustness—it contributes significantly to reducing the dependency on manual annotation of training data. Manual annotation, a laborious and costly process reliant on expert knowledge, introduces the possibility of inter-observer variability. By employing data augmentation to generate new samples, deep learning models can learn more effective medical image segmentation without being heavily reliant on vast quantities of manually annotated data. This is particularly advantageous in scenarios with limited annotated data availability, such as rare diseases or resource-constrained settings. Furthermore, data augmentation allows the simulation of diverse imaging conditions, including variations in image resolution, noise, or contrast, thereby enhancing the generalization capability of segmentation models when encountering new data. In summation, data augmentation emerges as a potent technique addressing challenges stemming from limited training data in medical image segmentation, with the potential to revolutionize clinical diagnosis and treatment accuracy and efficiency.

In this study, the data augmentation pipeline for a single-channel CT image with input intensity values in Hounsfield Units (HU) follows a sequential series of transformations:

1. Intensity Normalization and Clipping: The intensities within the tissue window are normalized to a range of (0,1), and any values outside the window are clipped.
2. Resampling: The image is resampled to match a pixel dimension of (0.8,0.8,0.8).
3. Foreground Cropping: Regions surrounding the foreground (i.e., areas of interest) are cropped.
4. Random Fixed-Sized Region Cropping: Three random fixed-sized regions of dimensions (192,192,64) are cropped. The centers of these regions are selected randomly from both foreground and background voxels, maintaining a 1:1 ratio.
5. Random Volume Rotations: The volume is randomly rotated.

6. Random Volume Zooming: Random zooming is applied to the volume.
7. Random Volume Smoothing: Gaussian kernels are randomly applied to smooth the volume.
8. Random Intensity Scaling: The intensity of the volume is randomly scaled.
9. Random Intensity Shifting: The intensity of the volume is randomly shifted.
10. Random Gaussian Noise Addition: Gaussian noise is randomly added to the volume.
11. Random Volume Flipping: The volume is randomly flipped in three directions.

It is duly noted that augmentations 4–11 are exclusively applied to the training data.

2.6 Evaluation Criteria

For medical image segmentation, the selection of the appropriate evaluation metric is crucial in order to assess the accuracy and efficacy of segmentation models. One of the widely adopted metrics, and the one utilized in this study, is the Dice score (also known as the Sørensen-Dice coefficient or F1 score). The Dice score is a robust and intuitive metric that quantifies the overlap between the predicted and ground truth segmentation masks. The Dice score is calculated using the following formula:

$$\text{Dice score} = \frac{2|A \cap B|}{|A| + |B|}. \tag{1}$$

where

- A represents the set of voxels in the predicted segmentation mask.
- B represents the set of voxels in the ground truth segmentation mask.
- value denotes$|\cdot|$ denotes the cardinality of a set, indicating the number of voxels in the respective masks.

The Dice score provides a value between 0 and 1, where a score of 1 indicates a perfect match between the predicted and ground truth masks, signifying a flawless segmentation. Conversely, a score of 0 represents no overlap between the masks, indicating a complete mismatch.

The Dice score is particularly suitable for medical image segmentation tasks due to its sensitivity to both false positives and false negatives. It rewards accurate localization and delineation of structures while penalizing both under-segmentation and over-segmentation errors. This sensitivity to boundary accuracy makes it a preferred choice in applications where precise anatomical structure delineation is of paramount importance, such as organ segmentation, tumor detection, and lesion delineation.

Throughout this study, the Dice score serves as the primary evaluation metric, providing a quantitative assessment of the model's segmentation performance. A higher Dice score corresponds to a more accurate and reliable segmentation, making it a pivotal metric in gauging the effectiveness of the proposed methodologies. In addition to the Dice score, other secondary metrics, such as sensitivity, specificity, and intersection over union (IoU), may also be considered to provide a comprehensive evaluation of the segmentation model's performance.

2.7 Training Pipeline

Creating an effective training strategy is fundamental to achieving a promising performance in medical image segmentation tasks. In this study, we have meticulously crafted a training pipeline (see Fig. 2) that combines cutting-edge optimization techniques and learning rate scheduling to maximize the performance of our neural network model.

To address the initial challenge of handling raw 3D images with varying dimensions, we employ a critical preprocessing step. The raw 3D images are resampled to a fixed size of small patches (Step 4 in the data augmentation pipeline, Sect. 2.4), which serve as the input for our model during training. This preprocessing step ensures uniformity in the training data, allowing our model to effectively learn intricate patterns and representations within these smaller, standardized patches.

It's important to note that for each image, we have set the number of patches at 4, and the batch size at 2, forming an effective batch size of 8. This configuration optimizes the training process by efficiently processing multiple patches concurrently, enhancing both computational efficiency and model convergence.

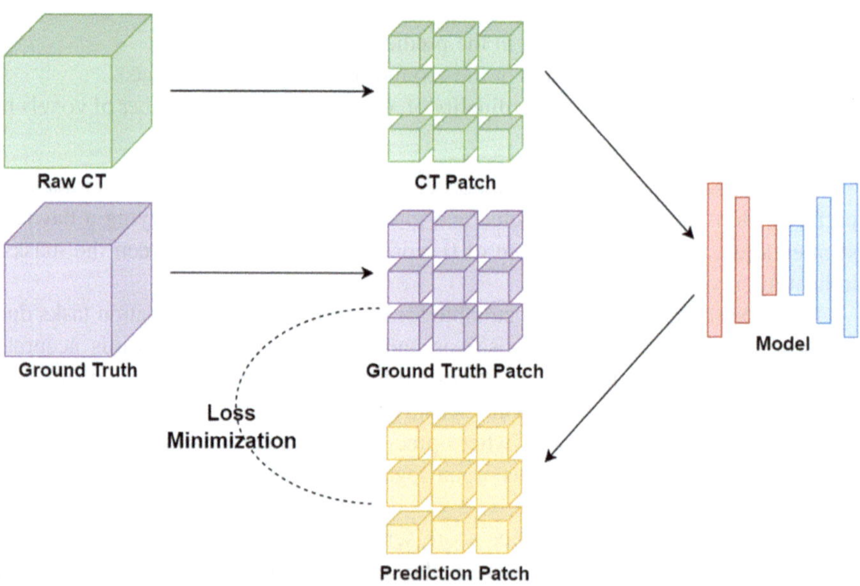

Fig. 2 The overall training pipeline in this study

Our training strategy is anchored by the utilization of the AdamW optimizer in conjunction with the DiceCELoss, a custom loss function tailored for the nuances of medical image segmentation. AdamW, known for its prowess in deep neural networks, plays a pivotal role in minimizing our specialized loss function and enhancing the accuracy and robustness of our segmentation model.

In addition to our choice of optimizer and loss function, we implement a cyclic learning rate scheduler to fine-tune the training process. This scheduler incorporates a dynamic range of learning rates, featuring a base learning rate of 1.0e-6 and a maximum learning rate of 5.0e-4, over a cyclic period spanning 100 epochs. This approach introduces adaptability into our training process, allowing our model to explore different learning rates during training, a crucial capability when dealing with the complex and diverse landscape of medical image data.

Our training regimen extends over 3000 epochs, affording our model ample opportunity to learn intricate patterns and representations within the standardized image patches. This extended duration underscores our commitment to achieving the highest possible segmentation accuracy and generalization capabilities.

2.8 *Inference with Sliding-Window*

For medical image segmentation, achieving high-precision results is of paramount importance, as accurate delineation of anatomical structures and pathology directly impacts clinical decision-making. While the training process, as discussed earlier, forms the bedrock for training robust segmentation models, the inference phase plays a critical role in applying these models to unseen data.

One common strategy employed during inference, particularly for volumetric data, is sliding window inference, which is also introduced in our inference pipeline (see Fig. 3). This technique is well-suited for processing large 3D medical volumes efficiently and ensuring that no regions are overlooked during segmentation.

In sliding window inference, a smaller, fixed-size "window" or "patch" is systematically moved across the entire 3D volume. At each position of the window, the trained segmentation model is applied to predict the segmentation mask for the contents of that window. The window is then shifted, overlapping with the previous one, until the entire volume has been processed. This approach allows the model to capture fine-grained details and ensures that even small or intricate structures within the data are not missed.

However, it's important to note that sliding window inference may introduce some challenges, such as handling overlapping predictions from neighboring windows and dealing with boundary artifacts. Strategies like weighted averaging or post-processing techniques can be employed to address these issues and ensure smooth and accurate segmentation results.

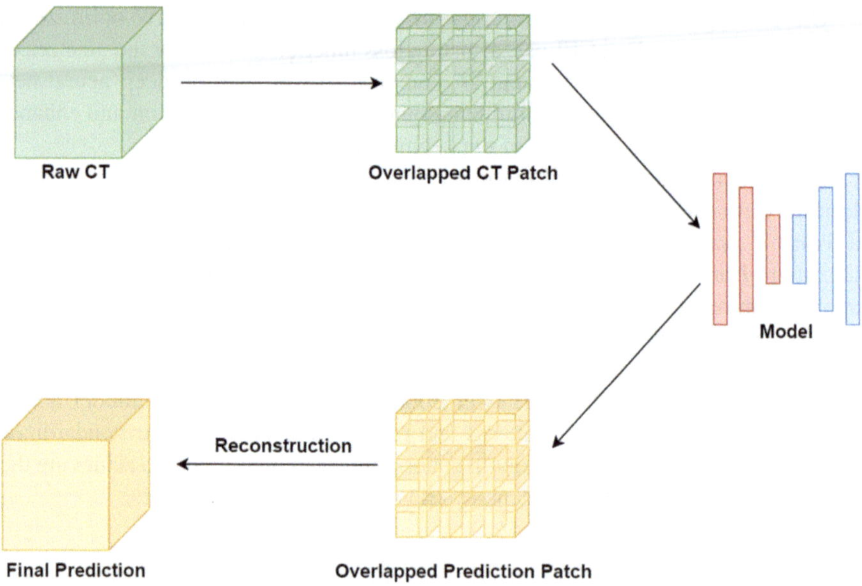

Fig. 3 Inference pipeline with sliding window inference

2.9 *Self-Training for Model Enhancement*

In the pursuit of enhancing the performance of our segmentation model, we employ a self-training methodology that allows the model to learn from unlabeled data. Self-training represents a powerful paradigm in semi-supervised learning, leveraging unlabeled data to bolster the model's capabilities. This self-training pipeline comprises a series of carefully designed steps, as outlined below.

Pseudo-labeling For each unlabeled image in our dataset, we initiate the process of pseudo-labeling. Voxels that are predicted to not belong to the background class, with prediction confidence exceeding a predefined threshold, receive a pseudo-label identical to their corresponding model prediction. Conversely, voxels with prediction confidence below this threshold are assigned the pseudo-label of background voxels.

Validation of Pseudo-Labels To ensure the quality and validity of the pseudo-labels, we calculate the ratio of each class within the pseudo-labeled voxels for each image. Images with pseudo-label ratios similar to those found in the labeled dataset are deemed to possess "valid" pseudo-labels. This step safeguards against potential noise introduced by mislabeled or ambiguous unlabeled data.

Training with Pseudo-Labels The core of our self-training approach lies in training the model using a combination of pseudo-labeled data and the labeled data from our dataset. This joint training process allows the model to refine its segmentation capabilities by incorporating insights gleaned from the unlabeled data. Notably, the set of pseudo-labeled data is periodically refreshed at specified intervals during training. This refreshing mechanism ensures that the model adapts to evolving patterns and variations in the unlabeled data, further enhancing its segmentation performance.

By integrating these self-training steps into our segmentation pipeline, we harness the power of semi-supervised learning, effectively leveraging unlabeled data to augment the model's ability to accurately classify and delineate anatomical structures and pathologies within medical images. This approach contributes to the development of a more robust and accurate segmentation model, with potential implications for improving clinical diagnosis and treatment planning.

3 Results and Discussion

The model achieved a Dice score of 0.67 for hepatic vessels and 0.65 for tumors in its overall performance on the testing set. Figures 4 and 5 exhibit two successful instances of model inference, both demonstrating highly promising results with prediction masks closely aligned with the ground truth. The model exhibits remarkable accuracy in identifying the position and size of tumors, showcasing its precision in anatomical structure delineation.

Furthermore, the figures illustrate the model's adeptness at mitigating noise introduced during the labeling process. This noise reduction is evident in the smoother vessel structures, which preserve the essential connectivity of vessel voxels. These observations underscore the model's ability to not only provide accurate segmentations but also enhance the overall visual quality of the segmented structures.

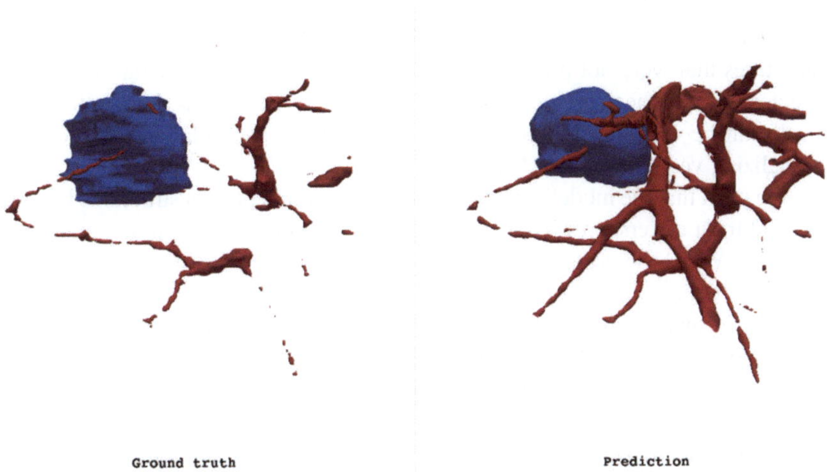

Fig. 4 Accurate segmentation of hepatic vessels and tumors

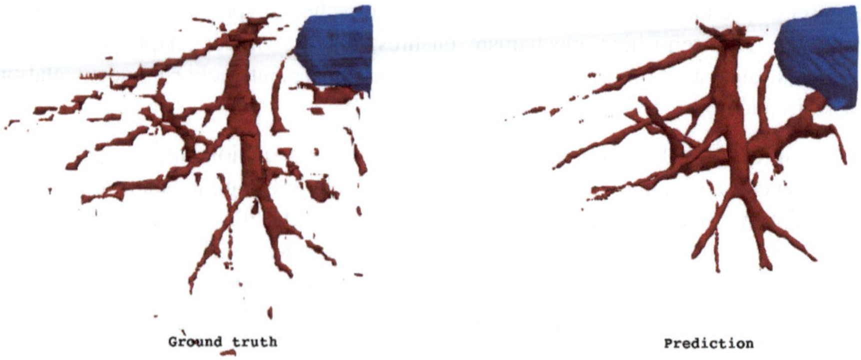

Fig. 5 Accurate segmentation of hepatic vessels and tumors

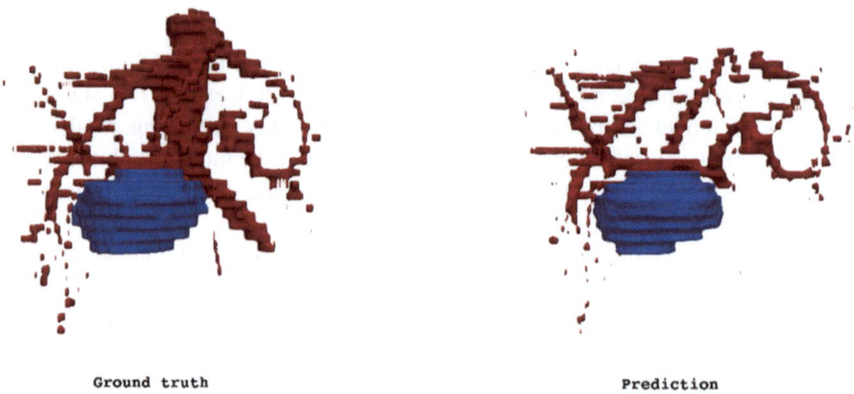

Fig. 6 Instances of unsuccessful hepatic vessel and tumor segmentation (IVC)

Additionally, it is worth noting that the model exhibits the capability to predict structures that were not present in the original ground truth annotations. Assessing the accuracy of these predictions poses a unique challenge, as the labeling process itself can be intricate and challenging for humans, particularly in the context of visualizing vessels within CT images.

We posit that the model's capacity to generalize and predict structures beyond the ground truth underscores its potential to offer valuable insights in the realm of data analysis. While the accuracy of these additional predictions may be challenging to ascertain definitively, they nonetheless highlight the model's ability to uncover previously unseen information within the data, potentially opening new avenues for exploration and understanding.

Nonetheless, it is crucial to acknowledge instances of prediction failure, as evidenced in Figs. 6 and 7. Notably, in Fig. 6, the model faces challenges in predicting the hepatic vessels, particularly the intricate structure of the inferior vena cava (IVC). It is important to consider the dataset's characteristics; typically,

Fig. 7 Instances of unsuccessful hepatic vessel and tumor segmentation (tumor)

IVC labeling is not consistently present in every image due to the specific nature of the data. This particular sample appears to be one of these special cases.

Given that the training data lack comprehensive information on IVC, it is reasonable to attribute the model's failure in this specific sample to the absence of training data pertaining to IVC structures. These observations emphasize the importance of data representation and distribution in influencing model performance, particularly in the context of specialized and less frequently labeled structures.

However, in Fig. 7, it becomes evident that the model faces challenges when it comes to the accurate prediction of the presence of a small tumor nestled within these complex vessel structures.

4 Conclusion

In conclusion, this study demonstrates the effectiveness of a semi-supervised deep learning approach for liver tumor and vessel segmentation in whole-body CT scans. The proposed model achieved a Dice score of 0.67 for hepatic vessels and 0.65 for tumors, showcasing remarkable accuracy in identifying the position and size of tumors and precision in anatomical structure delineation. It is worth noting that the model exhibits the capability to predict structures that were not present in the original ground truth annotations. The labeling process itself can be intricate and challenging for humans, particularly in the context of visualizing vessels within CT images. The inferior vena cava (IVC) was not considered to be a part of the liver in the current study, however, in some image data, IVC was included in the ground truth annotations. This highlights the importance of labeling in training the data. The study pipeline, which includes a meticulously crafted training strategy and a sliding-window inference pipeline, plays a critical role in achieving high-precision results. The Dice score serves as the primary evaluation metric, providing a quantitative assessment of the model's segmentation performance. Other secondary metrics, such as sensitivity, specificity, and intersection over union (IoU), may also be considered to provide a comprehensive evaluation of the segmentation model's performance. The potential implications of this research for medical professionals and patients are significant, as accurate delineation of anatomical structures and pathology directly impacts clinical decision-making. Overall, this study highlights the potential of deep learning techniques in medical image analysis and provides a promising avenue for future research in this field.

Acknowledgments Computational resources were provided by AI Biomedicine HPC, NHRI.

Conflict of Interest The authors have no relevant financial or non-financial interests to disclose.

References

1. Liu, H., Wang, H., Zhang, M.: Deep learning algorithm-based magnetic resonance imaging feature-guided serum bile acid profile and perinatal outcomes in intrahepatic cholestasis of pregnancy. Comput. Math. Methods Med. **2022**, 1–10 (2022)
2. Yang, B., Chang, Y., Liang, Y., Wang, Z., Pei, X., Xu, X.G., Qiu, J.: A comparison study between CNN-based deformed planning CT and CycleGAN-based synthetic CT methods for improving iCBCT image quality. Front. Oncol. **12**, 896795 (2022)
3. Zhou, H., Zhang, R., He, X., Li, N., Wang, Y., Shen, S.: MCEENet: multi-scale context enhancement and edge-assisted network for few-shot semantic segmentation. Sensors. **23**, 2922 (2023)
4. Kuang, H., Yang, Z., Zhang, X., Tan, J., Wang, X., Zhang, L.: Hepatic vein and arterial vessel segmentation in liver tumor patients. Comput. Intell. Neurosci. **2022**, 1–10 (2022)
5. Hatamizadeh, A., Tang, Y., Nath, V., Yang, D., Myronenko, A., Landman, B., Roth, H.R., Xu, D.: UNETR: transformers for 3D medical image segmentation. In: 2022 IEEE/CVF Winter Conference on Applications of Computer Vision (WACV), pp. 574–584. IEEE (2022)
6. Ronneberger, O., Fischer, P., Brox, T.: U-Net: convolutional networks for biomedical image segmentation. In: Navab, N., Hornegger, J., Wells, W., Frangi, A. (eds.) Medical Image Computing and Computer-Assisted Intervention – MICCAI 2015. MICCAI 2015 Lecture Notes in Computer Science, vol. 9351. Springer (2015)
7. Hussain, Z., Gimenez, F., Yi, D., Rubin, D.: Differential data augmentation techniques for medical imaging classification tasks. AMIA Annu. Symp. Proc. **2017**, 979–984 (2018)
8. Kebaili, A., Lapuyade-Lahorgue, J., Ruan, S.: Deep learning approaches for data augmentation in medical imaging: a review. J. Imag. **9**, 81 (2023)

Legacy Versus Algebraic Machine Learning: A Comparative Study

Imane M. Haidar, Layth Sliman, Issam W. Damaj, and Ali M. Haidar

1 Introduction

Artificial intelligence has two primary learning approaches: knowledge-based and statistical-based. The statistical-based approach is best implemented using neural networks (NNs), while the knowledge-based approach involves providing the machine with basic information and allowing it to conclude new examples. While the neural network is useful in solving problems, it has limitations. Humans do not need to see thousands of examples to learn because of the billions of neurons [1] in their brains that work together simultaneously with low power and high memory resources, allowing them to conceptualize data. The algebraic machine learning approach is introduced as a step toward accelerating learning through the conceptualization of datasets, which begins with small and automatically created features. In addition to the original work [2], other researches were done in the same context: In [3], the article explores the formal mathematical representation of finite atomized semilattices, a type of algebraic construction that is utilized in algebraic machine learning (AML) to establish and integrate models. The paper specifically investigates the comprehensive definition of these semilattices and highlights the formalization of crucial AML concepts such as the full crossing operator and pinning terms. The

I. M. Haidar (✉) · A. M. Haidar
Beirut Arab University Debbieh, Bakhqaoun, Lebanon
e-mail: i.haidar@bau.edu.lb; ari@bau.edu.lb

L. Sliman
EFREI, Paris-Pantheon-Assas University, Villejuif, France
e-mail: layth.sliman@efrei.fr

I. W. Damaj
Cardiff School of Technologies, Cardiff Metropolitan University, Cardiff, UK
e-mail: IDamaj@cardiffmet.ac.uk

M. N. Seyman (ed.), *2nd International Congress of Electrical and Computer Engineering*, EAI/Springer Innovations in Communication and Computing,
https://doi.org/10.1007/978-3-031-52760-9_13

analysis emphasizes the significance of these concepts within the AML framework. In addition, another article [4] presents a comprehensive literature review on the topic of semantic embeddings within the context of semilattices. The main focus is to explore the potential applications of semantic embeddings in addressing both machine learning and classic computer science problems. The authors provide a formal definition of semantic embeddings, which involve encoding problems as sentences in an algebraic theory that extends the theory of semilattices. The authors also introduce the concept of finite atomized semilattices as a formalism for studying the properties of these embeddings and their finite models. In the same context, this work discusses the benefits of the algebraic approach such as its less computational expense and ability to achieve high accuracy even for approximate problems with noise in the dataset. It represents a literature review of the state-of-the-art approaches in artificial intelligence beginning from the classical methods, passing through neural networks and fuzzy logic, and then gives a detailed explanation of algebraic machine learning. This section incorporates the algorithm steps: from the formation of the initial graph of the model to the clarification of the concept of atomization. The chapter continues by presenting the main model functions: first enforcing the so-called trace constraints for the positive and negative relations. Second, applying the crossing operation, then reduction before finishing with preparing the batch training for the next epoch, and so on. The chapter concludes by presenting a detailed comparison of knowledge propagation in each AI model and a chart of error rates on the MNIST handwrit-ten-digit database [5]. Future research directions are finally suggested.

2 Classical Machine Learning

Machine learning helps in solving many life challenges, especially in the field of image and sound detection, data analysis, and language processing, in addition to many other modern applications [6] like learning from biological sequences or email data in complex environments such as the internet. Also, many types of methods [7] are employed depending on the nature of the data and problem; they are divided into supervised learning [8], where data are labeled; unsupervised learning [9], where the model classifies input into groups without any pre-labeled step; and reinforcement learning [10], which needs some feedback from the environment to know if the decision made was true. There are myriads of handmade AI methods like linear regression [11], logistic regression [12], random forest [13], support vector machine [14], decision trees [15], and many others. They are based on some mathematical equations and discrete operations defined for a specific range of problems. This section will present a statistical approach [16] to recognize handwritten digits by determining known features and searching for these features in each input; the classification of a digit depends on how much the input matches the predefined characteristics. Features are combined into one vector and two methods are applied for classification: one is probabilistic, which deploys the majority vote concept to classify digits, i.e., the more the probability of a digit to have a specific feature, the more it maps to the right output. (Digit zero has the highest probability amongst other digits to have a hole in it.) The other one is based on the K-nearest-neighbors, which consists of determining a smaller number of features and determining how close a digit is to all these features. (Digit nine is close to features like a top hole with a vertical straight line.)

3 Neural Network

The neural network [17] concept is similar to classical methods, but with the distinction that it learns features on its own without human intervention. Although considered a black box algorithm, unlike transparent classical methods that explicitly inform users how outputs match inputs, neural networks are more useful as they can generalize to a broader array of classification and recognition problems without relying on handwritten rules. The neural network consists of interconnected nodes with weights assigned to each connection. During the learning process, the model adjusts the weights to predict the correct output through feedforward propagation and minimizes the error rate through backpropagation. An activation function is used to restrict values to a numerical range, such as between zero and one for probabilistic problems. Weights are multiplied by inputs and summed before entering the activation function. A bias value is appended to the summation to make the activation function more robust, for example, preventing an input of (0,0) from always producing a zero output after extensive training.

4 Continuously Constructive Neural Network

This updated version of the neural network, known as constructive NN [18], differs from the traditional neural network as it not only updates the weights but also alters the layers. While the neural network is considered to be a human-free learning method, it still needs human intervention to determine the number of layers where each contains a set of neurons. Constructive NN addresses this issue by utilizing two methods: tunnel networks and budding perceptions. Tunnel networks employ a parameter that determines whether a layer is active or inactive based on a nonlinear equation containing weight and bias. As more nonlinear layers are added, the network depth increases. In contrast, budding perceptions use a parameter associated with each layer, and during gradient-descent training, if more nonlinear processing is required, a new layer is created like a tree node. Both methods have the capability of adding and removing layers during training, with tunnel networks being better suited for pruning and budding perceptions for generating infinite NN depths.

5 Fuzzy Logic AI

Fuzzy logic is a rule-based method used for feature extraction that operates on a range of numbers instead of precise values. However, the data must be preprocessed to apply the rules. The feature extraction process involves two steps: feature detection, which identifies the features that preserve the essential information of the image, and feature selection, which determines the principal feature components to enable effective classification.

A great amount of research has been conducted in this field, particularly in the work by the author [19]. In this study, the author preprocessed the image and used 7 inputs, 1 output, and 57 rules. The author divided the image into seven segments by intersecting it with two horizontal and one vertical line, resulting in seven intersection points, which were used as the seven inputs. The rules were then formed to determine which digit corresponded to each combination of intersections. In the end, the author achieved an accuracy of 80% for handwritten digit recognition. In another more detailed work [20, 21], the authors work on a different type of inputs. They preprocess the image to standardize size, border ... then they use two groups of inputs to get the final output.

6 Neuro-Fuzzy AI

Neuro-fuzzy [22], known as hybrid fuzzy neural network, is an algorithm that combines the advantages of fuzzy logic and neural networks. This model imitates the composition of the human brain, with neural networks considered as the hardware and fuzzy logic representing the software. There are many types of structures used in Neuro-fuzzy, including sequential hybrid, which uses the output of one technology as the input for the other; auxiliary hybrid, where one technology serves as a sub-function for the other, with the principal technology invoking the other; and embedded hybrid, which merges both technologies so that one cannot work without the other. The hybrid method is considered the best one.

There are many ways of combining the two approaches, and each way reflects a different goal. The hybrid fuzzy neuro system is used in many modern applications using different architectures as in [23–27]. A classification algorithm [28] was made on images and digits using CIFAR and MNIST databases, respectively. The method has achieved 99.58% in MNIST which outperforms most of the AI models, while in CIPHAR 10 and 100, it achieves 88% and 63%, respectively, which is less than other methods.

7 Algebraic Machine Learning

Algebra has played an important role in AI [29] especially in queries and data representation [30]. The algebraic approach [2] is a parameter-free model that does not use function minimization and does not overfit. The training dataset is used to find the minimal algebraic characterization giving high accuracy in test data. The algebraic concept will be explained through the supervised vertical bar problem. Digit recognition is then introduced. First, the general steps will be listed then the detailed ones will be explained each in a subsection.

7.1 General Steps

1. Define terms, constants and atoms then draw a graph of M and M* with 0 (zero) and 0* atoms, respectively, as it is explained in sections: 1—Initial Graph and 2—Atomization. The terms are the examples from positive and negative classes, the constants are the elements of any term (like the pixels of an image) and the atoms are generated through training epochs, they encompass a combination of constants representing some features.
2. Add atoms to positive relations in M such that constant $v < T+$ and to negative relations in M* such that ($v < T-$), then verify that trace constraints are satisfied. If so, this ensures that NOT ALL the atoms of v are in T−.
3. Do full or sparse crossing to enforce ALL atoms of $v < T+$ with preservation of trace constraints.
4. Do reduction operation to minimize atom cardinality.
5. Do batch training that accumulates knowledge and preserves constraints then find the general formula.

7.2 The Vertical Bar Problem

In the vertical bar problem, the algebraic algorithm learns to classify a 2×2 image with black and white pixels into a positive class that has a vertical bar or a negative one that does not (Fig. 1), it begins training using five examples from the dataset. The first two (T1+, T2+) are positive and the other three (T1−, T2−, T3−) are negative. The goal is to build an algebra that classifies new images as positive or negative. After completing the training process, there will be four atoms. In order for a positive example to be considered valid, it must contain all four atoms, while a negative example must not contain at least one of the atoms. The term "contain" should be interpreted as having something in common with a given atom.

Fig. 1 Vertical bar problem. On the right left the positive class has a vertical bar. On the right, a negative example

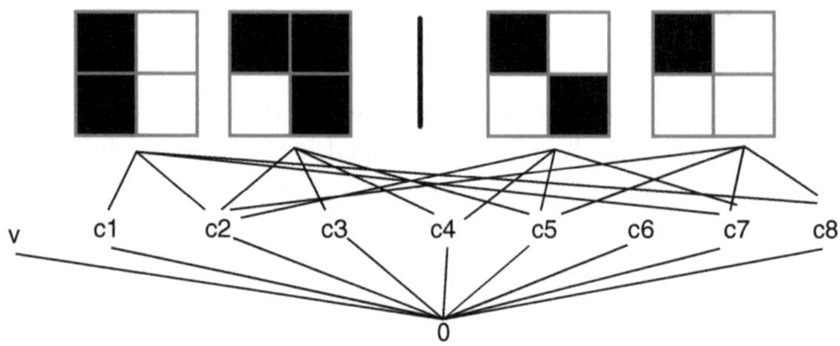

Fig. 2 Graph of M containing the positive and negative terms, constants and atom 0

7.3 *Initial Graph*

This algebra is represented using three main operations: the transitive relation, the partial order, and the unary operator in addition to three elements: constants, terms, and atoms. Constants represent the image pixels named c1–c8. The terms are the "merge" of the constants. Atoms (in Greek letters) are created through learning. Combining many constants using logic will lead to the formation of a single atom. A term is therefore also a merge of atoms. The merge operator allows to establishment of the inclusion relationship < between elements: a < b, a $\odot$ b = b.

In this example:

$$\mathrm{T1}+ = \mathrm{const1} \odot \mathrm{const2} \odot \mathrm{const7} \odot \mathrm{const8}. \tag{1}$$

Any of these constants, say c1, obeys c1 < T1+ because const1 $\odot$ T 1+ = T 1+. The training set consists of positive (v < T1+) and negative (v < T3−) relations where v is a constant that describes the algebra. The edge between elements a → b represents the inclusion operation. A "0" atom is added to all constants which will make exposition simpler. The constructed graph describes an algebra called M (Fig. 2); it evolves during the learning process until finding a good accuracy. The auxiliary algebra M* is a semilattice similar to the dual of M [6]. Each element in M has its dual in M* (Fig. 3). Additional edges are added in M* such that v < T1+. An atom 0* is included in M* to make the exposition simpler.

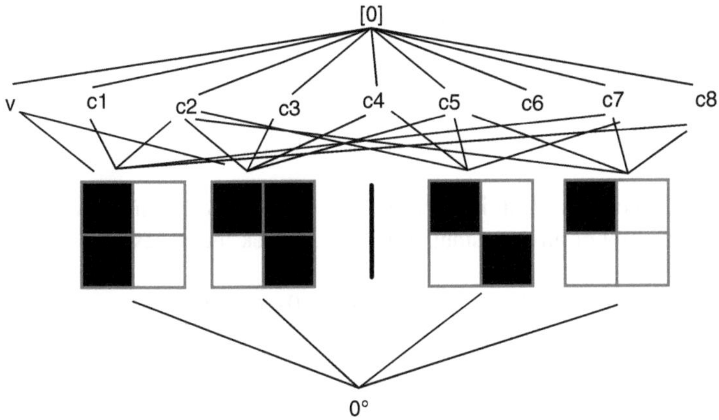

Fig. 3 Graph of M* containing: the dual atom [0], the constants (dual of constants, and dual of terms), and negative atom 0*

7.4 Atomization

The merge of any two elements is equal to the merge of their atoms

$$\mathrm{Gla}\left(a \odot b\right) = \mathrm{Gla}(a) \cup \mathrm{Gla}(b). \tag{2}$$

Gla(x) represents the lower graph of element x that contains all its atoms.

For the vertical bar problem, the goal is to find a set of atoms attached to v that are also included in the images T1+ and T2+ and not all included in the negative images.

7.5 Trace and Trace Constraints

The trace is an essential tool in learning; it tests if the transformation done to semilattice is acceptable or not, the trace should be kept invariant while transforming the algebra. The trace of an element represents the intersection of all the negative atoms edged to its dual in M*. For the vertical bar problem, trace constraints should be obeyed for the positive training examples, v < Ti+, which then needs to enforce the trace of positive term T1+ into the trace of constant v and at the same time enforce the fact that the trace of the negative image T1− is not included in Tr(v). Keep in mind that the goal is to find an atomization for M, but before that an atomization for M* should be calculated. In the vertical bar problem, for every negative example Ti−, an atom is added ζ I → [Ti]. As for the given case, three atoms are introduced $\zeta 1$, $\zeta 2$, and $\zeta 3$ in the graph of M*. The new atoms do not belong to gLa([v]) so the reverted negative relations are met. Now enforce the trace constraints for the negative examples. The trace of the negative image is

$$\mathrm{Tr}(T\,i-) = \mathrm{gLa}([\mathrm{zero}]) = \{0*, \zeta 1, \zeta 2, \zeta 3\} \tag{3}$$

and for constant v is also

$$\mathrm{Tr}(v) = \mathrm{gLa}([\mathrm{zero}]) = \{0*, \zeta 1, \zeta 2, \zeta 3\}. \tag{4}$$

The negative trace constraint is not achieved, so it should be enforced. For this, add an atom alpha to v or any constant equal to v. Check the traces,

$$\mathrm{Tr}(T\,i-) = \mathrm{Tr}(\mathrm{zero}) = \{0*, \zeta 1, \zeta 2, \zeta 3\} \tag{5}$$

and

$$\mathrm{Tr}(v) = \mathrm{Tr}(\mathrm{zero}) \cap \mathrm{Tr}(\mathrm{alpha}) = \{0*\} \tag{6}$$

Thus, obeying Tr(Ti−) not included in Tr(v), as required.

Now, for positive trace constraints v < Ti+, this condition Tr(Ti+) Tr(v) should be enforced. Calculate the trace,

$$\mathrm{Tr}(T\,i+) = \mathrm{Tr}(\mathrm{zero}) = \{0*, \zeta 1, \zeta 2, \zeta 3\} \tag{7}$$

and

$$\mathrm{Tr}(v) = \mathrm{Tr}(\mathrm{zero}) \cap \mathrm{Tr}(\mathrm{alpha}) = \mathrm{Tr}(\mathrm{alpha}) = 0* \tag{8}$$

Thus,

$$\mathrm{Tr}(\mathrm{Ti}+) \in /\mathrm{Tr}(v), \tag{9}$$

so, enforce it by adding atoms ϵi to the constants ci until

$$\mathrm{Tr}(T\,i+) = \mathrm{Tr}(\mathrm{zero}) \cap i\mathrm{Tr}(\epsilon i) \text{ equals } \mathrm{Tr}(v). \tag{10}$$

For the first term T1+, edge one atom to the first constant, $\epsilon 1 \rightarrow$ c1, so

$$\mathrm{Tr}(T\,1+) = \mathrm{Tr}(\mathrm{zero}) \cap \mathrm{Tr}(\epsilon 1) = \{0*, \zeta 1, \zeta 2, \zeta 3\} \cap \{0*\} = \{0*\}. \tag{11}$$

For the second training example, T2+, add one atom for each of its constants, $\epsilon 2 \rightarrow$ c3 and $\epsilon 3 \rightarrow$ c4.

Doing this,

$$\begin{aligned}\mathrm{Tr}(T\,2+) &= \mathrm{Tr}(0) \cap \mathrm{Tr}(\epsilon 2) \cap \mathrm{Tr}(\epsilon 3) = \{0*, \zeta 1, \zeta 2, \zeta 3\} \cap \{0*, \zeta 2\} \\ &\cap \{0*, \zeta 1, \zeta 3\} = \{0*\},\end{aligned} \tag{12}$$

as required.

7.6 Crossing Operations

After the trace enforcement, all negative relations are fulfilled. To satisfy the positive relations v < Ti+, crossing is employed. This operation replaces the atoms of v with other atoms that are included in Ti+ without affecting the previous relations. Consider two elements a and b that should be enforced with atoms:

$$\mathrm{GLa}(a) = \{\alpha, \beta, \chi\} \quad \text{and} \quad \mathrm{GLa}(b) = \{\chi, \delta, \varepsilon\}. \tag{13}$$

It holds that GLa(a), GLa(b). It is said that atoms α and β are "crossed" into b. In the vertical bar problem, the crossing is applied to the two positive terms. The crossing of v into Ti+ is done by creating new atoms in v and matching them to atoms in Ti+ such that the traces of all atoms do not change. After edging the new atoms (ϕ1, ϕ2 and ϕ3) to an atom alpha of v, its trace should be rechecked:

$$\mathrm{Tr}(\mathrm{alpha}) = \mathrm{Tr}(\phi 1) \cap \mathrm{Tr}(\phi 2) \cap \mathrm{Tr}(\phi 3). \tag{14}$$

The trace of alpha may change which is not acceptable. Before replacing alpha with the new atoms, make sure that its trace has not altered. Consider the set of atoms **Φ**1 in v which does not belong to the first positive example, T1+, and select one. Here, there is only **Φ**. Now select one of the atoms in T1+, which is ϵ1. Create a new atom ϕ1, and match it to ϕ and another to ϵ1. Check the atoms' traces: Tr(ϕ)i = {0∗}, after crossing Tr(ϕ) = Tr(ϕ1) = {0∗}.

Thus, it is not changed. For atom ϵ1, Tr(ϵ1)i = {0∗}, after the crossing Tr (ϵ1) = Tr(ϕ1) = {0∗}.

This also is not altered. Eliminate the original atoms ϕ and ϵ1. Do the crossing between v and T2+. Cross the atom in ϕ2 (the atoms in v which are not in T2+), ϕ1, with one of the atoms in T2+, like ϵ2. Create atom ϕ2 and edges ϕ2 → ϕ1 and ϕ2 → ϵ2. The trace of atom ϕ1 before crossing is Tr(ϕ1)i = {0∗}, but after the crossing is Tr(ϕ1) = {0∗, ζ2}. Select another atom ϵ3 in T2+ because the trace has changed. Generate a new atom ϕ3 and edges to ϕ1 and ϵ3. After appending ϕ3 with these new edges, the trace of ϕ1 becomes Tr(ϕ1) = Tr(ϕ2) ∩ Tr(ϕ3) = {0∗}, so now the trace invariance is fulfilled. The trace of atom ϵ2 also remains unchanged as Tr (ϵ2)i = {0*, ζ2} equals Tr(ϵ2) = Tr(ϕ2) = {0∗, ζ2}.

For the trace of ϵ3, Tr(ϵ3)i = Tr(ϵ3) = {0∗, ζ1, ζ3}, so it also remains unchanged. If there are more atoms in v, repeat. In this case, everything is done and the training examples all obey the constraints. The atoms of v are GLa(v) = {0, ϕ2, ϕ3} and the atoms for the positive training examples are also GLa(T1,2+) = {0, ϕ2, ϕ3}, while for the negative examples, GLa(T1,3−) = {0, ϕ3} and GLa(T2−) = {0, ϕ2}.

7.7 Reduction Operation

In crossing, the aim was to cross the atoms without changing the trace; for the reduction, it will be on constants. The reduction process finds the only atoms that preserve the trace and adds them to set Q before deleting the other atoms. Q is empty at the beginning, for each constant, select a subset of its atoms such that Tr(c) is preserved. The selected atoms are put in Q before moving onto another constant. Then continue by selecting atoms for the next constant c, beginning with the intersection between its atoms and Q, then add others such that

$$\mathrm{GLa(c)} \cap \mathrm{Q}, \text{equates } \mathrm{Tr(c)}. \tag{15}$$

In the vertical bar example,

$$\mathrm{Tr(c3)} = \mathrm{Tr(zero)} \cap \mathrm{Tr}(\phi 2) = \mathrm{Tr}(\phi 2), \tag{16}$$

so $\phi 2$ must be kept. Same thing for Tr(c4) = Tr($\phi 3$), so it cannot be deleted. Thus, this algebra cannot be reduced.

7.8 Batch Training

After learning the new atoms, the model accuracy should be checked using test data. If it is less than the desired value, training should continue with new examples. Many epochs will be applied. Assume it is epoch 1, for graph G(S) to be manageable, the elements of R0 and its examples' set should be removed in order to train the new one. However, this leads to the problem that some relations of R0 no longer hold. For that, the first epoch is replaced by a set of relations that represents its atoms: for each atom, create a term that contains all the complements of its constants. These terms are called "the pinning terms" of the algebra.

In the vertical bar example, two pinning terms are to be created. For atom $\phi 2$ and $\phi 3$, new pinning terms and relations are constructed after each epoch. They are not replaced, instead they accumulate and/or are discarded. After the batch training is completed, the algebra of this example will end up with four atoms from where a special formula is derived.

8 Qualitative and Quantitative Comparison

After explaining how each approach works, a detailed comparison will be conducted to draw some conclusions (refer to Table 1).

Table 1 Qualitative comparison of ML approaches

Approach	Architecture	Features	Classification
Classical	Separate graphs	Handcrafted	Using probability/knn
NN	Handcrafted	Crisp features auto extracted	Probability, weights, bias, activation function
CCNN	Auto built	Auto extracted	Probability, weights, bias, activation function
Fuzzy logic	Handcrafted	Handcrafted	Using if-then rules
Neuro-fuzzy	Handcrafted	Fuzzy features auto extracted	Using fuzzy operations, the activation function
Algebraic ML	Auto built	Auto extracted	Using atomization

Classical methods use handcrafted criteria to detect features, which are then fed into probabilistic or K-Nearest Neighbor algorithms for feature selection.

In neural networks, engineers construct an architecture that automatically discovers features using backpropagation, and the results are fed into an activation function for classification.

Continuous constructive neural networks (CCNN) are similar to neural networks, as they use parameters and activation functions, but they also update the architecture through backpropagation, adding and pruning layers until the optimal architecture is reached. Both neural networks and CCNN start with random or incorrect weights and attempt to minimize errors by comparing the calculated output to the target output.

Fuzzy logic AI uses handcrafted criteria and rules to detect and select features. Fuzzy neural networks merge the speed of fuzzy logic with the meaningfulness of the learning process in neural networks, reducing computational costs and producing desired results by omitting useless details.

Algebraic machine learning, on the other hand, begins with an empty example and enforces it to learn how to be a positive example and not a negative one simultaneously. This is achieved by giving the model a subset of positive and negative datasets and continuously growing them. Batch training is achieved by trying a new subset that accumulates knowledge, without suppressing old knowledge like in neural networks. This method changes the architecture of the model instead of updating parameters.

To compare the approaches, the classical method begins with true handcrafted features and maps them to the output, while neural networks begin with incorrect automated weights with a handcrafted architecture that iteratively moves toward the correct solution. CCNN begins with incorrect weights and architectures and iteratively improves both. Fuzzy logic begins with a handcrafted infrastructure and discovers the rules for each output class through training. Neuro-fuzzy also begins with incorrect weights but iteratively moves toward the correct class using fuzzy values instead of crisp ones. The neuro-fuzzy system has the advantage of giving meaning to weights and other parameters, such as the width and position of membership functions, and it is good at initializing a more appropriate set of weights. However, it has the drawback of dimensionality.

The algebraic approach, in contrast, begins with an empty example and extracts atoms by ensuring they are all in the positive class while not all being in the negative class. More atoms are added as more datasets are tried. Additionally, it preserves the negative example terms while training the new dataset and accumulating knowledge. Thus, algebraic learning starts with empty examples and adds true features iteratively, avoiding wrong temporal steps, unlike neural networks. It can also generalize and formalize problems. One advantage is that the model could be parallelized for big problems.

The table summarizes the qualitative differences among all approaches and shows the amount and position of human intervention (marked as “handcrafted”). It categorizes the algorithms by their architecture, features extraction method, and classification function. It is observed that classical and fuzzy methods are almost totally dependent on human decisions, while neural networks in both types (NN and neuro-fuzzy) are dependent on humans only in the architecture-building step. The programmer must try and error on the number of layers and neurons until getting a satisfying accuracy. In contrast, feature extraction and classification in neural networks are auto-calculated using mathematical equations. On the other hand, CCNN is better than the aforementioned neural networks since its architecture is auto-generated and updated based on a hyperparameter that decides when to add or prune layers. As for algebraic machine learning, it is clearly seen that the whole algorithm is totally independent of humans. This is caused by the nature of the model that employs a top-down approach beginning from data contained in the example (e.g., pixels in images) and then accumulating knowledge by simultaneously distinguishing the features of positive and negative sets.

In conclusion, algebraic machine learning offers a new horizon in artificial intelligence field by decreasing the amount of dependence on humans.

To conduct a quantitative comparison of the approaches, the error rate is chosen as the criteria for comparison as the other two criteria, false positive and negative error rates, were not found in all the algorithms. As shown in Fig. 4, the neuro-fuzzy approach has the lowest error rate of 0.42% since it omits useless details and generalizes some features into a wider range of values. The algebraic learning approach ranks second with an error rate of 1.07%, which is similar to fuzzy logic in widening the positive class. In this approach, positive examples must contain all the atoms, while each atom should not be entirely present in the example; it is sufficient for the tested example to have a common area with the atom. Moreover, multiple atomizations are used to achieve higher accuracy. Continuous constructive neural networks have an error rate of 1.84–2.83%, followed by neural networks with an error rate of 2%. The highest error rates were observed in classical methods (3.65–4.3%) and fuzzy logic AI (6%) since they use handcrafted features that might cause the algorithm to overlook some unseen features.

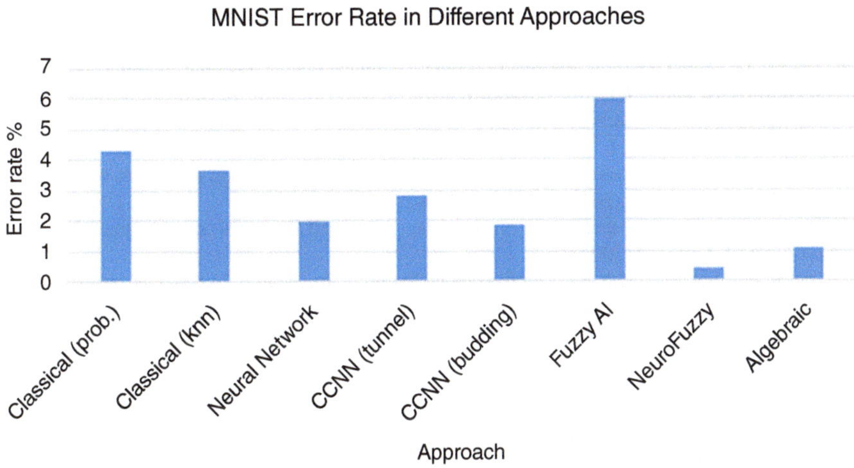

Fig. 4 MNIST error rate in different approaches

9 Conclusion

This chapter has presented various machine learning approaches, ranging from classical methods and fuzzy logic to neural, fuzzy neural, and constructive networks. All of these approaches were compared to the algebraic model, which has the advantage of eliminating error function minimization with its high computational cost and has been shown to generalize without overfitting. A qualitative and quantitative comparison was conducted, where the neuro-fuzzy model obtained the highest accuracy of 99.58% while benchmarking the MNIST dataset, followed by the algebraic model (98.93%) and then neural networks (98%). In future work, a combination of these approaches will be investigated to leverage the strengths of each one. A deeper study on the effectiveness of algebraic machine learning in a wider range of applications will be held.

Conflict of Interest The authors have no relevant financial or non-financial interests to disclose.

Data Availability Training and testing processes have been carried out using the MNIST dataset. MNIST dataset is publicly available at https://www.kaggle.com/datasets/hojjatk/mnist-dataset.

References

1. Zhang, J.: Basic neural units of the brain: neurons, synapses and action potential. Neurons Cognit. arXiv:1906.01703. (2019)
2. Martin-Maroto, F., de Polavieja, G.G.: Algebraic machine learning. arXiv:1803.05252. (2018)
3. Martin Maroto, F., de Polavieja, G.G.: Finite atomized semilattices. arXiv:2102.08050. (2021)
4. Martin Maroto, F., de Polavieja, G.G.: Semantic embeddings in semilattices. arXiv:2205.12618. (2022)

5. LeCun, Y., Cortes, C.: The MNIST database of handwritten digit. Retrieved from http://yann.lecun.com/exdb/mnist/ (1998)
6. Tzanis, G., Katakis, I., Vlahavas, I.: Modern applications of machine learning. J. Emerg. Technol. Web Intell. **1**(1), 10–22 (2006)
7. Muhamedyev, R.I.: Machine learning methods: an overview. Comput. Modell. New Technol. **19**(6), 14–29 (2015)
8. Liu, Q., Wu, Y.: Supervised learning. In: Data Mining and Knowledge Discovery for Big Data, pp. 451–471. Springer (2012)
9. Ghahramani, Z.: Unsupervised learning. In: Advanced Lectures on Machine Learning, pp. 72–112. Springer (2004)
10. Du, K.-L., Swamy, M.N.S.: Reinforcement learning. In: Neural Net-Works and Statistical Learning, pp. 547–561. Springer (2014)
11. Kumari, K., Yadav, S.: Linear regression analysis study. J. Pract. Cardiovasc. Sci. **4**, 33–36 (2018)
12. Sperandei, S.: Understanding logistic regression analysis. Biochem. Med. **24**(1), 12–18 (2014)
13. Biau, G., Scornet, E.: A random forest guided tour. Test. **24**(2), 165–192 (2015)
14. Hearst, M., Dumais, S.T., Osman, E., Platt, J., Scholkopf, B.: Support vector machines. IEEE Intell. Syst. **13**(4), 18–28 (1998)
15. Rokach, L., Maimon, O.: Decision trees. In: The Data Mining and Knowledge Discovery Handbook, pp. 165–192. Springer (2005)
16. Giuliodori, A., Lillo, R., Pena, D.: Handwritten digit classification. arXiv:1107.0579. (2011)
17. Van der Zwaag, B.J.: Handwritten digit recognition: a neural net-work demo. In: Artificial Neural Networks—ICANN, pp. 762–771. Springer (2001)
18. Irsoy, O., Alpaydin, E.: Continuously constructive deep neural networks. IEEE Trans. Neural Netw. Learn. Syst. **30**(4), 1158–1171 (2018)
19. Ferdinando, H.: Handwriting digit recognition with fuzzy logic. Jurnal Teknik Elektro. **3**, 84–87 (2003)
20. Jasim, M., Al-Saleh, A., Aljanaby, A.: A fuzzy based feature extraction approach for handwritten characters. Int. J. Comput. Sci. Issues. **10**(3), 208–215 (2013)
21. Jasim, M., Al-Saleh, A., Aljanaby, A.: A fuzzy logic based hand-written numeral recognition system. Int. J. Comput. Appl. **83**(12), 36–43 (2013)
22. Alavala, C.: Fuzzy Logic and Neural Networks: Basic Concepts and Applications. Springer (2009)
23. Marakhimov, A., Khudaybergenov, K.: Neuro-fuzzy identification of nonlinear dependencies. In: Proceedings of the 2019 9th International Conference on Computer Science and Information Technologies (CSIT), pp. 1–4. IEEE (2019)
24. Terziyska, M., Terziyski, Z.: Computationally efficient neuro-fuzzy predictive models. In: Proceedings of the 2020 6th International Conference on Control, Automation and Robotics (ICCAR), pp. 633–637. IEEE (2020)
25. Mishra, S.: Neuro-fuzzy models and applications. In: Fuzzy Systems and Data Mining III, pp. 81–117. Springer (2020)
26. Singh, H., Biswas, S.: Rule extraction from neuro-fuzzy system for classification using feature weights: neuro-fuzzy system for classification. Int. J. Fuzzy Syst. Adv. Appl. **9**(1), 59–79 (2020)
27. Khuzyatova, L., Galiullin, L.: Optimization of parameters of neuro-fuzzy model. Indones. J. Electr. Eng. Comput. Sci. **19**, 229–232 (2020)
28. Yazdanbakhsh, O., Dick, S.: A deep neuro-fuzzy network for image classification. arXiv:2001.01686. (2019)
29. Nilsson, N.J.: Logic and artificial intelligence. Artif. Intell. **47**, 31–56 (1991)
30. Pouly, M., Kohlas, J.: Generic Inference: A Unifying Theory for Automated Reasoning. Wiley, Hoboken., ISBN 9780470527016 (2011)

Comparison of Textual Data Augmentation Methods on SST-2 Dataset

Mustafa Çataltaş, Nurdan Akhan Baykan, and Ilyas Cicekli

1 Introduction

The field of natural language processing (NLP) has witnessed remarkable advancements in recent years, largely due to the surge in machine learning techniques and increased computational power [1]. As these machine learning methods often require large volumes of data, the significance of textual data augmentation has grown. This is because it helps generate diverse and high-quality data samples for training and testing machine learning models [2]. Basic textual data augmentation techniques have proven effective in enhancing the performance and generalization capabilities of NLP models. By introducing variations in sentence structure, semantics, and syntax, these techniques address challenges such as data scarcity, domain adaptation, and model robustness [3]. Each method presents a distinct way of augmenting textual data, offering researchers and practitioners invaluable tools for real-world language processing tasks.

Data augmentation (DA) is a technique widely used in the field of machine learning to increase the size and diversity of a training dataset [4]. General methods for data augmentation share the same path, which involves applying various transformations or modifications to the existing data and creating new samples that are similar but not identical to the original ones. Data augmentation has proven to be an effective method for improving the performance and robustness of machine learning models, particularly in scenarios where the available training data are limited or imbalanced [5]. Textual data augmentation methods can be grouped into three

M. Çataltaş (✉) · I. Cicekli
Hacettepe University, Ankara, Turkey
e-mail: mcataltas@ktun.edu.tr

N. A. Baykan
Konya Technical University, Konya, Turkey

M. N. Seyman (ed.), *2nd International Congress of Electrical and Computer Engineering*, EAI/Springer Innovations in Communication and Computing,
https://doi.org/10.1007/978-3-031-52760-9_14

categories based on the underlying techniques as proposed in [5]. These categories are noising-based DA, paraphrasing-based DA, and sampling-based DA.

Noise-based DA introduces random variations or perturbations to the original text to generate new samples [2]. The goal of this method is to simulate the noisy or imperfect data that a model might encounter in real-world scenarios. By familiarizing the model with various alterations of the same text, noise-based augmentation seeks to boost the model's robustness and generalization. This technique can be applied at three distinct levels: character-level noising is one such level, where noise is introduced by inserting, removing, or swapping characters within the text [6]. Word-level noising involves the substitution, deletion, or rearrangement of words. Finally, at the sentence level, entire sentences might be restructured, replaced, or shuffled within a paragraph to produce diverse contextual variations while maintaining the core message of the text.

Paraphrasing-based DA refers to the practice of modifying the original sentence to produce synthetic sentences that maintain the original's semantic meaning, albeit with variations in structure and wording [5]. A notable example of this technique is backtranslation [7], a method that leverages Sequence-to-Sequence (Seq2Seq) language models. The technique translates an original sentence between the source language and various target languages before reverting it to the source language. Although the back-translated data may not be identical to the original, it remains semantically similar. Utilizing this method, however, necessitates a reliable machine translation model to ensure the preservation of the sentence's semantic integrity.

Sampling-based DA involves generating new samples from an existing corpus using sampling techniques [5]. Unlike other methods that might concentrate on individual samples for each generation, it considers the entire corpus. As a result, the generated text emerges as a new variation that retains patterns consistent with the original corpus.

As a variety of new textual DA techniques have been proposed in recent years, the selection of DA method for enriching textual datasets has become a challenge. While lots of studies showcased individual results on specific tasks [6, 8–11] and some studies made reviews on them [2, 4, 5], an understanding of how they perform relative to each other on the same task is required. Addressing this gap, in this chapter, one sample DA method from each primary DA technique is selected, aiming to show their strengths and shortcomings through examination and comparison of text classification tasks. Our findings provide a starting point for researchers to make informed decisions on choosing DA methods, thus pushing the boundaries of current NLP practices and contributing to the field of NLP.

The organization of the chapter is as follows: Section 2 gives a brief description of related works, contextualizing our study within the broader landscape of NLP data augmentation literature. Section 3 outlines the datasets and specific augmentation methods examined, while Sect. 4 presents and analyzes the experimental results. Section 5 concludes the chapter by summarizing key findings and suggesting directions for future research.

2 Related Works

Textual data augmentation has received lots of interest in the field of NLP, especially in scenarios with limited or imbalanced training data. This interest led to an abundance of DA methods with different approaches ranging from adding random noise to sampling from corpus.

Noising-based augmentation of textual data has been done through simple manipulations of text such as insertion, deletion, and substitution. However, these operations might be applied in different levels of text which are character, word, and sentence. In [12], the insertion of sentences to legal documents was done randomly on legal documents. They selected sentences from other legal documents with the same label as the original document. Their thinking behind this strategy was that legal documents in the same category might tolerate sentences from each other. Another noising-based augmentation method was proposed in [6], where augmentation was done by inserting predefined punctuation characters randomly into text with the number of punctuations inserted proportional to document length.

Paraphrasing-based augmentation of textual data includes a variety of methods ranging from simple manipulation like synonym replacement [13] to paraphrasing with vector-space operations [10, 14]. Another method rooted in paraphrasing-based augmentation is introduced in [15]. In this approach, text data are augmented by manipulating sentences through cropping and rotating techniques, aided by dependency parsing. When applying these transformations, the dependency parser discerns interconnected words, ensuring their syntactic relationships are maintained. Moreover, the parser is adept at identifying vital sentence constituents, preventing them from being cropped out of context. Experimental results from [15] indicate that both the cropping and rotating techniques serve as effective data augmentation strategies, enhancing the accuracy of Part-of-Speech (PoS) taggers.

In the sampling-based DA technique, some rule-based methods have been proposed [16–18]. However, since large language models (LLMs) have been successful at understanding natural language and generating text within the boundaries of natural language [19], they are well suited for generating new data samples upon original data samples. There have been numerous works that utilize LLMs for sampling-based DA. In [9], a general method for using prompt-based LLMs for data augmentation is proposed. Before this study, there are studies using LLM's that are fine tuned with only text data, and it has been found to be effective. In this study, in addition to text, the model used for DA is also given the class to which the text belongs which provides conditionality for DA. In a format where prompt is composed of text tag, text, class tag, and class, an auto-encoder model (BERT) [20], an auto-regressive model (GPT-2) [19], and a Seq2Seq model (BART) [21] are fine-tuned for DA. In the experiments conducted for the text classification task, it was seen that the models trained with the datasets applied with BART led to better results. While BERT gives similar results to BART, it is seen that the GPT-2 model cannot preserve the target class information added to the text and therefore produces bad results. In the domain of sampling-based data augmentation, a significant contribution comes from the Knowledge Mixture Data Augmentation Model (KnowDA) [22]. Developed as a Sequence-to-Sequence (Seq2Seq) language model, KnowDA stands apart due to its unique pretraining strategy on a diverse array of NLP tasks. This strategy, named Knowledge Mixture Training (KoMT), aims to integrate task-specific insights from various NLP disciplines into a single model. The intention behind KoMT is to enable KnowDA to intuitively understand and synthesize the core principles of any target NLP task, even when confronted with a limited number of training instances. An impressive feature of KnowDA is its non-task-specific nature, implying its applicability across a wide spectrum of NLP challenges.

3 Methods

In this section, a general overview of the methods used in the study is given. In Fig. 1, a general flow of the study is shown.

3.1 Dataset

The Stanford Sentiment Treebank (SST-2) dataset [23] is a well-known dataset that is used as a baseline for sentiment analysis tasks in NLP. It consists of movie reviews labeled as either positive or negative sentiments. The dataset contains approximately 7791 sentences for training, 1800 sentences for validation, and 1821 sentences for testing. Each sentence is associated with a binary label indicating the sentiment

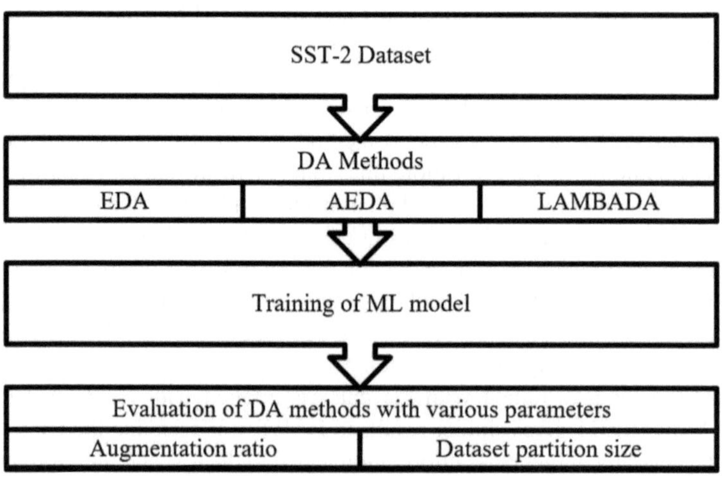

Fig. 1 A general overview of the study

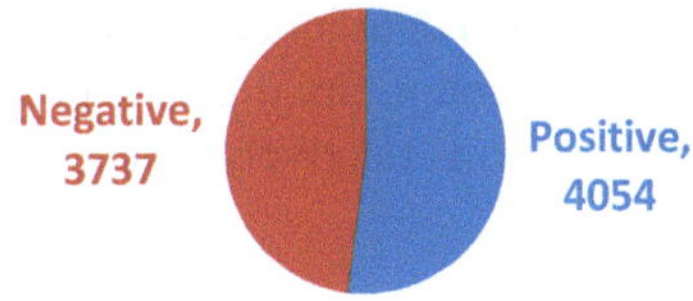

Fig. 2 Distribution of entries for target classes on the SST-2 dataset

expressed in the review. The SST-2 dataset provides a valuable benchmark for evaluating the performance of sentiment classification models and is a standard choice for researchers in the field.

The dataset consists of two target classes which are positive and negative. The distribution of entries in these classes is shown in Fig. 2. The distribution of entries is close to even which makes this dataset a better choice in this study since it would avoid any effect of imbalance on the performance of data augmentation methods.

3.2 Data Augmentation Methods

3.2.1 Noising-Based DA Method

Noising-based DA method used in this study is an easier data augmentation (AEDA) which is proposed in [6]. In this method, only random punctuation insertion is applied to augment textual data as noise. The random punctuation operation includes the determination of a number of punctuations for each sentence, and selecting the insertion position and punctuation randomly.

3.2.2 Paraphrasing-Based DA Method

Paraphrasing-based DA method used in this study is easy data augmentation (EDA) which is proposed in [11]. EDA combines different rule-based transformations to improve the text classification performance of machine learning models. DA methods used in EDA and their explanations are as follows:

- Synonym Replacement: A pre-determined number of non-stop words are chosen and replaced with their synonyms.
- Random Insertion: A random synonym of a random non-stop word in the sentence is inserted into a random position in the sentence.
- Random Swap: Randomly chosen two words from the sentence are swapped.
- Random Deletion: Randomly chosen words from the sentence are deleted from the sentence with probability p.

This approach requires a library like WordNet for synonym replacement and random insertion which makes it difficult to generalize for low-resource languages. Since the operations in this approach are mostly random, it is possible that the sentence loses its semantic or even sentimental coherence. To avoid excessive semantic and sentimental deterioration, the operation count for each sentence is limited by a ratio of token count. So, it is not suitable for text generation or machine translation tasks but performs relatively better in some classification tasks.

3.2.3 Sampling-Based DA Method

The sampling-based DA method used in this study is language-model-based data augmentation (LAMBADA), which is proposed in [8]. This method takes advantage of large language models (LLM) for data augmentation as the name of the method implies. In this method, an LLM is fine-tuned with the whole corpus and then used to generate new samples that resemble the original entries which makes LAMBADA a sampling-based DA method. The flow of the method includes multiple steps, which are shown in Fig. 3.

In the first step, a baseline model of choice is trained with existing data D_{train} in the dataset to be later used in the elimination of undesired synthetic data. The second step includes fine-tuning of an LLM using D_{train}. The input data D_{train} are defined in Eq. (1), where x is the sentence feature and y is the target feature. In the fine-tuning process, the input I to the LLM should follow the format in Eq. (2). In this format, SEP indicates a separator token and EOS indicates an end-of-sentence token.

$$D_{\text{train}} = \{x_i, y_i\}_{i=1}^{n} \tag{1}$$

$$I = y_1 \text{SEP} x_1 \text{EOS} y_2 \text{SEP} x_2 \text{EOS} \ldots y_n \text{SEP} x_n \text{EOS} \tag{2}$$

In the third step, auto-completion capabilities of LLM are used to generate synthetic data. The input I_{S} given to fine-tuned LLM should follow Eq. (3).

$$I_{\text{S}} = y_i \text{SEP} \tag{3}$$

By following these formats in Steps 2 and 3, the conditional generation of synthetic data based on the target class y is aimed.

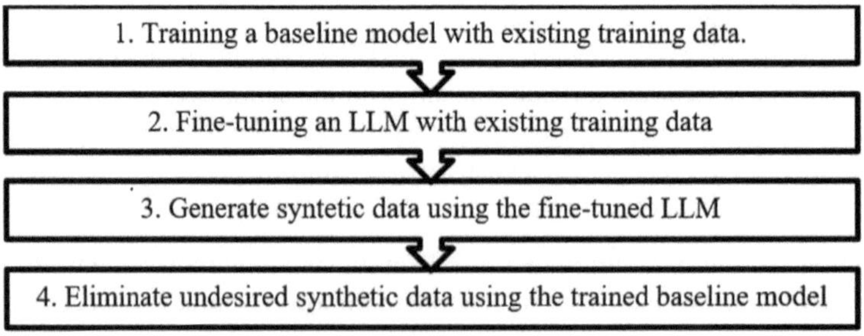

Fig. 3 The steps in the LAMBADA method [8]

In the last step, labels for generated sentences are predicted by the baseline model, and then these predictions are compared to the labels that are given to LLM when generating the sentence. If these two labels contradict each other, then the entry is eliminated from the generated sentences. After elimination, the top N generated sentence from each class is selected according to their confidence scores in baseline model predictions.

4 Experimental Results

4.1 General Parameters for DA Methods

To evaluate the performance of DA methods in different scenarios, two parameters are selected which are *augmentation ratio* and *data set size*. For each experiment final training dataset size $\#_{\text{training data}}$ is calculated using Eq. (4). For Eq. (4),

- *augmentation ratio* indicates how many synthetic sentences will be generated for each original sentence in the training dataset. This parameter is a good indicator of the quality of synthetic sentences generated by DA methods. As the augmentation ratio increases, the proportion of generated data in the dataset rises, which underscores the growing influence of the synthetic sentences relative to the original data.
- *data set size* indicates the size of the training dataset. The original training dataset is partitioned into smaller parts to test DA methods' performance in low-data regime scenarios.

$$\#_{\text{training data}} = \text{augmentation ratio} \times \text{data set size} + \text{data set size} \tag{4}$$

4.2 Classifier

To analyze the performance of DA methods, a fair classifier model is required. DistilBERT (a distilled version of BERT) [24] is chosen for experiments. DistilBERT is a lightweight variant of BERT [20], designed for low-resource setups with the performance close to BERT. It takes advantage of knowledge distillation [25], where the distilled model (DistilBERT) is trained to imitate the behavior of a larger teacher model (BERT), it retains about 97% of BERT's language understanding while being 40% smaller and 60% faster [24]. DistilBERT is particularly useful in real-time applications or environments where computational resources are limited. Like BERT, it is pre-trained on vast amounts of text data, making it capable of delivering robust performance even when fine-tuned on smaller datasets.

Table 1 Parameters used for the DistilBERT model

Parameter	Value
Optimizer	AdamW [26]
Learning rate	5e-05
Activation function	Gaussian Error Linear Units (GELUs) [27]
Batch size	8
Number of epochs	2
Pretrain checkpoint	DistilBERT-base-uncased
Dropout	0.2
Loss function	Cross entropy

The parameters used in the training process of DistilBERT in all experiments are given in Table 1.

4.3 Evaluation Metrics

The metrics utilized to evaluate the model's performance on classification tasks are accuracy and F1-score [27]. Accuracy is a metric that measures the proportion of correct classifications in relation to the total number of classifications made [27] and is generally selected for the evaluation of classification tasks because it provides a straightforward and intuitive understanding of a model's overall performance in classifying instances correctly. Eq. (5) shows the calculation of accuracy.

$$\text{Accuracy} = \frac{\text{Correct prediction}}{\text{Total number of predictions}} \tag{5}$$

The other metric F1-score is derived from recall and precision whose equations are shown in Eqs. (6) and (7), respectively. Equation (8) demonstrates the calculation of the F1-score. The F1-score metric is useful when the imbalance between classes exists since it considers false positives and false negatives, unlike accuracy.

$$\text{Recall} = \frac{\text{True positive}}{\text{True positive} + \text{false negative}} \tag{6}$$

$$\text{Precision} = \frac{\text{True positive}}{\text{True positive} + \text{false positive}} \tag{7}$$

$$\text{F1\_score} = 2 \times \frac{1}{\frac{1}{\text{precision}} + \frac{1}{\text{recall}}} \tag{8}$$

4.4 Classification Results

Results from the conducted experiments are presented in this section, categorized by two evaluation metrics accuracy and F1-score. In Fig. 4 and Table 2, the accuracies are displayed for all cases, while F1-scores for different scenarios are presented in Fig. 5 and Table 3.

As observed in Tables 2 and 3, when considering a dataset size of 100 samples, the model performs worse than a random predictor without data augmentation. However, with the introduction of all three DA methods, there's a significant boost in classification performance. Notably, as the augmentation ratio increases, LAMBADA consistently outperforms the other methods, while the latter tends to show decreased results as can be seen in Fig. 4.

In Table 2, it is observed that when no augmentation is applied to the training set, the DistilBERT classifier performs relatively better for low augmentation ratios for dataset sizes other than 100. On the other hand, when the augmentation ratio

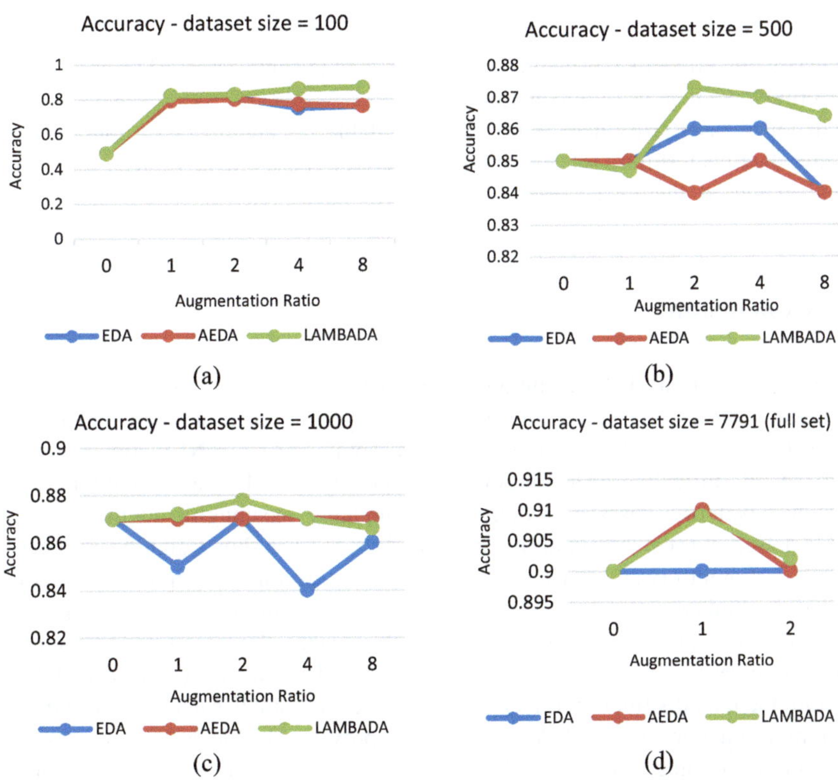

Fig. 4 Accuracies of each DA method for augmentation ratios 0, 1, 2, 4, and 8 for dataset sizes (**a**) 100, (**b**) 500, (**c**) 1000, and (**d**) 7791 (full training set)

Table 2 Accuracies of each DA method for augmentation ratios 0, 1, 2, 4, and 8 for dataset sizes 100, 500, 1000, and 7791 (full training set)

		Dataset size			
Augmentation method	Augmentation size	100	500	1000	Full dataset
No augmentation	0	0.49	0.85	0.87	0.90
EDA	1	0.79	0.85	0.85	0.90
	2	0.81	0.86	0.87	0.90
	4	0.75	0.86	0.84	–
	8	0.76	0.84	0.86	–
AEDA	1	0.79	0.85	0.87	0.91
	2	0.80	0.84	0.87	0.90
	4	0.77	0.85	0.87	–
	8	0.76	0.84	0.87	–
LAMBADA	1	0.82	0.84	0.87	0.90
	2	0.82	0.87	0.87	0.90
	4	0.85	0.87	0.87	–
	8	0.86	0.86	0.86	–

–: no results were obtained

increases, the effect of DA methods decreases, and in some cases, EDA and AEDA perform worse than no augmentation.

In full dataset size, there is no result available for augmentation ratios 4 and 8 since the final resulting sets exceed our resources. Therefore, in Tables 2 and 3 and Figs. 4d and 5d, there is no result available for augmentation ratios 4 and 8.

5 Conclusion

Data augmentation (DA) techniques for NLP have proven to be effective in addressing data scarcity and improving model performance in small datasets for easier tasks like text classification. However, applying DA in NLP comes with its own set of difficulties. The linguistic complexity of NLP tasks, with intricate language patterns, poses a challenge in generating meaningful augmented data while preserving semantic integrity. Nevertheless, DA has shown significant success in more straightforward tasks such as text classification and sentiment analysis, where the limited set of features allows for effective augmentation strategies. On the other hand, in experiments with larger dataset sizes, the effects of DA methods tend to be relatively small, as the abundance of data already enables models to capture the data distribution effectively. However, for smaller datasets, where data scarcity is a major concern, DA methods have proven valuable, as they increase the diversity of training samples, leading to improved generalization and better model performance.

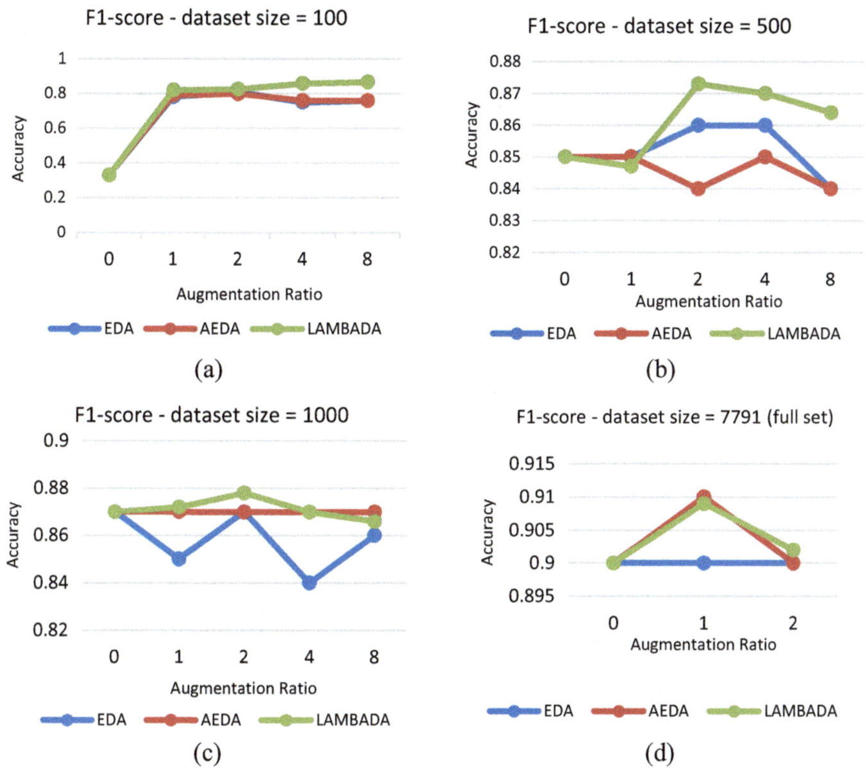

Fig. 5 F1-scores of each DA method for augmentation ratios 0, 1, 2, 4, and 8 for dataset sizes (**a**) 100, (**b**) 500, (**c**) 1000, and (**d**) 7791 (full training set)

Table 3 F1-scores of each DA method for augmentation ratios 0, 1, 2, 4, and 8 for dataset sizes 100, 500, 1000, and 7791 (full training set)

		Dataset size			
Augmentation method	Augmentation size	100	500	1000	Full dataset
No augmentation	0	0.33	0.85	0.87	0.90
EDA	1	0.78	0.85	0.85	0.90
	2	0.81	0.86	0.87	0.90
	4	0.75	0.86	0.84	–
	8	0.76	0.84	0.86	–
AEDA	1	0.79	0.85	0.87	0.91
	2	0.80	0.84	0.87	0.90
	4	0.76	0.85	0.87	–
	8	0.76	0.84	0.87	–
LAMBADA	1	0.82	0.84	0.87	0.90
	2	0.82	0.87	0.87	0.90
	4	0.85	0.87	0.87	–
	8	0.86	0.86	0.86	–

–: no results were obtained

In this study, we compared data augmentation (DA) methods across three distinct categories: noising-based, paraphrasing-based, and sampling-based DA. From each category, a top-performing DA method was chosen; AEDA for noising-based, EDA for paraphrasing-based, and LAMBADA for sampling-based. To evaluate the performance of these methods, we utilized an experimental setup accounting for parameters such as augmentation size and dataset size. Results from our experiments revealed that the sampling-based method, LAMBADA, consistently delivers superior results in low-data scenarios, even with a high augmentation rate. While all methods enhance the classifier's performance when the augmentation rate is low, both EDA and AEDA sometimes have no impact or even a detrimental effect when compared to cases without any augmentation. In conclusion, while each technique excels in specific scenarios, LAMBADA generally outperforms the other methods.

References

1. Vaswani, A., Shazeer, N., Parmar, N., Uszkoreit, J., Jones, L., Gomez, A.N., Kaiser, Ł., Polosukhin, I.: Attention is all you need. Adv. Neural Inf. Proces. Syst. **30**, 1–11 (2017)
2. Chen, J., Tam, D., Raffel, C., Bansal, M., Yang, D.: An empirical survey of data augmentation for limited data learning in NLP. Trans. Assoc. Comput. Ling. **11**, 191–211 (2023)
3. Liu, P., Wang, X., Xiang, C., Meng, W.: A survey of text data augmentation. In: 2020 International Conference on Computer Communication and Network Security (CCNS), pp. 191–195. IEEE (2020)
4. Feng, S.Y., Gangal, V., Wei, J., Chandar, S., Vosoughi, S., Mitamura, T., Hovy, E.: A survey of data augmentation approaches for NLP. arXiv preprint arXiv:2105.03075. (2021)
5. Li, B., Hou, Y., Che, W.: Data augmentation approaches in natural language processing: a survey. AI Open. **3**, 71–90 (2022)
6. Karimi, A., Rossi, L., Prati, A.: AEDA: an easier data augmentation technique for text classification. arXiv preprint arXiv:2108.13230. (2021)
7. Sennrich, R., Haddow, B., Birch, A.: Improving neural machine translation models with monolingual data. arXiv preprint arXiv:1511.06709. (2015)
8. Anaby-Tavor, A., Carmeli, B., Goldbraich, E., Kantor, A., Kour, G., Shlomov, S., Tepper, N., Zwerdling, N.: Do not have enough data? Deep learning to the rescue! In: Proceedings of the AAAI Conference on Artificial Intelligence, pp. 7383–7390. AAAI (2020)
9. Kumar, V., Choudhary, A., Cho, E.: Data augmentation using pre-trained transformer models. arXiv preprint arXiv:2003.02245. (2020)
10. Li, G., Wang, H., Ding, Y., Zhou, K., Yan, X.: Data augmentation for aspect-based sentiment analysis. Int. J. Mach. Learn. Cybern. **14**, 125–133 (2023)
11. Wei, J., Zou, K.: EDA: easy data augmentation techniques for boosting performance on text classification tasks. arXiv preprint arXiv:1901.11196. (2019)
12. Yan, G., Li, Y., Zhang, S., Chen, Z.: Data augmentation for deep learning of judgment documents. In: Intelligence Science and Big Data Engineering. Big Data and Machine Learning: 9th International Conference, IScIDE, pp. 232–242. Springer, Nanjing (2019)
13. Zhang, X., Zhao, J., LeCun, Y.: Character-level convolutional networks for text classification. In: Advances in Neural Information Processing Systems, vol. 28. Curran Associates, Inc. (2015)
14. Wu, X., Lv, S., Zang, L., Han, J., Hu, S.: Conditional bert contextual augmentation. In: Computational Science–ICCS 2019: 19th International Conference, pp. 84–95. Springer (2019)

15. Şahin, G.G., Steedman, M.: Data augmentation via dependency tree morphing for low-resource languages. arXiv preprint arXiv:1903.09460. (2019)
16. Chen, Y., Kedzie, C., Nair, S., Galuščáková, P., Zhang, R., Oard, D.W., McKeown, K.: Cross-language sentence selection via data augmentation and rationale training. arXiv preprint arXiv:2106.02293. (2021)
17. Kober, T., Weeds, J., Bertolini, L., Weir, D.: Data augmentation for hypernymy detection. arXiv preprint arXiv:2005.01854. (2020)
18. Shi, H., Livescu, K., Gimpel, K.: Substructure substitution: structured data augmentation for NLP. arXiv preprint arXiv:2101.00411. (2021)
19. Radford, A., Wu, J., Child, R., Luan, D., Amodei, D., Sutskever, I.: Language models are unsupervised multitask learners. OpenAI Blog. **1**, 9 (2019)
20. Devlin, J., Chang, M.-W., Lee, K., Toutanova, K.: BERT: pre-training of deep bidirectional transformers for language understanding. arXiv preprint arXiv:1810.04805. (2018)
21. Lewis, M., Liu, Y., Goyal, N., Ghazvininejad, M., Mohamed, A., Levy, O., Stoyanov, V., Zettlemoyer, L.: BART: denoising sequence-to-sequence pre-training for natural language generation, translation, and comprehension. arXiv preprint arXiv:1910.13461. (2019)
22. Wang, Y., Zheng, J., Xu, C., Geng, X., Shen, T., Tao, C., Jiang, D.: KnowDA: all-in-one knowledge mixture model for data augmentation in few-shot NLP. arXiv preprint arXiv:2206.10265. (2022)
23. Socher, R., Perelygin, A., Wu, J., Chuang, J., Manning, C.D., Ng, A.Y., Potts, C.: Recursive deep models for semantic compositionality over a sentiment treebank. In: Proceedings of the 2013 Conference on Empirical Methods in Natural Language Processing, pp. 1631–1642. Association for Computational Linguistics, Seattle (2013)
24. Sanh, V., Debut, L., Chaumond, J., Wolf, T.: DistilBERT, a distilled version of BERT: smaller, faster, cheaper and lighter. arXiv preprint arXiv:1910.01108. (2019)
25. Hinton, G., Vinyals, O., Dean, J.: Distilling the knowledge in a neural network. arXiv preprint arXiv:1503.02531. (2015)
26. Loshchilov, I., Hutter, F.: Decoupled weight decay regularization. arXiv preprint arXiv:1711.05101. (2017)
27. Powers, D.M.: Evaluation: from precision, recall and F-measure to ROC, informedness, markedness and correlation. arXiv preprint arXiv:2010.16061. (2020)

Machine Learning-Based Malware Detection System for Android Operating Systems

Rana Irem Eser, Hazal Nur Marim, Sevban Duran, and Seyma Dogru

1 Introduction

The increasing trend of mobile devices has become a prominent aspect of modern society. Mobile devices, such as smartphones and tablets, have witnessed a remarkable surge in popularity and usage in recent years. Their portability, convenience, and multifunctionality have made them an integral part of people's daily lives. Mobile devices have revolutionized communication, enabling instant connectivity with others across the globe through voice calls, text messages, and various messaging applications. They have also become powerful tools for accessing information, browsing the internet, managing social media, and conducting online transactions.

Mobile devices cannot use traditional operating systems primarily due to the differences in hardware architecture and user experience requirements. Traditional operating systems, such as those found on desktop or laptop computers, are designed to work with specific hardware configurations, including ×86 processors and larger displays. Mobile devices, on the other hand, typically utilize ARM-based processors and have smaller screens. Additionally, mobile devices have unique user experience needs, such as touch-based interactions, mobile-specific applications, and optimized power consumption. Traditional operating systems are not tailored to meet these requirements out of the box.

The two most popular mobile device operating systems are Android and iOS. Android was introduced in 2008 as a mobile operating system that is rapidly growing in popularity. Android, developed by Google, was released as an open-source operating system and offered to mobile device manufacturers. Android is the most

R. I. Eser (✉) · H. N. Marim · S. Duran · S. Dogru
Computer Engineering Department, Biruni University, Istanbul, Turkey
e-mail: ranaeser77@gmail.com

M. N. Seyman (ed.), *2nd International Congress of Electrical and Computer Engineering*, EAI/Springer Innovations in Communication and Computing,
https://doi.org/10.1007/978-3-031-52760-9_15

widely used mobile operating system with a market share of approximately 72% worldwide [1].

Therefore, malware attacks specifically target Android devices. It aims to affect Android devices as much as possible and maximize their negative effects. It damages computer systems, networks, or devices and defines any software or code designed with malicious intent to gain unauthorized access to them. Malware is a type of software designed to execute malicious code on smart technological devices [2, 3]. Malware refers to harmful software that performs unwanted or risky activities on target machines. Malware variations can effectively intercept personal data, launch distributed denial of service attacks, and cause disruption to such systems or technological gadgets.

Malware is always a concern in the digital world, and mobile devices are particularly vulnerable [4, 5]. The stakes in the field of information security have never been higher, but this persistent threat has put a long shadow over it. Once upon a time, when the Android platform was still young, it was easy to spot and counteract such harmful apps. In essence, the Android operating system was a place where new ideas could flourish alongside opportunities for commercial gain. The study of API calls within a safe and regulated sandbox environment has become a standard technique for discovering and mitigating low-level forms of malware [6]. Malware has evolved over time, becoming more sophisticated. As the environment changed, so did the sophistication with which enemies hid their malice. Once a simple undertaking, keeping up with hackers has become a complex and ever-changing game of cat and mouse for the cybersecurity community. Therefore, it became crucial to develop sophisticated and diverse methods for mobile security.

Cutting-edge technologies, machine learning (ML) algorithms, behavior-based analysis, and a dogged determination to remain ahead of the attackers characterize today's fight against mobile malware. In light of this ongoing conflict, it is more important than ever to adapt and innovate in the cybersecurity industry in order to safeguard the rapidly growing ecosystem of mobile devices. Machine learning has emerged as a method to overcome these challenges and develop a more effective malware detection system. Using data obtained from real-time monitoring, machine learning algorithms can analyze malware patterns and behaviors. This makes it possible to detect complex and unpredictable malicious activities.

Malware detection has come a long way, and contemporary methods leveraging machine learning (ML) techniques have proven to be significantly more efficient than traditional approaches, such as statistical and knowledge-based solutions [7–10]. Recognizing the potential of ML in enhancing security, this study undertook the development of a Malware Detection System, harnessing the power of both static and dynamic analysis techniques, coupled with state-of-the-art deep learning algorithms. In this research endeavor, the DREBIN dataset served as the foundation, providing a diverse and comprehensive collection of malicious and benign mobile applications. The choice of the Python programming language facilitated the development of a robust and versatile system.

The static analysis phase of the project employed sophisticated techniques, including Random Forest and Decision Trees. These methods were instrumental in examining the initial state of the dataset, helping to identify the most crucial attributes and formulate predictions. Through this analysis, the system gained a deeper understanding of the structural characteristics of potential threats, enabling it to classify new applications effectively. In the dynamic analysis portion, the study delved into the realm of deep learning, leveraging powerful algorithms such as Artificial Neural Networks (ANN) and Deep Artificial Neural Networks (DNN). These neural networks were well-suited for analyzing the behavioral patterns and nuances of applications. Their ability to learn complex relationships within the data allowed the system to detect subtle and evolving malware characteristics, adapting to the ever-changing landscape of threats in the mobile ecosystem.

By combining dynamic and static analysis methods with the capabilities of Deep Learning, this Malware Detection System emerged as a formidable guardian against the evolving landscape of malicious applications. The research not only contributes to the ever-evolving field of cybersecurity but also underscores the potency of machine learning in bolstering our defenses against digital threats.

The remaining sections of the chapter are structured as follows: In the subsequent section, an examination of the literature about the subject matter is conducted. This is followed by an explanation of the methodology employed for the proposed model in Sect. 3. Section 4 shows the experimental results, with a focus on highlighting the noteworthy outcomes. Finally, the chapter concludes by discussing prospects and potential avenues for further research in this domain.

2 Literature Review

In the ever-expanding realm of Android-based cybersecurity, the study by Bayazit et al. stands out as a significant contribution, addressing the growing concern associated with Android malware [11]. Their research endeavors culminated in the development of an advanced system, a robust amalgamation of diverse machine and deep learning models, designed with the sole purpose of identifying and combating malware threats that target the Android operating system. The researchers adopted a multifaceted approach, combining dynamic and static analysis techniques to create a comprehensive defense mechanism against Android malware. In dynamic analysis, suspected malware samples were executed within a controlled environment, allowing for the observation of their behavior without exposing real Android devices to potential harm. On the other hand, static analysis focused on scrutinizing the characteristics and attributes of malware files without the need for actual execution, reducing potential risks significantly.

One of the key strengths of their study lies in the meticulous comparative analysis of various models and analysis methods employed. This comparative approach shed light on the strengths and weaknesses of each technique, providing a holistic understanding of their performance. Through rigorous experimentation, Bayazit and the research team discovered that their proposed models surpassed existing methodologies, achieving remarkable accuracy rates. Notably, the Long

Short-Term Memory (LSTM) model, when applied in the static analysis, demonstrated an outstanding accuracy of 0.988, while the Convolutional Neural Network-Long Short-Term Memory (CNN-LSTM) model, in the dynamic analysis, achieved an impressive accuracy of 0.953.

These findings not only underline the effectiveness of machine and deep learning in the context of Android malware detection but also offer a beacon of hope for the ongoing battle against these ever-evolving digital threats. Bayazit et al.'s work not only contributes to the field of cybersecurity but also showcases the potential for innovative solutions in countering the Android malware epidemic.

The efficiency of the proposed model for detecting real malware was demonstrated in a pioneering manner, setting a precedent for future research endeavors in this domain [12]. This groundbreaking work served as a tangible testament to the viability of the method, showcasing its potential to identify and mitigate real-world malware threats. Furthermore, in a study conducted by Sasisharan [13], an innovative approach for Android malware detection was put forward, emphasizing a behavior-based methodology. This approach heralded a shift towards understanding and combating malware not solely through static attributes but by observing how they behaved in real-world scenarios. The foundation of this method lay in the organization and coding of a malicious dataset, enabling the system to pinpoint suspicious API classes. By scrutinizing the behaviors and actions of these classes, intricate patterns were generated, providing valuable insight into the characteristics and tactics of malware. This behavioral analysis approach marked a departure from traditional, signature-based detection methods, acknowledging the dynamic and evolving nature of modern malware threats. The study by Sasisharan illuminated the potential of observing and understanding the behavioral traits of malware, and it opened up new avenues for the development of advanced detection systems that could adapt to the ever-changing tactics of malicious actors. This behavior-based approach represents a significant leap forward in the ongoing battle against Android malware, as it offers a more resilient and adaptable line of defense against these constantly evolving digital adversaries.

Multiple flow alignments and hidden applied Markov model (HMM) profile for different application families. An accuracy of 94.5% was achieved in classification. Wu et al. developed DroidMat, which uses character types such as authorizations and permissions, distribution of components, intent messages, and UPA (Application Programming Interface) calls [14]. They collected malware studies from the "Contagio Mobile" website [15] and benign application studies from the Google Play store. The datasets contain 238 malware samples and 1500 benign application samples. They used different clustering algorithms for classification and obtained a 97.87% accuracy rate in their tests. In 2014, a comprehensive review of malware detection on Android phones was conducted [16].

The researchers who developed the "Drebin" dataset proposed a detailed static analysis for malware detection to avoid a rapid consumption of smartphone system resources. Therefore, it was decided to keep the character types obtained from APK files as comprehensive as possible. A total of eight different attribute types were recorded, including attributes such as file and application authorizations, UPA calls, and network addresses. All these attributes were then combined into a common vector space to automatically identify malicious communication objects. Extensive experiments were conducted using 5560 malware and 123,453 innocuous software samples [16]. By using the SVM (support vector machine) algorithm in the classification stage with machine learning detection of malware, a success rate of 0.94 and a false positive rate of about 0.01 were reached.

An unsupervised single-class Support Vector Machine approach for the detection of Android malware was implemented in a different paper in [17]. However, this method focused exclusively on non-malignant (benign) samples, making it less suitable for addressing imbalanced classification problems, where the number of benign samples far exceeds that of malicious ones. While this approach brought an interesting perspective to the field, it underscored the need for more comprehensive solutions that could effectively handle the imbalance in real-world datasets, where malicious applications are a minority.

Moreover, in another recent study supported by machine learning, researchers conducted an in-depth analysis of the source code of 200 malicious and 200 benign Android applications. The results of this meticulous investigation were impressive, with an F1 score of 95.1% achieved [18]. This high F1 score underscores the effectiveness of their methodology in accurately distinguishing between harmful and benign applications. The use of machine learning in this context not only showcases the potential for fine-grained analysis but also raises the bar for the accuracy and reliability of Android malware detection systems.

These studies collectively emphasize the evolving landscape of Android malware detection, where researchers are continuously exploring diverse approaches, from unsupervised SVMs to detailed source code analysis, to enhance the efficacy and precision of their systems. As the mobile security domain matures, the integration of machine learning techniques becomes increasingly vital in the ongoing battle to safeguard Android devices from an ever-adapting array of malicious threats.

A notable recent contribution in this domain is the EveDroid model [19]. This model demonstrated remarkable performance with an exceptionally high F1 score of 99.8%. The evaluation was conducted on datasets comprising diverse benign app samples collected from PlayDrone and the Google Play Store, as well as malicious app samples obtained from VirusShare [19].

Furthermore, Wenbo et al. [20] introduced a malware detection system for Android devices that leverages multiple deep-learning techniques. Their approach incorporates the analysis of permissions, API calls, hardware components, and purpose properties of applications. The deep-learning-based algorithms were categorized into three groups: DNN, CNN, and CNN-GRU. The experiments involved the examination of 5560 malware samples and 16,666 benign samples, resulting in an impressive accuracy rate of 98.74%.

3 Methodology

In this study, a malware detection system using static and dynamic analysis methods and Deep Learning algorithms is developed. For the study, the DREBIN dataset was used. Python Programming language was used for data preprocessing and ML and deep learning models for malware detection in Android systems. Python-specific Integrated Development Environments (IDE) PyCharm, Spyder IDE, and Jupyter Notebook were used in the study process. Random Forest Algorithm, XGBoost (eXtreme Gradient Boosting) Algorithm, Gaussian Naive Bayes (GNB), AdaBoost (AB) Algorithm, Decision Tree Algorithm, and Support Vector Machine Algorithm were used in static analysis. On the other side, artificial neural networks (ANN) and deep artificial neural networks (DNN) were used for dynamic analysis.

3.1 Dataset

The Drebin dataset consists of 123,453 different Android applications collected between 2010 and 2012. The apps used in this study consist of 5560 for malware and 9476 for goodware. There are 216 features. There is one target variable that contains two different values, whether there is an attack or not. 0 indicates no attack; 1 indicates an attack. It was collected between 2010 and 2012, and therefore, it represents older malware and threats in the dataset. While mobile threats are evolving rapidly, the timeliness of the dataset is limited. This dataset represents data collected in a specific mobile environment. Therefore, issues such as the ability to generalize across different mobile environments or for different types of malware are other issues discussed.

3.2 Algorithms

The algorithms used in this study are: GaussianNB is a Naive Bayes classification method based on the Bayes theorem and used to determine the classes of data points using probability calculations. K Nearest Neighbors the KNN algorithm classifies based on the labels of the data points depending on the number of neighbors. KNN is accepted as one of the simplest algorithms for machine learning. For prediction, the nearest neighbors in the dataset are searched. Random Forest or random decision forest algorithm is an ensemble method for classification and regression for a collection of data sets. One of the important aspects of using Random Forest in Python is overfitting. The risk of overfitting is low, so it can be more generalized. One of the important features of using Random Forest in Python is that it is (overfitting). The risk of overfitting is low, so more generalizable models are obtained.

AdaBoost: The most important feature of AdaBoost that distinguishes it from other algorithms is that it creates a strong learner by combining weak learners and corrects errors by weighting. In this way, it can better model the complexity of the data set and achieve higher performance.

Logistic Regression used in binary classification problems is a probability-based approach. Decision Tree: The most important features that distinguish the Decision Tree algorithm from other algorithms are its simple structure, the fact that it is not dependent on the distribution of the data, that it shows the importance of attributes, and that it is adaptable to the data. These features make Decision Tree an effective classification algorithm that can be used in a wide range of applications. SVM uses a classification method that aims for maximum marginal separation. SVM aims to find the best separation between classes and build decision boundaries around the support vectors.

XGBoost, which stands for eXtreme Gradient Boosting, is a powerful algorithm for machine learning that excels in making accurate forecasts through the technique of gradient boosting. It achieves this by combining a collection of weak forecasting models, often decision trees, into a single, robust forecasting model. Gradient boosting is a technique that incrementally builds AE estimation model by improving upon the weaknesses of the previous models, thereby creating a stronger overall model. What sets XGBoost apart is its optimization of the gradient boosting method and its focus on scalability, making it a highly efficient and adaptable framework for various ML tasks.

Artificial Neural Network (ANN) gathers information by detecting patterns and relationships in data and learning or training independently of programming. An ANN consists of hundreds of independent units, artificial neural cells, or processing elements, which form the neural structure and are organized in layers. These connected units are associated with coefficients called weights. Deep Neural Network DNNs have made a huge impact in the field of deep learning. DNNs are an algorithm capable of highly automated feature learning. Multi-Layer Perceptron (MLP) Algorithm is an algorithm that takes inputs on a sample, processes these inputs in hidden layers, and finally produces a result in the output layer. The nodes in each layer perform operations on the inputs, which are multiplied by weights and passed through an activation function. In this way, the MLP learning algorithm aims to optimize the weights and gain the ability to produce a desired output. Optimizer Adam Algorithm Adam is an algorithm that adaptively adjusts the learning rate over iterations to ensure that the weights converge to the best values during the training process.

3.3 Performance Calculation

In this chapter, Accuracy, recall, F1 score, precision, ROC, and AUC metrics were used and Confusion matrix (CM) shows the details to measure the performance of the models. The confusion matrix is a fundamental tool for evaluating the performance of classification algorithms. It provides a clear breakdown of how well a model is at distinguishing between different classes. The key elements of a confusion matrix include:

- True Positives (TP): These are instances where the model correctly predicts positive cases. In other words, it accurately identifies instances that actually belong to the positive class.
- True Negatives (TN): These are instances where the model correctly predicts negative cases. It accurately identifies instances that truly belong to the negative class.
- False Positives (FP): These are instances where the model incorrectly predicts positive cases. It wrongly classifies instances from the negative class as belonging to the positive class.
- False Negatives (FN): These are instances where the model incorrectly predicts negative cases. It wrongly classifies instances from the positive class as belonging to the negative class.

With these elements in the CM, various evaluation metrics can be calculated, including:

- Accuracy: This is the proportion of correct predictions out of the total predictions, and it is calculated as Eq. (1). It provides an overall measure of the model's correctness.
- Precision: Also known as Positive Predictive Value, precision is the proportion of true positive predictions out of all positive predictions, calculated as Eq. (2). It assesses how well the model performs when it predicts positive.
- Recall: Also known as Sensitivity or True Positive Rate, recall is the proportion of true positive predictions out of all actual positive instances, calculated as Eq. (3). It measures the model's ability to capture all positive instances.
- F1 Score: The F1 score is the harmonic mean of precision and recall, providing a balance between these two metrics calculated as Eq. (4).

Specificity: Specificity, also known as True Negative Rate, measures the proportion of true negative predictions out of all actual negative instances. It is calculated as TN/(TN + FP).

$$\text{Accuracy} = \frac{(\text{TP} + \text{TN})}{\text{TP} + \text{FN} + \text{TN} + \text{FP}} \times 100 \tag{1}$$

$$\text{Precision} = \frac{\text{TP}}{\text{TP} + \text{FP}} \times 100 \tag{2}$$

$$\text{Recall} = \frac{\text{TN}}{\text{FN} + \text{TN}} \times 100 \tag{3}$$

$$\text{F1 score} = 2 \times \frac{\text{Precision} \times \text{recall}}{\text{Precision} + \text{recall}} \tag{4}$$

These metrics hold significant importance in the assessment of classification algorithms across various domains, including applications in healthcare, finance, and machine learning. They play a critical role in the context of distinguishing between different classes, which is essential for informed decision-making. The ROC (Receiver Operating Characteristic) curve is a valuable graphical tool utilized to evaluate the performance of classification models. It provides a visual representation of the sensitivity (True Positive Rate) and specificity (1 − False Positive Rate) of the classification model. The Area Under the Curve (AUC) of the ROC curve serves as a quantitative measure of the overall performance of the model. A higher AUC value, closer to 1, indicates superior model performance.

3.4 Model Building and Data Preprocessing

Data preprocessing is a critical step in the field of data science and machine learning that cannot be overstated. It serves as the foundation upon which accurate, reliable, and meaningful insights are built. The process of data preprocessing involves tasks like cleaning, transforming, and organizing raw data into a usable format, which is essential for accurate analysis and model training. By addressing issues such as missing values, outliers, and inconsistencies, data preprocessing enhances the quality of the data, making it more robust and suitable for modeling. It also involves feature scaling, selection, and engineering, all of which contribute to the performance of machine learning algorithms. Furthermore, data preprocessing can improve the interpretability of results, reduce overfitting, and speed up the training process. In essence, data preprocessing is the bedrock of effective data analysis and modeling, ensuring that the output is both trustworthy and valuable for making informed decisions and predictions.

After examining the data set for outliers and null values, null values were encountered. These null values were filled according to the median of the values in the column where they were found. No outliers were found. After these procedures, GaussianNB, K Nearest Neighbor, Random Forest, AdaBoost, Logistic Regression, Decision Trees, and Decision Support Machines algorithms were applied to the dataset.

4 Result and Discussion

In this section, it is aimed to serve as a pivotal component of this academic or scientific chapter, acting as the culmination of extensive research and experimentation. In this section, researchers present the findings and observations gleaned from the investigations during the work. We design and test our model on the computer structure as depicted in Table 1.

Table 1 Computer properties

Component	Component name
Processor	Intel® Core™ i5-1035G1 CPU @ 1.00 GHz 1.19 GHz
RAM	8 GB RAM
Graphics card	Intel UHD Graphics
Operation system	Windows 10 Pro 64-bit

Table 2 Confusion matrix

Model	TP	TN	FP	FN
GNB	1133	999	864	12
KNN	1096	1836	27	49
RF	1115	1852	11	30
AB	1082	1802	61	63
LogR	1096	1836	27	49
DT	1113	1827	36	32
GNB	1133	999	864	12
KNN	1096	1836	27	49

Table 3 Model value comparison

Model	Accuracy	Recall	Precision	F1-score
GNB	70.87	56.73	98.95	72.11
KNN	97.80	97.36	96.85	97.11
RF	98.67	99.1	97.37	98.23
AB	95.87	94.66	94.49	94.58
LogR	97.47	97.59	95.72	96.64
DT	97.80	96.95	97.29	97.12
XGBoost	97.81	97.45	96.77	97.11
SVM	97.80	97.36	96.85	97.11

After cleaning and organizing the data set, GaussianNB, K-Nearest Neighborhood (KNN), Random Forest, AdaBoost, Logistic Regression, Decision Tree, Support Vector Machines (SVM), XGBoost, ANN, and DNN algorithms were applied. The test size was set as 0.2 for all algorithms. The TP, TN, FP, and FN values obtained are shown in Table 2.

When the values are compared, it is seen that all algorithms except GaussianNB give high values (Table 3). The Random Forest algorithm gave the best result with an accuracy of 98.67%. This model has a usable reliability. The performances of the algorithms are compared as shown in Fig. 1.

As seen in this figure, the best performance is reached with Random Forest Algorithm while the worst case is reached with Adaboost (Fig. 2). Random Forest and AdaBoost are both powerful ensemble methods, but they have distinct characteristics and are suited to different types of problems. Random Forest is often favored for its simplicity, stability, and robustness, while AdaBoost can be effective for boosting the performance of weaker models, particularly when interpretability is

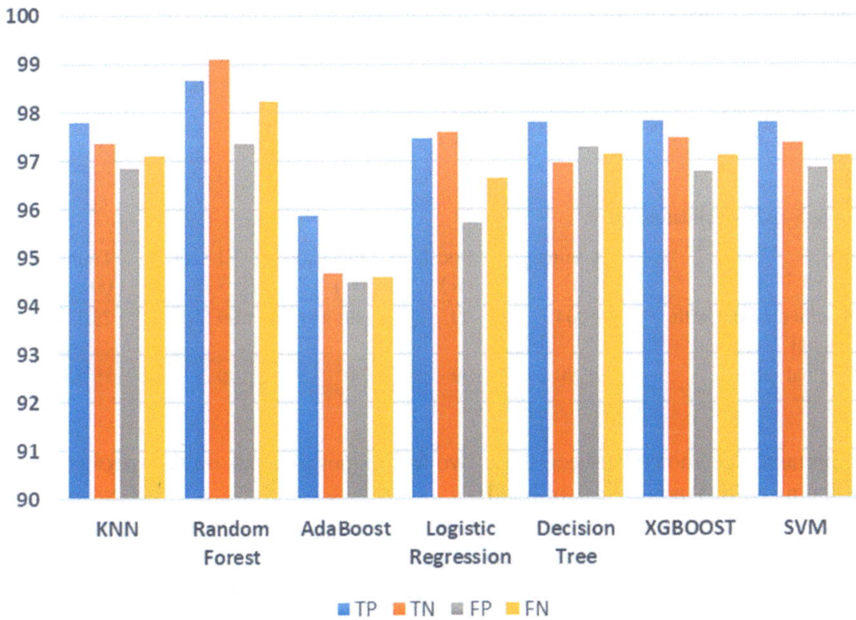

Fig. 1 Performance evaluation

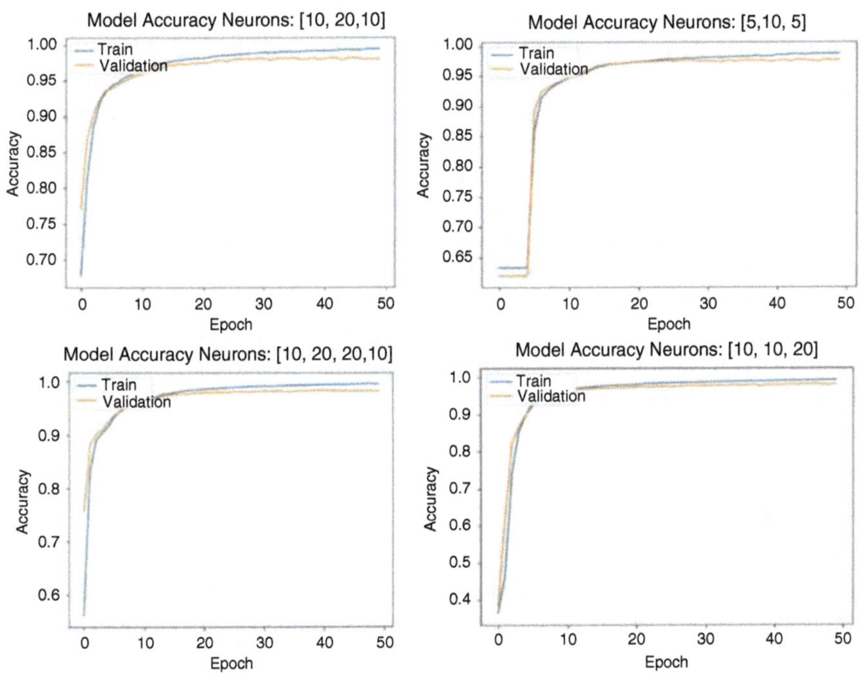

Fig. 2 Model accuracy according to different DNN designs

Table 4 ANN validation graph

Epochs	Train accuracy (%)	Train loss (%)	Validation accuracy (%)	Validation loss (%)	Train duration (%)
10	95.43	17.79	96.11	17.12	2.07
50	97.76	7.63	97.87	8.08	6.83

Table 5 ANN validation graph

Model	Accuracy (%)	Recall (%)	Precision (%)	F1-score (%)	Train time (s)
Modified model with dropout	99.04	99	99	99	23.11
Original model	98.90	99	98	99	19.55

important. The choice between these two techniques depends on the specific needs of the machine learning task at hand.

Deep learning methods were then applied. The table of values obtained by changing the iteration (epoch) values is as follows.

According to Table 4, the training accuracy of the 10–10-iteration (epoch) model is approximately 95.43% and the validation accuracy is approximately 96.11%. Similarly, the training accuracy of the model with 50 iterations (epochs) is approximately 97.76% and the validation accuracy is approximately 97.87%. These results show that the model achieves a higher accuracy with more iterations (epochs). The longer training period allowed the model to learn the patterns in the dataset better. Considering the training loss values, the model with 50 iterations (epochs) has a lower loss value. This indicates that the model has a better generalization ability.

These results show that more iterations (epochs) can improve the model performance and better results can be obtained. However, there should be a balance between the performance of the model and the computational cost. The appropriate number of iterations (epochs) should be chosen according to the needs and resources. The train time of the models is depicted in Table 5.

In the deep learning method, the best performance result is the model performance modified with Dropout. The value obtained is the best result with 99.04%. If we make a comparison between machine learning and deep learning results, the best result is the deep learning results.

5 Conclusion

In a digitalized world where Android devices have a large economic market share, secure and efficient use is becoming increasingly important. As the popularity of smart devices has increased, their use cases have diversified. This diversity has attracted the attention of malware developers and led them to create malware for various purposes. According to research, the most widely used mobile operating system worldwide is Android with a rate of 82.8%. In a digitalized world where Android devices have a large economic market share, secure and effective use is becoming increasingly important. The reliability and usability of smart mobile devices are negatively affected when healthy interaction between users and devices cannot be established. Therefore, providing a secure service against malware is a fundamental requirement for malware detection to effectively use these devices.

The number of cyber-attacks that specifically target valuable information has been the primary driver behind the huge growth in the number of security concerns that the cyber community has with Android systems. Malware is used by attackers as a technique to penetrate Android devices, with the goal of gaining control of various components of the device and abusing them. Users with Android devices are at a greater risk of being attacked, mostly as a result of the open-source nature of the Android platform and the simplicity with which its applications may be accessed. The development of algorithms for machine learning has been an essential component in the effort to handle these various security concerns. These algorithms make it possible to speed up the learning process, which in turn leads to operations that are more efficient in the product, technology, and service domains. This study conducts an in-depth literature assessment, with a particular emphasis on the utilization of machine learning strategies for the identification of malicious software on Android-based platforms. It offers a variety of recently published datasets and research papers, in addition to summary tables and an examination of several detection algorithms applicable to this setting. Experimental studies have shown that the values of machine learning detection models varied according to the algorithms applied. According to the results, the Random Forest algorithm gave the best result with an accuracy of 98.67%. In the deep learning method, the best performance result is the model performance modified with Dropout. The value obtained is the best result with 99.04% accuracy.

As a future work or some, new deep learning models can be used as mentioned in [21, 22]. The performance of the system can be increased by the use of feature selection methods as in [23, 24] and to increase the efficiency of the systems, the processing power of GPU technologies can be taken into consideration. By the way, not only the training time but also the execution time can be decreased as mentioned in [25, 26].

Conflict of Interest The authors have no relevant financial or non-financial interests to disclose.

Data Availability Training and testing processes have been carried out using the DREBIN dataset.

DREBIN dataset can be reachable at https://www.sec.cs.tu-bs.de/~danarp/drebin/download.html.

References

1. Statcounter: Desktop vs mobile vs tablet market share worldwide. https://gs.statcounter.com/platform-market-share/desktop-mobile-tablet. Last accessed 2023/06/10
2. Bayazit, E.C., Sahingoz, O.K., Dogan, B.: Malware detection in Android systems with traditional machine learning models: a survey. In: 2020 International Congress on Human-Computer Interaction, Optimization and Robotic Applications (HORA), pp. 1–8. IEEE, Ankara (2020)
3. Kumar, P., Gupta, G.P., Tripathi, R.: Toward design of an intelligent cyber attack detection system using hybrid feature reduced approach for IoT networks. Arab. J. Sci. Eng. **46**, 3749–3778 (2021)
4. Karbab, E.B., Debbabi, M., Derhab, A., Mouheb, D.: MalDozer: automatic framework for Android malware detection using deep learning. Digit. Investig. **24**, 48–59 (2018)
5. Kumar, P., Gupta, G.P., Tripathi, R.: A cyber attack detection framework focused on federated learning and fog-cloud architecture for IoT networks. Comput. Commun. **166**, 110–124 (2021)
6. Arslan, R.S., Doğru, İ.A., Barışçı, N.: Permission comparison-based malware detection system for Android mobile applications. J. Polytech. **20**(1), 175–189 (2017)
7. Almomani, I., Qaddoura, R., Habib, M., Alsoghyer, S., Al Khayer, A., Aljarah, I., Faris, H.: Android ransomware detection based on a hybrid evolutionary approach in the context of highly imbalanced data. IEEE Access. **9**, 57674–57691 (2021)
8. Agrawal, P., Trivedi, B.: Machine learning classifiers for Android malware detection. In: Data Management, Analytics and Innovation, pp. 311–322. Springer, Berlin/Heidelberg (2021)
9. Amouri, A., Alaparthy, V.T., Morgera, S.D.: A machine learning-based intrusion detection system for mobile Internet of Things. Sensors. **20**, 461 (2020)
10. Hussain, M.S., Khan, K.U.R.: A survey of IDS techniques in MANETs using machine learning. In: Proceedings of the Third International Conference on Computational Intelligence and Informatics, pp. 743–751. Springer, Singapore (2020)
11. Bayazit, E.C., Sahingoz, O.K., Dogan, B.: Deep learning-based malware detection for Android systems: a comparative analysis. Tehnički Vjesnik. **30**(3), 787–796 (2023)
12. Dini, G., Martinelli, F., Saracino, A., Sgandurra, D.: MADAM: a multi-level anomaly detector for Android malware. In: Computer Network Security, pp. 240–253. Springer, Berlin (2012)
13. Sasidharan, S.K., Thomas, C.: ProDroid – an Android malware detection framework based on a profile hidden Markov model. Pervasive Mob. Comput. **72**, 1–16 (2021)
14. Wu, D., Mao, C., Wei, T., Lee, H., Wu, K.: DroidMat: Android malware detection through manifest and API calls tracing. In: Seventh Asia Joint Conference on Information Security, pp. 62–69. IEEE, Tokyo (2012)
15. Contagio Mobile: Android Fakebank samples. http://contagiominidump.blogspot.com. Last accessed 28/09/2023
16. Arp, D., Spreitzenbarth, M., Hübner, M., Gascon, H., Rieck, K., Siemens, C.: Drebin: effective and explainable detection of Android malware in your pocket. In: 21st Annual Symposium on Network and Distributed System Security. Internet Society, San Diego (2014)
17. Sahs, J., Khan, L.: A machine learning approach to Android malware detection. In: Proceedings of the 2012 European Intelligence and Security Informatics Conference, pp. 141–147. IEEE, Odense (2012)
18. Milosevic, N., Dehghantanha, A., Choo, K.-K.R.: Machine learning-aided Android malware classification. Comput. Electr. Eng. **61**, 266–274 (2017)
19. Lei, T., Qin, Z., Wang, Z., Li, Q., Ye, D.: EveDroid: event-aware Android malware detection against model degrading for IoT devices. IEEE Internet Things J. **6**(4), 6668–6680 (2019)
20. Wenbo, F., Linlin, Z., Chenyue, W., Yingjie, H., Yuaner, Y., Kai, Z.: AMC-MDL: a novel approach to Android malware classification using multimodal deep learning. In: 2020 IEEE Intl Conf on Dependable, Autonomic and Secure Computing, Intl Conf on Pervasive Intelligence and Computing, Intl Conf on Cloud and Big Data Computing, Intl Conf on Cyber Science and Technology Congress (DASC/PiCom/CBDCom/CyberSciTech), pp. 251–256. IEEE (2020)

21. Sismanoglu, G., Onde, M.A., Kocer, F., Sahingoz, O.K.: Deep learning based forecasting in stock market with big data analytics. In: 2019 Scientific Meeting on Electrical-Electronics and Biomedical Engineering and Computer Science (EBBT), pp. 1–4. IEEE, Istanbul (2019)
22. Prakash, A., Chauhan, S.: A comprehensive survey of trending tools and techniques in deep learning. In: 2023 International Conference on Disruptive Technologies (ICDT), pp. 289–292. IEEE, Greater Noida (2023)
23. Korkmaz, M., Sahingoz, O.K., Diri, B.: Feature selections for the classification of webpages to detect phishing attacks: a survey. In: 2020 International Congress on Human-Computer Interaction, Optimization and Robotic Applications (HORA), pp. 1–9. IEEE, Ankara (2020)
24. Kumar, K.R., Nakkeeran, R.: A comprehensive study on denial of service (DoS) based on feature selection of a given set datasets in Internet of Things (IoT). In: 2023 International Conference on Signal Processing, Computation, Electronics, Power and Telecommunication (IConSCEPT), pp. 1–8. IEEE, Karaikal (2023)
25. Baykal, S.I., Bulut, D., Sahingoz, O.K.: Comparing deep learning performance on BigData by using CPUs and GPUs. In: 2018 Electric Electronics, Computer Science, Biomedical Engineerings' Meeting (EBBT). IEEE, Istanbul (2018)
26. Ghioldi, F., Piscaglia, F.: Acceleration of supersonic/hypersonic reactive CFD simulations via heterogeneous CPU-GPU supercomputing. Comput. Fluids. **266**, 106041 (2023)

A Comparative Study of Malicious URL Detection: Regular Expression Analysis, Machine Learning, and VirusTotal API

Jason Misquitta and Anusha Kannan

1 Introduction

The Internet has become a crucial part of our lives, and it offers many benefits. However, with these benefits come threats, including viruses, malware, and other malicious activities. One of the most significant threats is through malicious URLs, which are links that appear legitimate but are designed to harm your computer or steal your personal information.

To counter this threat, antivirus software has become a crucial tool in safeguarding computers and networks against malicious URLs. Its primary purpose is to identify and eliminate malicious software from a computer or network. Additionally, antivirus software has the capability to detect malicious URLs and proactively prevent users from accessing them. Identifying malicious URLs is a pivotal aspect of cybersecurity, given the increasing sophistication and frequency of cyberattacks. Numerous methods and strategies have been suggested for detecting malicious URLs in recent times. This paper seeks to assess and contrast the efficiency of various approaches employed within this domain.

In this study, we employed three distinct approaches for detecting malicious URLs. The first method involved the utilization of custom-written code, which disassembled URLs into various components and applied functions and checkers to determine their malicious nature. The second approach utilized a pre-existing CSV dataset containing URLs from over 11,000 websites. Each entry in the dataset contained 30 parameters describing the websites, along with a class label indicating whether they were classified as phishing websites (1) or not (−1). We compared the

J. Misquitta · A. Kannan (✉)
School of Computer Science and Engineering, Vellore Institute of Technology, Chennai, Tamil Nadu, India
e-mail: anusha.k@vit.ac.in

M. N. Seyman (ed.), *2nd International Congress of Electrical and Computer Engineering*, EAI/Springer Innovations in Communication and Computing,
https://doi.org/10.1007/978-3-031-52760-9_16

performance of nine machine learning models on this dataset to identify the models with the highest accuracy and F1-scores. The third method employed a VirusTotal API Key, allowing users to input a website's URL for classification as malicious or not. Additionally, if a URL was detected as malicious, a report summarizing the analysis by various security vendors was generated.

2 Literature Review

Reference [1] presents an in-depth exploration of machine learning's application in identifying malicious URLs. The survey encompasses diverse methodologies, algorithms, and techniques employed in this domain. It investigates the challenges and advancements in URL-based threat detection, serving as a valuable resource for researchers and practitioners working on cybersecurity and threat mitigation. The authors explore the changing nature of cybersecurity risks and the contribution of machine learning in enhancing online security.

Reference [2] focuses on employing machine learning to identify malicious URLs through lexical features. The study explores the utilization of various linguistic and contextual attributes present in URLs for effective detection. The authors propose a methodology that involves feature extraction and classification using machine learning algorithms. The research aims to enhance cybersecurity by offering a technique to swiftly recognize potentially harmful URLs. The findings contribute to the ongoing efforts in devising robust tools for safeguarding online environments against evolving cyber threats.

Reference [3] presents a comprehensive exploration of malicious URL detection strategies, employing a synergistic blend of machine learning and heuristic methodologies. The research investigates the effectiveness of these combined techniques in accurately identifying potentially harmful URLs. Leveraging machine learning algorithms for intricate pattern recognition and incorporating rule-based heuristics for deeper analysis, the proposed approach aims to achieve enhanced precision in malicious URL identification. The study significantly contributes to the ongoing discourse on cybersecurity. By addressing the evolving challenges posed by cyber threats, this research offers valuable insights and practical implications for safeguarding digital environments.

Reference [4] addresses the opacity surrounding the labeling process of online scan engines like VirusTotal, which researchers heavily rely on to categorize malicious URLs and files. The paper aims to unravel the intricacies of the labeling generation process and assess the reliability of scanning outcomes. Focusing on VirusTotal and its 68 third-party vendors, the study centers on the labeling process for phishing URLs. Through a comprehensive investigation involving the establishment of mimic PayPal and IRS phishing websites and subsequent URL submissions for scanning, the authors analyze the network traffic and dynamic label changes within VirusTotal. The study uncovers significant insights into VirusTotal's operational mechanisms and label accuracy. Findings reveal challenges faced by vendors in identifying all phishing sites, with even the most proficient vendors failing to detect 30% of phishing sites. Moreover, inconsistencies emerge between VirusTotal scans and some vendors' in-house scanners, shedding light on the need for more robust methodologies to evaluate and effectively utilize VirusTotal's labels.

Phishing represents a type of online social manipulation aimed at stealing digital identities by posing as legitimate entities. This entails sending malicious content through channels like emails, chats, or blogs, often embedded with URLs leading to deceptive websites designed to extract private information from victims. The main objective of [5] is to construct a system capable of scrutinizing and categorizing URLs, primarily to identify phishing attempts. Rather than resorting to traditional methods involving website visits and subsequent feature extraction, the proposed strategy focuses on URL analysis. This not only ensures a level of distance between the attacker and the target but also demonstrates faster performance when compared to alternative methods like conducting internet searches, fetching content from destination websites, and utilizing network-level characteristics as observed in prior research. The research extensively explores various facets of URL analysis. This encompasses performance assessments conducted on balanced and imbalanced datasets, covering both controlled and real-time experimental scenarios. Additionally, the study explores the distinctions between online and batch learning approaches.

Reference [6] examines a lightweight strategy for detecting and categorizing malicious URLs based on their specific attack types. The efficacy and efficiency of using lexical analysis as a proactive means of identifying these URLs is demonstrated. The study identifies a comprehensive set of essential features required for precise categorization. The accuracy of this approach is evaluated using a dataset comprising over 110,000 URLs. An examination is conducted to assess the effects of obfuscation methods.

Due to mobile screens' limited size, manually verifying phishing URLs sent via text messages becomes challenging. Clicking a smishing URL can either direct users to phishing sites or attempt to implant harmful software. Reference [7] aims to counter Smishing attacks on Android devices through a novel application integrating established phishing APIs. This background-running app validates URL authenticity in received text messages. Five APIs were tested on a 1500 URL dataset, revealing the VirusTotal API's 99.27% accuracy but slower response (12–15 s) and the Safe-Browsing API's 87% accuracy with swift (0.15 ms) response. Depending on the application's urgency or security emphasis, the choice of API varies.

Reference [8] introduces "Obfuscapk," an open-source tool designed for enhancing the security of Android apps through obfuscation. Obfuscation is crucial to protect apps from reverse engineering. The tool operates as a "black-box," obfuscating apps without requiring their original source code. It applies various techniques, including renaming classes, methods, variables, and encrypting strings, making the code harder to decipher. As a response to the vulnerability of APK files to reverse engineering, Obfuscapk contributes significantly to mobile app security. The paper outlines the tool's functionalities, emphasizes its role in safeguarding intellectual property, and showcases how it mitigates potential security vulnerabilities. Obfuscapk's approach serves as a valuable contribution to the field of Android app security.

Due to a lack of security awareness, numerous web applications are vulnerable to web attacks. Addressing this, there is a pressing need to bolster web application reliability by effectively identifying malicious URLs. However past efforts have predominantly employed keyword matching for this purpose, such an approach lacks adaptability. Reference [9] introduces a groundbreaking technique that merges statistical analysis via gradient learning with the extraction of features using a sigmoidal threshold. The study utilizes naïve Bayes, decision tree, and SVM classifiers to evaluate the effectiveness and efficiency of this method. The empirical outcomes emphasize its strong ability to detect patterns, yielding an accuracy rate that exceeds 98.7%. Notably, the method has been deployed online for large-scale detection, successfully analyzing around 2 TB of data daily.

The surge in internet-connected devices like smartphones, IoT, and cloud networks has led to increased phishing attacks exploiting human vulnerabilities. Unlike system-focused attacks, phishing deceives users into revealing personal data. Addressing this, a streamlined phishing detection approach utilizing only nine lexical features has been introduced in [10]. Unlike resource-intensive methods, this approach is suitable for constrained devices. It applied Random Forest and was tested on the ISCXURL-2016 dataset with 11,964 instances.

3 Methodology

In this paper, malicious URL detection was done in three separate unique ways. Each method has been described below.

3.1 Segmenting and Analyzing URL

In the first method, the following self-written code in Fig. 1 was used for breaking up the URL into different fragments and then evaluating it using various functions. The code checks if the URL belongs to a Malicious or Safe Domain. It checks for suspicious keywords [11]. Parsing is done through regular expression to see if the URL matches the correct format. The path and file name within the URL was checked for additional clues about the content. For instance, URLs with file extensions like .exe, .zip, or .js might indicate potential malware downloads [12].

```python
import re
import requests

# Define a list of known malicious domains
MALICIOUS_DOMAINS = [
    'badurl.com',
    'malwaredomain.com',
    'evilhost.com',
    # Add more malicious domains as needed
]
# Define a list of known safe domains
SAFE_DOMAINS = [
    'google.com',
    'microsoft.com',
    'apple.com',
    # Add more safe domains as needed
]
# Define a list of suspicious keywords
SUSPICIOUS_KEYWORDS = [
    'download',
    'execute',
    'install',
    'update',
    # Add more suspicious keywords as needed
]

def is_malicious_url(url):
    # Extract the domain from the URL using regular expressions
    domain_regex = r'https?://(?:www\.)?([\w\.-]+)'
    match = re.search(domain_regex, url)
    if not match:
        # If the URL doesn't match the expected format, assume it's malicious
        return True
    domain = match.group(1)

    # Check if the domain is in the list of known malicious domains
    if domain in MALICIOUS_DOMAINS:
        return True
    # Check if the domain is in the list of known safe domains
    if domain in SAFE_DOMAINS:
        return False

    # Check if the URL contains any suspicious keywords
    for keyword in SUSPICIOUS_KEYWORDS:
        if keyword in url.lower():
            return True

    # If the URL has not been identified as either safe or malicious, assume it's safe
    return False
# Prompt user for URL to check
url = input("Enter a URL to check: ")

if is_malicious_url(url):
    print('Warning: This URL may be malicious!')
else:
    print('This URL seems safe.')
```

Fig. 1 Code for analyzing URL

3.2 Comparing Nine Machine Learning Models

In the second method, a dataset from Kaggle was used: https://www.kaggle.com/eswarchandt/phishing-website-detector. A small part of the dataset is shown in Fig. 2. It has 11,054 samples with 32 features.

The final attribute serves as the class label, indicating whether it categorizes the website as a phishing site (1) or not (−1). The following models were trained on this dataset and then compared on evaluation metrics.

- *Gradient Boosting Classifier*: Gradient Boosting Classifier serves as an ensemble learning method that merges multiple weak models (decision trees) to establish a more robust classifier. It follows a sequential training process, wherein each tree aims to rectify the mistakes made by the preceding trees. Renowned for its exceptional accuracy and proficiency in handling extensive datasets, this algorithm finds valuable application in malicious URL detection and other domains where precision and scalability are essential.
- *CatBoost Classifier*: CatBoost Classifier is a gradient boosting algorithm that is optimized for categorical data. It handles categorical variables efficiently by encoding them as numerical variables and reducing the impact of noisy variables. It also performs well on datasets with missing values, without requiring imputation [13].
- *Random Forest*: Random Forest represents an ensemble learning algorithm utilized for malicious URL detection, which constructs numerous decision trees and amalgamates their predictions. To mitigate overfitting, each tree is created using a random subset of the features. Known for its straightforwardness, impressive accuracy, and capacity to manage extensive datasets, Random Forest serves as a valuable tool in the field of security to effectively identify potentially harmful URLs [14].

R	S	T	U	V	W	X	Y	Z	AA	AB	AC	AD	AE	AF
InfoEmail	Abnormall	WebsiteFo	StatusBar(	DisableRig	UsingPopu	IframeRed	AgeofDom	DNSRecor	WebsiteTr	PageRank	GoogleInd	LinksPoint	StatsRepo	class
1	1	0	1	1	1	1	-1	-1	0	-1	1	1	1	-1
-1	-1	0	1	1	1	1	1	-1	1	-1	1	0	-1	-1
1	1	0	1	1	1	1	-1	-1	1	-1	1	-1	1	-1
1	1	0	-1	1	-1	1	-1	-1	0	-1	1	1	1	1
-1	-1	0	1	1	1	1	1	1	1	-1	1	-1	-1	1
-1	-1	0	1	1	1	1	1	-1	-1	-1	1	0	-1	-1
1	1	0	1	1	1	1	-1	-1	0	-1	1	0	1	-1
1	1	0	1	1	1	1	1	-1	1	1	1	0	1	1
1	1	0	1	1	1	1	1	-1	0	-1	1	0	1	-1
-1	-1	0	1	1	1	1	-1	1	1	1	1	-1	-1	1
-1	-1	0	1	1	1	1	-1	-1	-1	-1	1	0	-1	-1
1	1	0	-1	1	-1	1	1	-1	-1	-1	1	0	1	-1
1	1	0	1	1	1	1	-1	-1	0	-1	1	1	1	-1
1	1	0	1	1	1	1	1	-1	1	-1	1	-1	1	1
1	1	0	1	1	1	1	1	-1	-1	-1	1	0	1	-1
-1	-1	0	1	1	1	1	1	-1	0	-1	1	1	-1	-1
-1	-1	0	1	1	1	1	-1	1	1	-1	1	1	-1	-1
-1	-1	0	1	1	1	1	1	-1	-1	1	1	-1	-1	1
-1	-1	0	-1	-1	-1	-1	1	-1	0	-1	1	0	-1	1
-1	-1	0	-1	1	-1	1	-1	1	1	-1	1	-1	-1	1
-1	-1	0	1	1	1	1	-1	1	-1	-1	1	0	-1	1
1	1	0	1	1	1	1	1	1	0	-1	1	-1	1	1

Fig. 2 A small section of the dataset

- *Multi-layer Perceptron*: Multi-layer Perceptron is a neural network structure comprising multiple layers of perceptrons. It is capable of learning complex non-linear relationships between inputs and outputs. It is a versatile algorithm that can be used for classification, regression, and other machine learning tasks [15].
- *Support Vector Machine*: Support Vector Machine (SVM) represents a binary classification technique that seeks to identify the best hyperplane to separate two classes while maximizing the margin between them. SVM is particularly renowned for its proficiency in handling datasets with numerous dimensions and those that are not linearly separable.
- *Decision Tree*: Decision Tree is a classification algorithm that builds a model resembling a tree to make decisions and determine their potential outcomes. This straightforward and easy-to-understand algorithm is capable of handling both categorical and continuous data. Its versatility makes it a valuable tool in various domains [16].
- *K-Nearest Neighbors*: K-Nearest Neighbors (KNN) is a non-parametric classification technique that assigns a class label to an instance based on the class labels of its k-closest neighbors within the training dataset. This straightforward algorithm does not involve any training phase and is capable of addressing multi-class problems. Nonetheless, its performance may be hindered by its relatively slow and memory-intensive nature, particularly when applied to large datasets [17].
- *Logistic Regression*: For the purpose of malicious URL detection, Logistic Regression can be employed as a linear classification algorithm. By utilizing a logistic function and input features, it can calculate the probability of a URL being malicious or benign. Due to its simplicity and interpretability, Logistic Regression can be a valuable tool for identifying potentially harmful URLs in various security applications [18].
- *Naive Bayes Classifier*: The Naive Bayes Classifier is a classification method that relies on probabilities and makes an assumption that the input features are conditionally independent when given the class. It employs Bayes' theorem to compute the posterior probability for each class, taking into account the provided features. Naive Bayes is renowned for its simplicity and efficiency, making it suitable for handling high-dimensional data. This algorithm finds application in various fields like spam filtering, sentiment analysis, and document classification.

The main aim of the second method is to see which machine learning models are best suited to malicious URL classification.

3.3 Using VirusTotal API Key

The third method involved the use of VirusTotal API key to check whether the URL is malicious or not.

VirusTotal is a no-cost web-based platform that examines files and web addresses for viruses, worms, trojans, and other types of malicious software. Users can submit a file or URL to VirusTotal, and the service will scan it using more than 70 different antivirus engines and other malware detection tools. VirusTotal also offers an API that allows developers to integrate its malware scanning capabilities into their own applications [19]. By using a VirusTotal API key, developers can submit files or URLs programmatically and retrieve the results for further analysis or processing. The VirusTotal API key can also be used to retrieve additional information about malware samples, such as their behavior and network activity. VirusTotal's API is widely used in the cybersecurity industry for threat intelligence, malware analysis, and incident response [20].

4 Results

4.1 Results of Self-Written Code

As seen in Fig. 3, the first method gave the following outputs for various URLs.

The outputs were highly accurate (close to 90%) in classifying the URLS. Thus, the self-written code proved to be quite effective in detecting malicious URLs.

4.2 Results of Comparison Between Nine Machine Learning Models

Table 1 was obtained after running all the machine learning models mentioned in the second method. The table displays the evaluation metrics (accuracy, F1-score, recall, and precision) of all the nine models. It is evident that the accuracy, F1-score, and recall are extremely high for all the models except for Naive Bayes Classifier. This

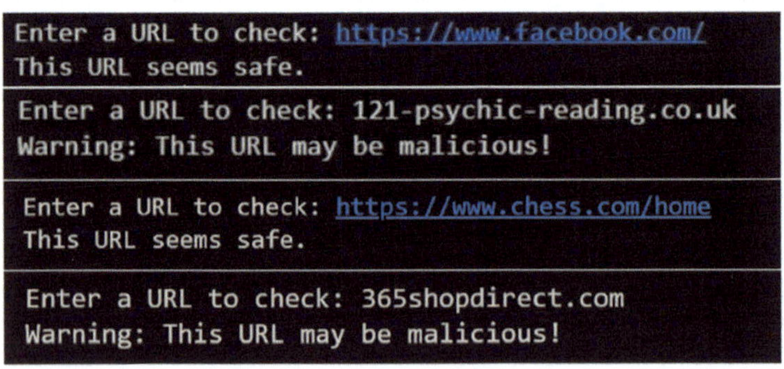

Fig. 3 Output of code

Table 1 Evaluation metrics comparison of nine models

Models	Accuracy	F1 score	Recall	Precision
Gradient Boosting Classifier	0.974	0.977	0.994	0.986
CatBoost Classifier	0.972	0.975	0.994	0.989
Random Forest	0.966	0.969	0.993	0.989
Multi-layer Perceptron	0.966	0.969	0.977	0.993
Support Vector Machine	0.964	0.968	0.980	0.965
Decision Tree	0.960	0.964	0.991	0.993
K-Nearest Neighbors	0.956	0.961	0.991	0.989
Logistic Regression	0.934	0.941	0.943	0.927
Naive Bayes Classifier	0.605	0.454	0.292	0.997

Fig. 4 Plotting accuracy of gradient boosting classifier

indicates that all these models (except Naive Bayes) are equally useful and suited to malicious URL detection.

Training and test accuracy was plotted for each model. In Fig. 4, the accuracy curves for the best model (Gradient Boosting Classifier) are shown.

To understand the pairwise relationships between different features in the dataset, we applied the Seaborn Pairplot as seen in Fig. 5. The Pairplot gave us an understanding of the amount of similarity between five key features. PrefixSuffix, SubDomains, and WebsiteTraffic all show high correlation with respect to each other.

A heatmap in Fig. 6. was also plotted to visualize the positive and negative correlations between the features. We found that the feature HTTPS had the highest positive correlation (0.71) in determining the class of the URL, followed by AnchorURL (0.69). There were three features: Favicon, UsingPopUpWindow, and IframeRedirection that had zero correlation with respect to the class of the URL.

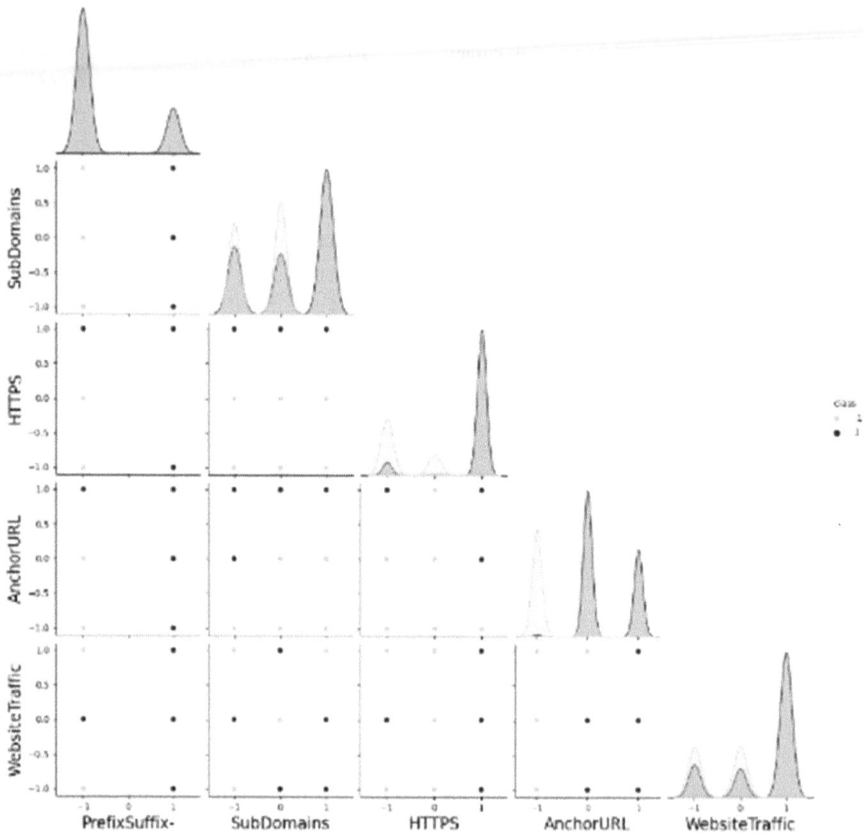

Fig. 5 Pairplot

4.3 Results Obtained Using VirusTotal API Key

In Fig. 7, we see outputs from the third method where VirusTotal API Key is used. The total detections are the number of antivirus engines that have analyzed that particular URL. The positive detections are the number of antivirus engines that have classified it as malicious. Even if only one antivirus engine detects the URL as malicious, it is declared dangerous and an analysis report is issued as displayed in Fig. 8.

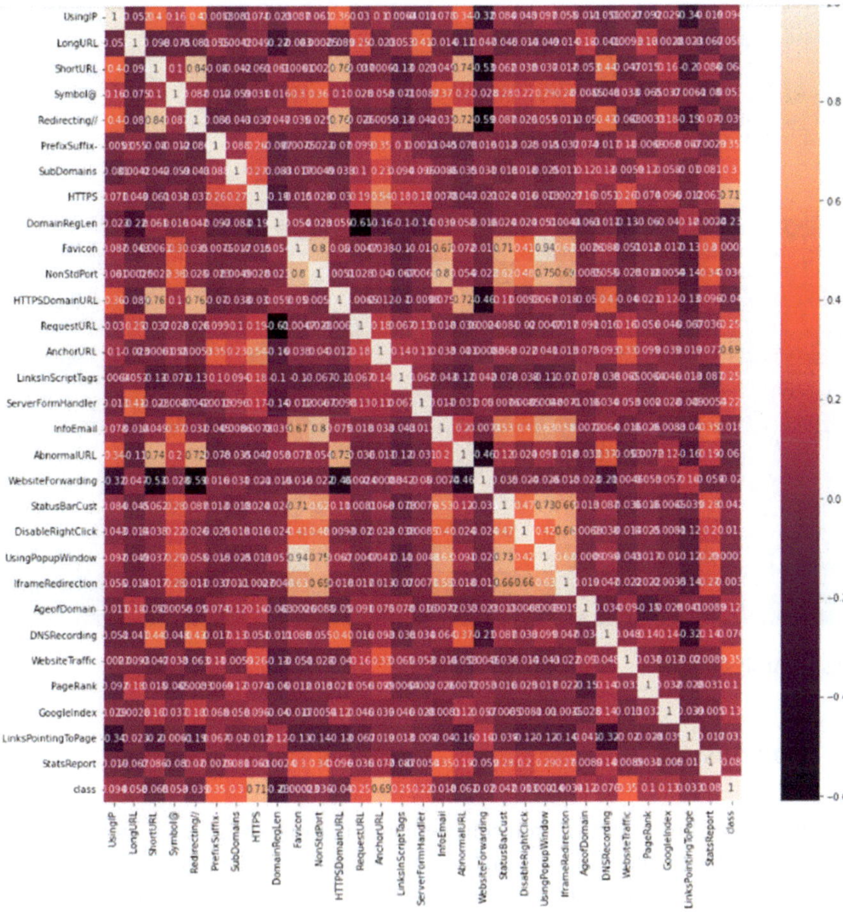

Fig. 6 Heatmap

```
Enter the URL to scan (or type 'exit' to quit): https://11011020.web.fc2.com/
The URL https://11011020.web.fc2.com/ is malicious.
Positive detections: 1
Total detections: 65
A detailed report has been saved in the same directory as this script: /content/h1
Scan another URL? (y/n): y
Enter the URL to scan (or type 'exit' to quit): https://028cdxyk.com/
The URL https://028cdxyk.com/ is malicious.
Positive detections: 11
Total detections: 89
A detailed report has been saved in the same directory as this script: /content/h1
Scan another URL? (y/n):
```

Fig. 7 Checking whether website is malicious or not

Fig. 8 Analysis report

5 Conclusion

The main conclusion drawn from this research is to investigate different machine learning models, conduct Data Analysis on the phishing dataset, and comprehend their characteristics. Some of the following observations are concluded:

- Gradient Boosting Classifier gave the best accuracy of 97.4% and also gave the best F1-score of 0.977. Hence, it was the best model to reduce the chance of malicious attachments.
- Naive Bayes Classifier gave the highest precision of 0.997; however, it was the worst model as all the other evaluation metrics were low.
- In the exploration of the Phishing dataset, it was revealed that attributes like "AnchorURL," "HTTPS" and "WebsiteTraffic" hold even greater paramount importance in the classification of a URL as either a phishing URL or a legitimate one.
- "Favicon," "UsingPopUpWindow" and "IframeRedirection" were the features that had zero correlation in the URL classification.
- The results of the self-written code were verified using the VirusTotal API Key. The code proved to be 90% successful in correctly identifying malicious URLs.

Future work based on our study could involve exploring hybrid approaches that combine machine learning and deep learning techniques for enhanced malicious URL detection, and investigating the integration of real-time threat intelligence feeds to improve accuracy and response speed.

Statements and Declaration

Author Contributions: J.M. and A.K. chose the dataset; J.M. wrote the code for segmenting and analyzing URL; J.M. and A.K. implemented the different machine learning models and compared them; J.M. verified the results of the code using VirusTotal API key; J.M. prepared the original draft submission; and A.K. conducted a thorough review and made edits to the manuscript. All authors have reviewed and approved the final published version of the manuscript.

Funding: No external funding was provided to this research.

Competing Interests: The authors state that they have no conflicts of interest.

Acknowledgments: Not applicable.

References

1. Sahoo, D., Chenghao L., Steven, C.H.: Malicious URL detection using machine learning: a survey. arXiv preprint arXiv:1701.07179 (2017)
2. Raja, A.S., Vinodini, R., Kavitha, A.: Lexical features based malicious URL detection using machine learning techniques. Mater. Today: Proc. **47**, 163–166 (2021)
3. Begum, A., Srinivasu, B.: A study of malicious url detection using machine learning and heuristic approaches. In: Advances in Decision Sciences, Image Processing, Security and Computer Vision: International Conference on Emerging Trends in Engineering (ICETE), vol. 2. Springer International Publishing, Cham (2020)
4. Peng, P., Yang, L., Song, L., Wang, G.: Opening the blackbox of virustotal: analyzing online phishing scan engines. In: Proceedings of the Internet Measurement Conference, pp. 478–485. Association for Computing Machinery, New York (2019)
5. Verma, R., Das, A.: What's in a url: fast feature extraction and malicious url detection. In: Proceedings of the 3rd ACM on International Workshop on Security and Privacy Analytics, pp. 55–63. Association for Computing Machinery, New York (2017)
6. Mamun, M.S.I., et al.: Detecting malicious urls using lexical analysis. In: Network and System Security: 10th International Conference, NSS, Taipei, Taiwan, 28–30 Sept 2016
7. Phadke, P., Christina, T.: Analysis of API driven application to detect smishing attacks. In: European Conference on Cyber Warfare and Security, Chester, Cham (2021)
8. Aonzo, S., Georgiu, G.C., Verderame, L., Merlo, A.: Obfuscapk: an open-source black-box obfuscation tool for android apps. SoftwareX. **11**, 100403 (2020)
9. Cui, B., He, S., Yao, X., Shi, P.: Malicious url detection with feature extraction based on machine learning. Int. J. High Perform. Comput. Netw. **12**(2), 166–178 (2018)
10. Gupta, B.B., Yadav, K., Razzak, I., Psannis, K., Castiglione, A., Chang, X.: A novel approach for phishing urls detection using lexical based machine learning in a real-time environment. Comput. Commun. **175**, 47–57 (2021)
11. Janet, B., Joshua, A.K.R.: Malicious URL detection: a comparative study. In: 2021 International Conference on Artificial Intelligence and Smart Systems (ICAIS). IEEE, Coimbatore, Cham (2021)
12. Liu, C., et al.: Finding effective classifier for malicious URL detection. In: Proceedings of the 2018 2nd International Conference on Management Engineering, Software Engineering and Service Sciences. Association for Computing Machinery, New York (2018)
13. Ibrahim, A.A., Ridwan, R.L., Muhammed, M.M., Abdulaziz, R.O., Saheed, G.A.: Comparison of the catboost classifier with other machine learning methods. Int. J. Adv. Comput. Sci. Appl. **11**(11), 738–748 (2020)
14. Weedon, M.D.T., James, D.-P.: Random forest explorations for URL classification. In: 2017 International Conference on Cyber Situational Awareness, Data Analytics and Assessment (Cyber SA). IEEE, London, Cham (2017)

15. Odeh, A.J., Keshta, I., Abdelfattah, E.: Efficient detection of phishing websites using multilayer perceptron. Int. J. Interact. Mob. Technol. **14**(11), 22–31 (2020)
16. Patil, D.R., Patil, J.B., et al.: Malicious urls detection using decision tree classifiers and majority voting technique. Cybern. Inf. Technol. **18**(1), 11–29 (2018)
17. Assegie, T.A.: K-nearest neighbor based url identification model for phishing attack detection. Indian J. Artif. Intell. Neural Netw. **1**, 18–21 (2021)
18. Rupa, C., et al.: Malicious url detection using logistic regression. In: 2021 IEEE International Conference on Omni-Layer Intelligent Systems (COINS). IEEE, Barcelona, Cham (2021)
19. Prasad, S.K., Budhathoki, D.R., Dasgupta, D.: Forensic analysis of ransomware families using static and dynamic analysis. In: 2018 IEEE Security and Privacy Workshops (SPW). IEEE, San Francisco, Cham (2018)
20. Deng, K.C., Juremi, J.: BEsafe-validating URLs and domains with the aid of indicator of compromise. In: 2023 15th International Conference on Developments in eSystems Engineering (DeSE). IEEE, Baghdad & Anbar, Cham (2023)

An Efficient Technique Based on Deep Learning for Automatic Focusing in Microscopic System

Fatma Tuana Dogu, Hulya Dogan, Ramazan Ozgur Dogan, Ilyas Ay, and Sena F. Sezen

1 Introduction

Microscope is one of the most useful imaging tools in the preliminary diagnosis of diseases caused by various pathogens such as viruses, bacteria, and fungi. The microscopic examination process determines the infection degree of patient and disease severity during the preliminary diagnosis of diseases. The microscopic system has a limited field of view. Therefore, during the microscopic examination process, the expert must position the microscope stage in the 3D axis (*X*-*Y*-*Z*) and

F. T. Dogu (✉)
Department of Electrical and Electronics Engineering, Faculty of Engineering, Karadeniz Technical University, Trabzon, Turkiye

H. Dogan
Department of Software Engineering, Faculty of Engineering, Karadeniz Technical University, Trabzon, Turkiye

Drug and Pharmaceutical Technology Application & Research Center, Karadeniz Technical University, Trabzon, Turkiye

R. O. Dogan
Department of Software Engineering, Faculty of Engineering and Natural Sciences, Gumushane University, Gumushane, Turkiye

I. Ay
Drug and Pharmaceutical Technology Application & Research Center, Karadeniz Technical University, Trabzon, Turkiye

S. F. Sezen
Drug and Pharmaceutical Technology Application & Research Center, Karadeniz Technical University, Trabzon, Turkiye

Department of Pharmacology, Faculty of Pharmacy, Karadeniz Technical University, Trabzon, Turkiye

M. N. Seyman (ed.), *2nd International Congress of Electrical and Computer Engineering*, EAI/Springer Innovations in Communication and Computing,
https://doi.org/10.1007/978-3-031-52760-9_17

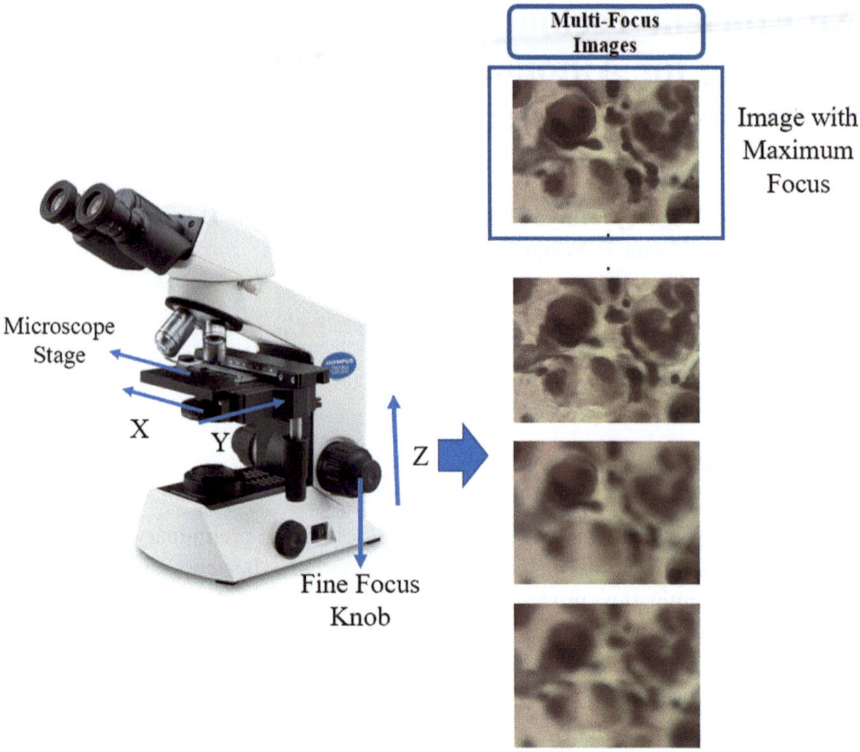

Fig. 1 Visual representation of manual optimum focusing achievement by expert

scan without losing focus with hand-eye coordination in order to analyze the entire area of the sample. This process takes a lot of time. Microscopic imaging quality is expected to be constantly maintained in this process. Therefore, while performing the examination, the expert may lose concentration and the probability of making mistakes is high. When any sample area is examined without paying attention or in a very short time, viruses, bacteria, and fungi on the sample may be overlooked, and incorrect diagnoses and findings may occur.

In the microscopic examination process, which requires scanning in the 3D axis (*X*-*Y*-*Z*), the primary phase is to achieve optimum focusing. Figure 1 shows visual representation of manual optimum focusing achievement by expert. To achieve optimum focusing as given in Fig. 1, experts move the microscope stage in the *Z* axis with the fine focus knob and find the image with the maximum focus between multi-focus images.

Automatic focusing is defined as the automation of finding the image with maximum focus by the computer [1]. Automatic focusing is performed on image sequence consisting of 2D images (multi-focus images) with the same field of view and different focuses, which are converted into digital form by the camera placed on the microscope and transferred to the computer [2]. This process provides higher quality and more effective imaging by reducing human dependency in microscopic systems [3, 4]. It is one of the important requirements for minimizing various noise and distortions in image sequences. In the literature, there are two different types of automatic focusing: active and passive [5]. Active automatic focusing systems include an additional material that measures distance between the object and the lens. They are quite expensive but can work in real time. On the other hand, passive automatic focusing systems, which are based on image processing, do not require any additional materials and greatly simplify the installation. These systems work by analyzing multi-focus images [6, 7]. The steps of this automatic focusing system are as follows:

1. Creating multi-focus images: In this step, a sequence consisting of images with the same field of view and different focuses is obtained by moving the microscope stage in the Z axis.
2. Extracting image focus values: In this step, focus information is extracted for each multi-focus image in the sequence using the focus function. In the literature, focus functions are divided into six groups [4]. (1) Gradient-based focus functions: Assuming that focused images have sharper edges than blurred images, these functions use variance or first-order derivatives to calculate the image focus value. Examples of gradient-based focus functions can be given as follows: Tenengrad [8], Quadratic Gradient [8], Gaussian Derivative [9], Thresholded Absolute Gradient [8], 3D Gradient [10], Gradient Energy [11], and Tenengrad Variance [8]. (2) Laplacian-based focus functions: Assuming that focused images have sharper edges than blurry images, these functions use Laplacian or second-order derivatives to calculate the image focus value. Some of the Laplacian-based focus functions are as follows: Modified Laplacian [12], Laplacian Energy [10], Laplacian Variance [13], Diagonal Laplacian [14], 3D Laplacian [15], and Multidirectional Modified Laplacian [16]. (3) Statistics-based focus functions: These focus functions use various image statistics to calculate the focus information of images. Examples of the statistics-based focus functions can be given as follows: Chebyshev Moments [17], Variance [8], Histogram Range [8], Modified Variance [8], Local Variance [13], Eigenvalue [18], Normalized Variance [8], and Histogram Entropy [8]. (4) Discrete Cosine Transform (DCT)-based focus functions: These focus functions use DCT coefficients to calculate the focusing levels of images from their frequency content. Some of the DCT-based focus functions are as follows: Modified DCT [19], Reduced DCT Energy Ratio [20], and DCT Energy Ratio [21]. (5) Discrete Wavelet Transform (DWT)-based focus functions: Assuming that focused images have higher frequencies, these focus functions use DWT coefficients to obtain the frequencies and positional information of the pixels for extracting the image focus values. Examples of the DWT-based focus functions can be given as follows: 3D DWT [22], Variance of DWT Coefficients [23], and Ratio of DWT Coefficients [23]. (6) Other focus functions: These focus functions extract the focusing information of the images using different features than the previous groups. Some of the focus functions are as follows: Brenner [8], Autocorrelation [8], Spatial Frequency [8], Absolute

Central Moment [24], Image Curve [25], Gabor Features [26], Image Contrast [27], Helmli and Scherer's Average Method [25], Ratio of Discrete Curvelet Transform Coefficients [28], Local Binary Pattern [29], 3D Steerable Filters [30], and 2D Steerable Filters [31].
3. Generating probability density function of image focus values: In this step, probability density function is created using the focus function values of the multi-focus images.
4. Determining the image with maximum focus value: In this step, the image with the highest value in the probability density function is determined and considered as the focused image.

As we mentioned above, many researchers have studied automatic focusing systems and proposed many functions to extract focus information from multi-focus images. Nevertheless, these focusing systems still consist of several significant limitations. The limitations of the literature studies that presented automatic focusing systems and proposed focus function can be summarized as follows:

1. Active automatic focusing systems can offer excellent accuracy and performance in real time. However, they are quite costly and require a supplementary material that measures the object's distance from the lens.
2. Although the literature contains numerous studies proposing various focus functions, passive automatic focusing systems are uncommon despite their lower cost and lack of equipment requirements.
3. Recent functions based on Gabor, 2D, and 3D steerable filters provide a more precise degree of focus for multi-focus images than conventional focus functions (Laplacian, Gradient). Literature studies reveal, however, that these functions are incapable of coping with various constraints, such as insufficiently precise characterization of curves and borders in images and longer execution time.
4. The literature contains an extensive number of focus functions. Nevertheless, their efficacy varies depending on the microscope and sample type. In order to determine the optimal focus function in their field, the researchers conduct the studies [3, 7, 32].
5. Convolutional neural networks (CNN) and deep learning have evolved significantly and swiftly in recent years [33, 34] in the research disciplines of image processing, computer vision, and medical image analysis. However, only just a few of CNN-based automatic focusing systems have been reported in the literature.

The purpose of this study is to develop an efficient deep learning–based technique to overcome the limitations of previous research presenting automatic focusing systems and proposing a focus function. The following are the primary contributions of the proposed study:

1. According to our review of the relevant literature, this is the first study to propose a deep learning–based technique for automatic focusing of microscopic systems.

2. The proposed technique acquires the pixel's focus degrees using deep features, which produces a more pronounced variation in images compared to previous research that utilized only the gray levels of multi-focus images.
3. This study provides more efficient and sample-free automatic focusing, in contrast to previous studies with various limitations, such as imprecise characterization of curves and edges in images, longer execution time, and performance variation depending on the sample and microscope.
4. In order to provide a comprehensive and sample-free analysis of the performance of automatic focusing techniques, novel multi-focus image sequences are generated in this study using various samples and magnification objectives.

The rest of this study has been designed as follows: The structure and overview of deep learning model, which forms the foundation for the proposed automatic focusing technique, is given in Sect. 2. Section 3 presents the experimental results and discussion. Finally, Sect. 4 summarizes the conclusion.

2 Methodology

The presented study has mainly focused on generating novel multi-focus image sequences and suggesting a deep learning–based technique for automatic focusing. These subdivisions of study are defined as follows:

2.1 Multi-focus Image Sequences

In order to provide a comprehensive and sample-free analysis of the automatic focusing techniques performance, multi-focus image sequences are created in this study. In order to generate these multi-focus image sequences, firstly five tissue samples of mouse, which are liver, intestine, heart, kidney, and lung, stained with hematoxylene eosin (HE), are prepared, and then multi-focus images are acquired scanning microscope stage on the *Z* axis. Examples of multi-focus image sequences – 1, 2, 3, 4, and 5, details of which are listed in Table 1, are illustrated in Fig. 2. However, 10× and 40× magnification objectives are utilized in the Zeiss Primo

Table 1 Details of multi-focus image sequences

Sequence	Tissue	Mag. objective	Image resolution	Number of images
Image sequence – 1	Liver	10×	1920 × 1080	104
Image sequence – 2	Intestine	40×	1920 × 1080	52
Image sequence – 3	Heart	40×	1920 × 1080	69
Image sequence – 4	Kidney	40×	1920 × 1080	61
Image sequence – 5	Lung	10×	1920 × 1080	62

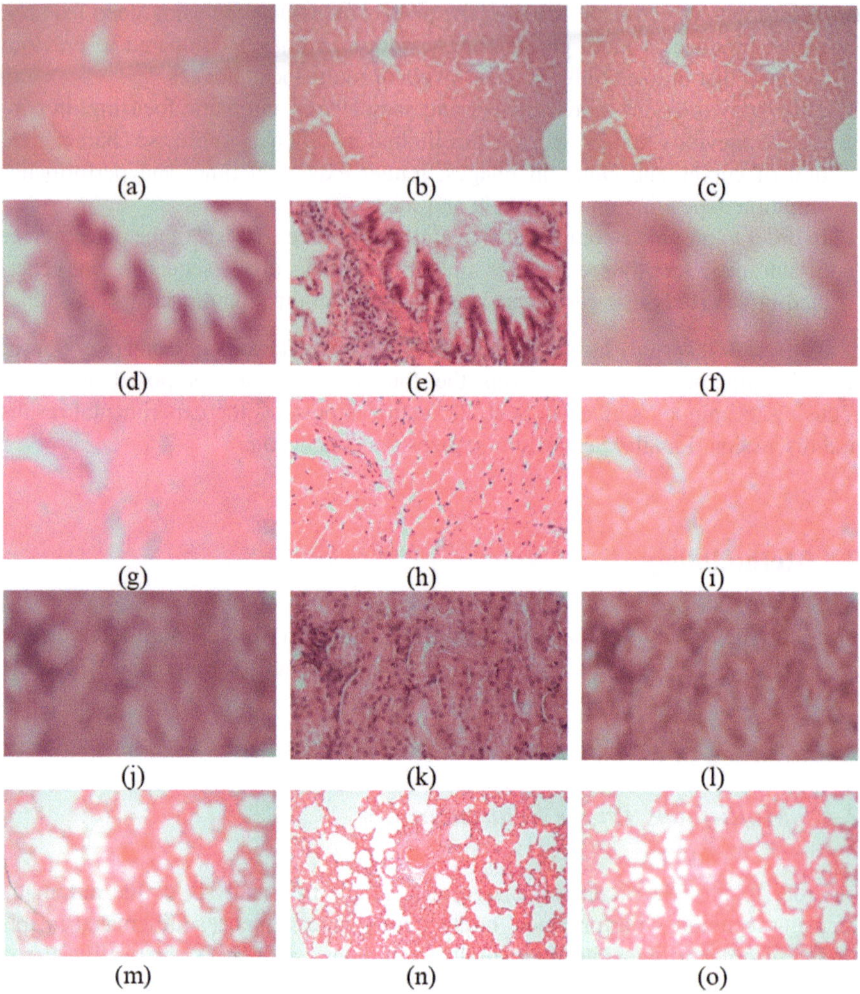

Fig. 2 Examples of multi-focus image sequences – 1 (**a**–**c**), 2 (**d**–f), 3 (**g**–**i**), 4 (**j**–**l**), and 5 (**m**–**o**)

microscope. The multi-focus images of all sequences are captured using the Zeiss Axiocam microscope camera. They have same pixel resolution, which is 1920 × 1080, and are saved in PNG file format. The image numbers in sequences are 104, 52, 69, 61, and 62.

2.2 Deep Learning–Based Techniques for Automatic Focusing

In this study, an efficient technique based on deep learning is developed for automatic focusing on microscopic system. This technique selects the image with maximum focus value between multi-focus image sequence. The schematic diagram of deep learning–based automatic focusing technique is displayed in Fig. 3. As seen in Fig. 3, the suggested technique comprises of five basic steps. These steps are described as follows:

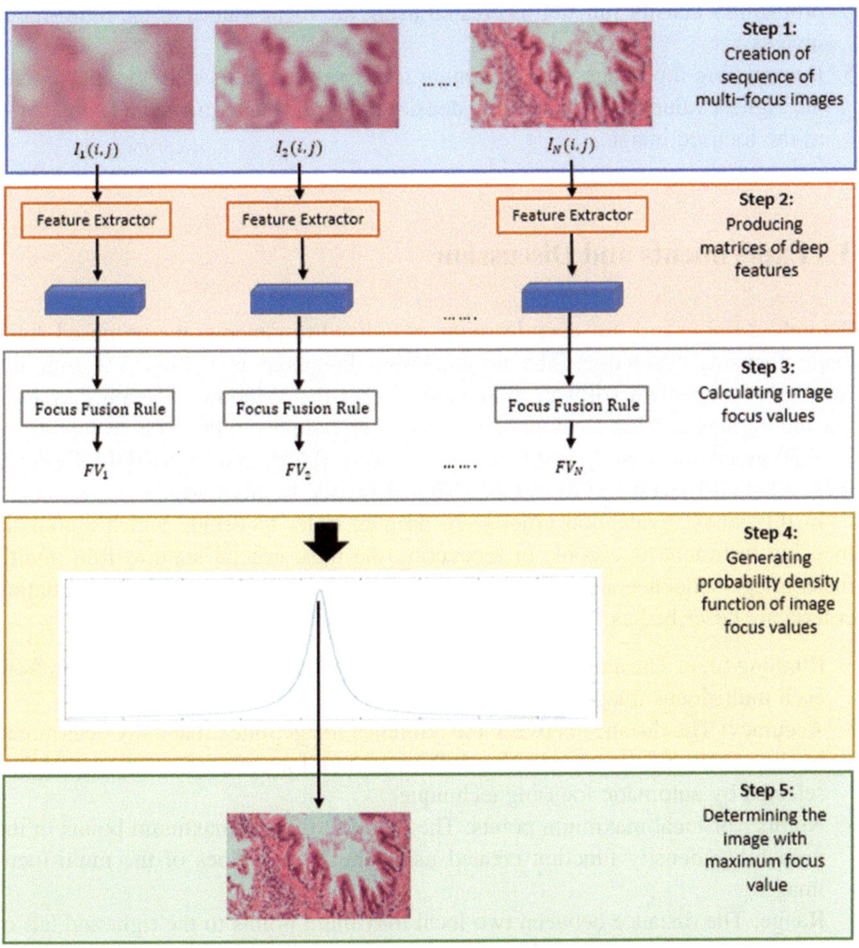

Fig. 3 Schematic diagram of deep learning–based automatic focusing system: (1) Creation of sequence of multi-focus images. (2) Producing matrices of deep features. (3) Calculating image focus values. (4) Generating probability density function of image focus values. (5) Determining the image with maximum focus value

1. Creation of sequence of multi-focus images: In the first step, a sequence of multi-focus images ($I_1(i, j)$, $I_2(i, j)$, $I_3(i, j)$, …, $I_N(i, j)$) are obtained by moving the microscope stage in the Z axis, where i and j are pixel resolution, N represents the image indexes.
2. Producing matrices of deep features: In the second step, matrices of deep features ($7 \times 7 \times 512$) are generated by utilizing the feature extractor of VGG16 structure which are pretrained with ImageNet weights.
3. Calculating image focus values: In the third step, the focus values (FV_1, FV_2, FV_3,…, FV_N) of multi-focus images are calculated with focus fusion rule. The focus fusion rule computes the average values on matrices of deep features.
4. Generating probability density function of image focus values: In this step, probability density function is created using the focus values of the multi-focus images.
5. Determining the image with maximum focus value: In this step, the image with the highest value in the probability density function is determined and considered as the focused image.

3 Experiments and Discussion

Pytorch, a framework for deep learning, is utilized to generate the proposed automatic focusing technique. The programming language is Python 3.9 with the package management software miniconda 3 and the Windows operating system. For the implementation of automatic focusing techniques, a PC with an Intel Core i7-9750 processor running at 2.60 GHz, 32 GB of RAM, and an NVIDIA GeForce RTX 3060 GPU with 12 GB of GPU VRAM is used in this study.

In this study, evaluation criteria are used in order to decide which automatic focusing technique is capable of recovering the most crucial features from multi-focus image sequence and provides more effective performance. These evaluation criteria are described as follows:

- Running time: The amount of time required to extract the focus information from each multi-focus image in the sequence.
- Accuracy: The distance between the reference image index manually determined by the expert and the image index with maximum focusing information, which is selected by automatic focusing technique.
- Number of local maximum points: The number of local maximum points in the probability density function created using the focus values of the multi-focus images.
- Range: The distance between two local maximum points to the right and left of the global maximum point in the probability density function created using the focus values of the multi-focus images.

Table 2 Ideal values of evaluation criteria for multi-focus image sequences

Sequence	Running time	Accuracy	Number of local maximum points	Range	Noise level (0.001)	Noise level (0.003)	Noise level (0.005)
Image sequence – 1	Minimum	0	0	104	0	0	0
Image sequence – 2	Minimum	0	0	52	0	0	0
Image sequence – 3	Minimum	0	0	69	0	0	0
Image sequence – 4	Minimum	0	0	61	0	0	0
Image sequence – 5	Minimum	0	0	62	0	0	0

- Noise level: The distance between the maximum focused image index of the sequence added with Gaussian noise and the maximum focused image index of the original sequence.

Table 2 shows the ideal values of the evaluation criteria for multi-focus image sequences – 1, 2, 3, 4, and 5. The running time of ideal automatic focusing technique is expected to be minimum for each image sequence. It is thought that the most focused image selected using automatic focusing technique is the same as the reference image determined by the expert. In this case, the accuracy of ideal automatic focusing technique should be 0. Similarly, the number of local maximum points in the probability density function created with ideal automatic focusing technique should be close to 0. The distance between the two local maximum points to the right and left of the focused image index, which is at the global maximum point, is expected to be equal to the number of multi-focus images in sequence. In this study, new sequences are created by adding noise at different standard deviations (0.001 – 0.003 – 0.005) to the multi-focus image sequences and compared with the originals. It is ideal for an automatic focusing technique that the focused image of the sequence with added Gaussian noise and the focused image of the original image sequence have the same index in the probability density functions. For this reason, the noise level criterion is expected to be close to 0.

In order to evaluate the effectiveness of our automatic focusing technique, a total 15 techniques from six groups are applied to the multi-focus image sequences. Tables 3, 4, 5, 6, and 7 indicate the evaluation criteria results of different automatic focusing techniques for all multi-focus image sequences. As stated previously, an ideal automatic focusing technique is expected to have minimum running time, accuracy, number of local maximum points and noise levels. As shown in Tables 3,

Table 3 Evaluation criteria results of different automatic focusing techniques for multi-focus image sequence – 1

Technique	Running time	Accuracy	Number of local maximum points	Range	Noise level (0.001)	Noise level (0.003)	Noise level (0.005)
Tenengrad	17.74	5	21	53	9	12	7
Thresholded Gradient	19.75	6	22	43	9	15	12
Quadratic gradient	22.67	5	36	42	8	13	10
Laplacian energy	21.98	6	28	35	11	14	15
Modified energy	22.33	7	38	40	13	13	14
Variance	18.96	5	30	37	15	8	10
Normalized variance	25.86	7	48	32	11	14	10
Histogram entropy	37.75	10	40	15	13	9	12
Modified DCT	40.76	9	27	5	7	14	9
3D DWT	35.86	5	31	24	14	13	7
Variance of DWT coefficients	33.75	7	41	25	9	14	9
Spatial frequency	20.74	6	40	25	15	7	10
Autocorrelation	15.63	8	46	9	13	13	7
2D Steerable Filters	38.53	6	31	36	10	13	8
Helmli	20.44	5	31	38	10	12	9
Deep learning	**21.94**	**3**	**10**	**92**	**3**	**3**	**3**

4, 5, 6, and 7, the performances of the automatic techniques vary according to the tissue type and magnification objective. For example, Tenengrad provides better performance for multi-focus image sequences – 1 and 2, while Variance provides better performance for multi-focus image sequence – 3. The results of evaluation criteria indicate that the recommended technique performs better than the state-of-art techniques developed automatic focusing system in the literature. Contrary to literature techniques with various constraints such as low precise characterization of curves and edges in images and performance variation according to the sample, this study provides more efficient and sample-free automatic focusing technique. The gray scales of multi-focus images are inappropriate for the automatic focusing, as given in Tables 3, 4, 5, 6, and 7. The proposed technique based on deep learning can provide adequate efficiency to transmit crucial details from images.

Table 4 Evaluation criteria results of different automatic focusing techniques for multi-focus image sequence – 2

Technique	Running time	Accuracy	Number of local maximum points	Range	Noise level (0.001)	Noise level (0.003)	Noise level (0.005)
Tenengrad	9.75	4	13	26	4	5	7
Thresholded gradient	10.98	3	5	34	6	6	3
Quadratic gradient	10.79	10	13	26	6	4	3
Laplacian energy	10.63	2	8	30	7	4	6
Modified energy	12.65	1	5	32	3	4	6
Variance	13.45	9	7	22	7	3	3
Normalized variance	12.33	4	12	28	3	4	3
Histogram entropy	14.9	10	15	29	4	6	6
Modified DCT	15.54	9	13	23	4	7	6
3D DWT	18.53	7	15	39	6	7	3
Variance of DWT coefficients	15.43	7	15	27	5	6	6
Spatial frequency	11.24	2	15	38	5	3	7
Autocorrelation	13.54	6	9	38	4	3	3
2D Steerable Filters	18.87	8	12	29	3	6	6
Helmli	14.67	6	6	21	5	4	4
Deep learning	**10.53**	**0**	**2**	**45**	**0**	**2**	**1**

Table 5 Evaluation criteria results of different automatic focusing techniques for multi-focus image sequence – 3

Technique	Running time	Accuracy	Number of local maximum points	Range	Noise level (0.001)	Noise level (0.003)	Noise level (0.005)
Tenengrad	9.65	6	16	53	4	8	8
Thresholded gradient	10.57	8	14	57	4	9	8
Quadratic gradient	11.32	8	13	46	5	6	5
Laplacian energy	9.87	3	11	52	10	9	7
Modified energy	11.55	4	11	41	10	8	10

(continued)

Table 5 (continued)

Technique	Running time	Accuracy	Number of local maximum points	Range	Noise level (0.001)	Noise level (0.003)	Noise level (0.005)
Variance	13.56	3	13	55	6	7	6
Normalized variance	14.53	9	11	41	9	5	5
Histogram entropy	15.68	8	16	57	7	8	10
Modified DCT	18.9	5	12	55	10	6	10
3D DWT	19.47	10	13	58	10	10	10
Variance of DWT coefficients	20.47	3	11	43	10	9	6
Spatial frequency	16.89	6	12	49	6	5	6
Autocorrelation	21.52	3	10	45	5	10	5
2D Steerable filters	22.33	7	11	47	6	7	4
Helmli	17.58	6	16	55	7	10	6
Deep learning	**14.45**	**1**	**8**	**63**	**1**	**2**	**1**

Table 6 Evaluation criteria results of different automatic focusing techniques for multi-focus image sequence – 4

Technique	Running time	Accuracy	Number of local maximum points	Range	Noise level (0.001)	Noise level (0.003)	Noise level (0.005)
Tenengrad	8.54	2	15	36	4	5	5
Thresholded gradient	9.76	4	21	31	2	5	5
Quadratic gradient	10.47	2	22	34	5	4	4
Laplacian energy	9.96	3	23	34	3	4	2
Modified energy	10.95	3	23	41	2	5	5
Variance	13.87	2	18	40	3	2	3
Normalized variance	13.76	5	18	34	3	2	3
Histogram entropy	15.11	2	22	43	3	3	3
Modified DCT	17.09	2	16	41	5	3	4
3D DWT	18.92	2	13	32	3	5	5
Variance of DWT coefficients	19.54	1	22	36	4	4	3

(continued)

Table 6 (continued)

Technique	Running time	Accuracy	Number of local maximum points	Range	Noise level (0.001)	Noise level (0.003)	Noise level (0.005)
Spatial frequency	14.99	1	17	31	4	2	3
Autocorrelation	20.54	1	13	41	2	2	2
2D Steerable filters	21.58	4	13	35	4	2	5
Helmli	16.98	1	21	40	3	4	4
Deep learning	**13.64**	**0**	**10**	**48**	**1**	**1**	**2**

Table 7 Evaluation criteria results of different automatic focusing techniques for multi-focus image sequence – 5

Technique	Running time	Accuracy	Number of local maximum points	Range	Noise level (0.001)	Noise level (0.003)	Noise level (0.005)
Tenengrad	8.62	3	17	38	4	3	6
Thresholded gradient	9.57	3	19	38	2	2	3
Quadratic gradient	10.56	3	16	43	6	3	3
Laplacian energy	10.06	4	18	36	3	6	2
Modified energy	11.11	1	15	42	2	3	4
Variance	13.99	3	15	40	5	6	4
Normalized variance	13.84	5	14	38	3	3	4
Histogram entropy	15.3	2	15	45	5	6	2
Modified DCT	17.27	3	18	44	6	6	5
3D DWT	19.16	4	13	37	5	6	5
Variance of DWT coefficients	19.89	4	15	36	6	4	2
Spatial frequency	15.26	4	12	37	5	2	6
Autocorrelation	21.64	4	19	41	4	2	2
2D Steerable filters	22.9	2	18	41	6	6	4
Helmli	18.93	4	15	45	4	2	3
Deep learning	**14.35**	**1**	**9**	**48**	**1**	**2**	**2**

4 Conclusion

In order to automatically determine the image with maximum focus value, this study improves an efficient technique based on deep learning. In contrast to studies in the literature, our suggested technique offers better focus representation and higher performance. It has minimal computational costs and complexity because no pre- or post-processing algorithm is needed. Moreover, the suggested automatic focusing technique acquires the image focus values from deep features, in contrast to previous studies employing only gray levels of original images, which can provide more sharp variation regarding images. Our proposed technique is evaluated theoretically and practically using a data set comprising of real microscope image sequences. To investigate which automatic focusing technique can extract more vital characteristics of multi-focus images, evaluation criteria are used, which are running time, accuracy, number of local maximum points, range, and noise level. The results of these evaluation criteria for the multi-focus image sequences show that our suggested automatic focusing technique is more effective than the other techniques.

Acknowledgments We thank Karadeniz Technical University Drug and Pharmaceutical Technology Application & Research Center for their support. This study was supported by a grant from The Scientific and Technological Research Council of Turkiye (TUBITAK) (Project no. 1919B012203634).

References

1. Dogan, H., Ekinci, M.: Automatic panorama with auto-focusing based on image fusion for microscopic imaging system. Signal Image Video Process. **8**, 5–20 (2014)
2. Wang, C., Huang, Q., Cheng, M., Ma, Z., Brady, D.J.: Intelligent autofocus. arXiv preprint arXiv:2002 12389 (2020)
3. Doğan, H., Baykal, E., Ekinci, M., Ercin, M.E., Ersöz, Ş.: Determination of optimum auto focusing function for cytopathological assessment processes. In: Medical Technologies National Congress (TIPTEKNO), pp. 1–4. IEEE, Trabzon (2017)
4. Shi, H., Shi, Y., Li, X.: Study on auto-focus methods of optical microscope. In: 2nd Int. Conf. on Circuits, System and Simulation (ICCSS 2012), vol. 46. IPCSIT (2012)
5. Santos, A., Ortiz De Solorzano, C., Vaquero, J.J., Pena, J.M., Malpica, N., del Pozo, F.: Evaluation of autofocus functions in molecular cytogenetic analysis. J. Microsc. **188**(3), 264–272 (1997)
6. Rudnaya, M.E., Mattheij, R.M.M., Maubach, J.M.L.: Evaluating sharpness functions for automated scanning electron microscopy. J. Microsc. **240**(1), 38–49 (2010)
7. Saini, G., Panicker, R.O., Soman, B., Rajan, J.: A comparative study of different auto-focus methods for mycobacterium tuberculosis detection from brightfield microscopic images. In: IEEE Distributed Computing, VLSI, Electrical Circuits and Robotics (DISCOVER), pp. 95–100, Mangalore (2016)
8. Pertuz, S., Puig, D., Garcia, M.A.: Analysis of focus measure operators for shape-from-focus. Pattern Recogn. **46**(5), 1415–1432 (2013)
9. Geusebroek, J.M., Cornelissen, F., Smeulders, A.W., Geerts, H.: Robust autofocusing in microscopy. Cytom.: J. Int. Soc. Anal. Cytol. **39**, 1–9 (2000)

10. Ahmad, M.B., Choi, T.S.: Application of three-dimensional shape from image focus in LCD/TFT displays manufacturing. IEEE Trans. Consum. Electron. **53**(1), 1–4 (2007)
11. Huang, W., Jing, Z.: Evaluation of focus measures in multi-focus image fusion. Pattern Recogn. Lett. **28**(4), 493–500 (2007)
12. Nayar, S.K.: Shape from focus system for rough surfaces. In: Physics-Based Vision: Principles and Practice: Radiometry, vol. 1, pp. 347–360. CRC Press, New York (1993)
13. Pech-Pacheco, J.L., Cristóbal, G., Chamorro-Martinez, J., Fernández-Valdivia, J.: Diatom autofocusing in brightfield microscopy: a comparative study. In: Proceedings 15th International Conference on Pattern Recognition. ICPR-2000, vol. 3, pp. 314–317. IEEE, Barcelona (2000)
14. Thelen, A., Frey, S., Hirsch, S., Hering, P.: Improvements in shape-from-focus for holographic reconstructions with regard to focus operators, neighborhood-size, and height value interpolation. IEEE Trans. Image Process. **18**(1), 151–157 (2008)
15. An, Y., Kang, G., Kim, I.J., Chung, H.S., Park, J.: Shape from focus through Laplacian using 3D window. In: In 2008 Second International Conference on Future Generation Communication and Networking, vol. 2, pp. 46–50. IEEE, Hainan (2008)
16. Yan, T., Hu, Z., Qian, Y., Qiao, Z., Zhang, L.: 3D shape reconstruction from multifocus image fusion using a multidirectional modified Laplacian operator. Pattern Recogn. **98**, 107065 (2020)
17. Yap, P.T., Raveendran, P.: Image focus measure based on Chebyshev moments. IEE Proc.-Vision, Image Signal Process. **151**(2), 128–136 (2004)
18. Wee, C.Y., Paramesran, R.: Measure of image sharpness using eigenvalues. Inf. Sci. **177**(12), 2533–2552 (2007)
19. Lee, S.Y., Kumar, Y., Cho, J.M., Lee, S.W., Kim, S.W.: Enhanced autofocus algorithm using robust focus measure and fuzzy reasoning. IEEE Trans. Circuits Syst. Video Technol. **18**(9), 1237–1246 (2008)
20. Lee, S.Y., Yoo, J.T., Kumar, Y., Kim, S.W.: Reduced energy-ratio measure for robust autofocusing in digital camera. IEEE Signal Process. Lett. **16**(2), 133–136 (2009)
21. Shen, C.H., Chen, H.H.: Robust focus measure for low-contrast images. In: 2006 Digest of Technical Papers International Conference on Consumer Electronics, pp. 69–70. IEEE, Las Vegas (2006)
22. Ali, U., Mahmood, M.T.: 3D shape recovery by aggregating 3D wavelet transform-based image focus volumes through 3D weighted least squares. J. Math. Imaging Vis. **62**, 54–72 (2020)
23. Xie, H., Rong, W., Sun, L.: Construction and evaluation of a wavelet-based focus measure for microscopy imaging. Microsc. Res. Tech. **70**(11), 987–995 (2007)
24. Shirvaikar, M.V.: An optimal measure for camera focus and exposure. In: Thirty-Sixth Southeastern Symposium on System Theory, pp. 472–475. IEEE, Atlanta (2004)
25. Helmli, F.S., Scherer, S.: Adaptive shape from focus with an error estimation in light microscopy. In: 2nd International Symposium on Image and Signal Processing and Analysis, pp. 188–193. IEEE Cat, Pula (2001)
26. Mahmood, F., Mahmood, J., Zeb, A., Iqbal, J.: 3D shape recovery from image focus using Gabor features. In: Tenth International Conference on Machine Vision, pp. 368–375 (2018). https://doi.org/10.1117/12.2309440
27. Nanda, H., Cutler, R.: Practical calibrations for a real-time digital omnidirectional camera. CVPR Tech. Sketch. **20**(2), 1–4 (2001)
28. Minhas, R., Mohammed, A.A., Wu, Q.J.: Shape from focus using fast discrete curvelet transform. Pattern Recogn. **44**(4), 839–853 (2011)
29. Lorenzo, J., Castrillon, M., Méndez, J., Deniz, O.: Exploring the use of local binary patterns as focus measure. In: International Conference on Computational Intelligence for Modelling Control & Automation, pp. 855–860, Vienna (2008)
30. Fan, T., Yu, H.: A novel shape from focus method based on 3D steerable filters for improved performance on treating textureless region. Opt. Commun. **410**, 254–261 (2018)
31. Minhas, R., Mohammed, A.A., Wu, Q.M., Sid-Ahmed, M.A.: 3D shape from focus and depth map computation using steerable filter. In: International Conference Image Analysis and Recognition, pp. 573–583. Springer, Berlin, Heidelberg (2009)

32. Xia, X., Yao, Y., Liang, J., Fang, S., Yang, Z., Cui, D.: Evaluation of focus measures for the autofocus of line scan cameras. Optik. **127**(19), 7762–7775 (2016)
33. Liu, Z., Lv, Q., Yang, Z., Li, Y., Lee, C.H., Shen, L.: Recent progress in transformer-based medical image analysis. Comput. Biol. Med. **164**, 107268 (2023)
34. Li, J., Chen, J., Tang, Y., Wang, C., Landman, B.A., Zhou, S.K.: Transforming medical imaging with transformers? A comparative review of key properties, current progresses, and future perspectives. Med. Image Anal. **85**, 102762 (2023)

Part II
Computing

Parallel Modeling of Complex Dynamic Systems in the Vector-Matrix Form

Volodymyr Svjatnij and Artem Liubymov

1 Introduction

One of the aspects of the problem of friendliness of parallel computing systems to subject area experts (users) is the transition from programming parallel simulators of complex dynamical systems (CDS) to their construction by means of modeling languages. The concept of development of parallel modeling languages (PMLs) proposed in [1] is based on the analogy between MIMD-processes and the main functional elements of object-oriented (OO), equation-oriented (EO), and block-oriented (BO) sequential languages (SL) of modeling. In [1], we called block diagrams for solving differential equations using functional SL-elements (BO-, OO-, EO-, or SL-simulators) as specifications of sequential simulators. A virtual parallel MIMD simulator shall be the name for the structure of MIMD-processes, which is built on the basis of analogies between SL-specifications and MIMD-parallelism in terms of SL-element – process ("block diagram – process," "object – process," and "operator – process") relation. In general, a SL-simulator of a CDS consists of *n* SL-elements (SLE), to which in a MIMD simulator, *n* analog process should correspond. Theoretically, all SLEs and corresponding processes have interconnections according to the logic of solving the equations of the CDS simulation model. The complete set of connections of the virtual process T_i, which has one output and *n* inputs, can be expressed as follows:

$$\mathrm{VST}_i = (S_{i1}T_1S_{i2}T_2 \ldots S_{ik}T_k \ldots S_{in}T_n) \tag{1}$$

V. Svjatnij · A. Liubymov (✉)
Donetsk National Technical University, Pokrovsk, Ukraine
e-mail: vsvjatnyi@gmail.com; artemliubymov@gmail.com

M. N. Seyman (ed.), *2nd International Congress of Electrical and Computer Engineering*, EAI/Springer Innovations in Communication and Computing,
https://doi.org/10.1007/978-3-031-52760-9_18

where i is the process number, $i = 1, 2, ..., n$; S_{ik} is the switching parameter: $S_{ik} = 1$ if there is a connection between the processes $T_i \leftrightarrow T_k$ and $S_{ik} = 0$ if there is no $T_i \leftrightarrow T_k$ connection. In the case of $k = i$, we have $S_{ii} = 1$, because the intermediate result of the i-process is used in it for further calculations. In set (1), we assume that the virtual number of inputs of process Ti is equal to the number of processes. Under this assumption, we introduce the virtual switching matrix SM:

$$\text{VSM} = \begin{pmatrix} S_{11} & S_{12}\ldots & S_{1n} \\ S_{21} & S_{22}\ldots & S_{2n} \\ & \ldots & \\ S_{k1} & S_{k2}\ldots & S_{kn} \\ & \ldots & \\ S_{n1} & S_{22}\ldots & S_{nn} \end{pmatrix} \quad (2)$$

The interrelationships between processes are characterized by the matrix of the simulator's state:

$$\text{SM} = \text{KM} * \text{DT} = \begin{pmatrix} S_{11}T_1 & S_{12}T_2\ldots & S_{1k}T_k\ldots S_{1n}T_n \\ S_{21}T_1 & S_{22}T_2\ldots & S_{2k}T_k\ldots S_{2n}T_n \\ & \ldots & \\ S_{k1}T_1 & S_{k2}T_2\ldots & S_{kk}T_k\ldots S_{kn}T_n \\ & \ldots & \\ S_{n1}T_1 & S_{n2}T_2\ldots & S_{nk}T_k\ldots S_{nn}T_n \end{pmatrix} \quad (3)$$

where DT is the diagonal matrix of processes.

The simulation model of the dynamical system under study is described by a system of equations, each of which is an implicit function that defines an unknown variable and is solved by the corresponding SLE structure and, by analogy, by MIMD-processes Ti. The solution of VARTi, being the output value of the process Ti, is the result of certain operations on the set of variables that are fed to the inputs of the process Ti according to the equations of the simulation model. In general, the variables of VARTk are the output values of all other processes; therefore, the specification of the virtual MIMD simulator with all possible connections between processes is the following set of variables [1]:

$$\begin{cases} \text{VART}_1 = \text{FUNT}_1(S_{11}\text{VART}_1\ S_{12}\text{VART}_2\ldots S_{1n}\text{VART}_n); \\ \text{VART}_2 = \text{FUNT}_2(S_{21}\text{VART}_1\ S_{22}\text{VART}_2\ldots S_{2n}\text{VART}_n); \\ \ldots \\ \text{VART}_k = \text{FUNT}_k(S_{k1}\text{VART}_1\ S_{k2}\text{VART}_2\ldots S_{kn}\text{VART}_n); \\ \ldots \\ \text{VART}_n = \text{FUNT}_n(S_{n1}\text{VART}_1\ S_{n2}\text{VART}_2\ldots S_{nn}\text{VART}_n), \end{cases} \quad (4)$$

where FUNTi – operations of Ti processes on input variables that are fed from the outputs of all other processes participating in solving the equation system of the simulation model.

Specification (4) summarizes the structures of virtual MIMD simulators at two possible levels: the basic level: "SL-element – process" provides for MIMD-processes of "shallow granulation" [3], during the execution of which there is an uneven load of processes and an unfavorable ratio between the volume of computational operations and data exchange operations between processes; the level "Group of SL-elements – process" corresponds to the stage of simulators development, at which the specifications use the composition of SL-elements according to a certain principle, for example, "One equation of the simulation model – Group of SL-elements for solving the equation – Process of coarse granulation."

Statement of the problem: To define the processes of coarse granulation and indicators of their a priori analysis within the framework of the tasks of devitalization of virtual MIMD simulators [1].

2 Types of CDS Simulation Models and Specifications of SL-Simulators

2.1 Systems of Nonlinear Differential Equations of the First-Order

General view of the system:

$$\frac{d\overline{x}}{dt} = \overline{F}\left(\overline{S}, \overline{Y}, \overline{f}(t), \varphi(\overline{X}), \varphi(\overline{Y})\right) \tag{5}$$

where $\overline{X} = (x_1, x_2, \ldots, x_n)^{\mathrm{T}}$ – vector of unknown variables; F – symbol of operations of the right sides for $x_1, x_2, \ldots, x_n$; $f(t)$ – known time functions. The right side of (5) contains dependencies on vector X components (cross-connections), where x_1 depends on x_j, where $j \neq 1$, $j = (2, 3,\ldots, n)$, on some vector Y (external influences), on nonlinear functions $\varphi(\overline{X})$ and $\varphi(\overline{Y})$, which can be physical phenomena (nonlinear electrical resistances, turbulent and quasi-turbulent process flows of liquids and gases, power characteristics of generators, pumps, fans, motors, compressors, technological characteristics of chemical reactors, hysteresis phenomena, etc.).

The specification of the SL-simulator provides for one integrator and SL-elements that implement the operations of the right sides for each system equation (Fig. 1).

In a number of problems, the problem of generating a formal description in the form of (5), that is, a system of first-order equations, is relevant. The block diagram can be started by placing all the components of the right sides on the block F (Fig. 1). It is necessary to draw the block diagram in such a way that it may be possible to "superimpose" the structure of the corresponding MIMD-processes. Figure 1 "hints" at balancing the loads of processes F_1, ..., F_n and integrators int 1 ... int n.

It is necessary to pay attention to the functional blocks $\varphi(X)$ and $\varphi(Y)$. It may be expedient to place them in groups F_1, ..., F_n. In BO language, this is realistic, but we need to analyze if it is so in MIMD.

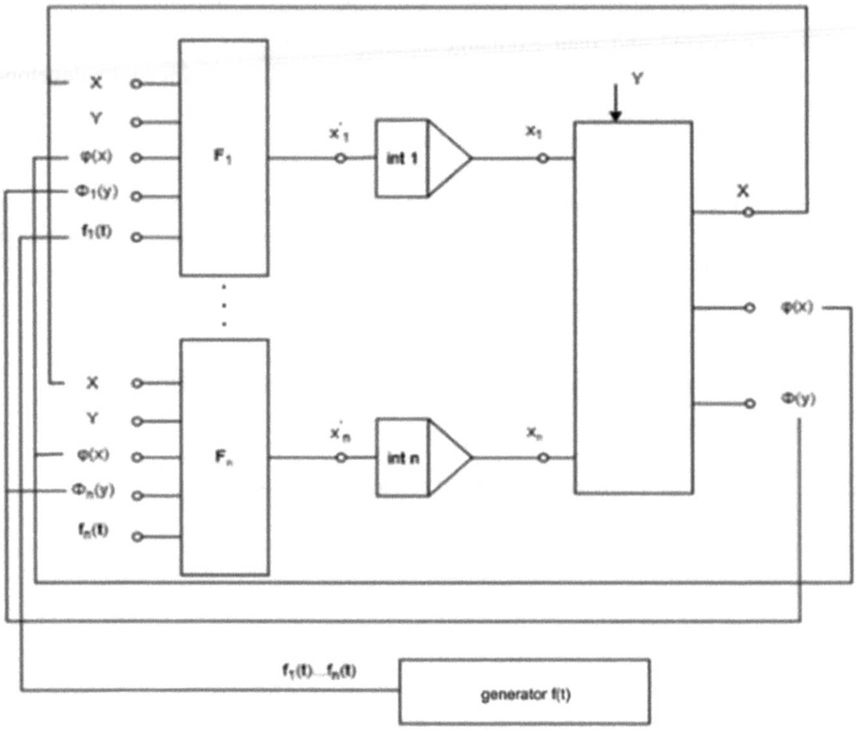

Fig. 1 SL-simulator based on a system of equations of the first-order

2.2 Systems of Nonlinear Differential Equations of Different Orders

This form of the system appears when describing the structures of automation systems (links of different orders – 1st, 2nd, etc.) and technological systems (reactors, pipelines, distillation columns, crystallizers, and other devices) by components. Usually, such systems are reduced to first-order systems, that is, to option 2.1, using ancillary variables.

However, from the point of view of structural correspondence between the components of the SL-simulator and the object under study, the following methodology is appropriate:

1. Equations are solved with respect to the highest derivative, that is, the simulation model will have left sides of different orders, and the right sides may have derivatives of corresponding lower orders.
2. A chain of blocks with the number of integrators equal to the order of the equation corresponds to each equation.

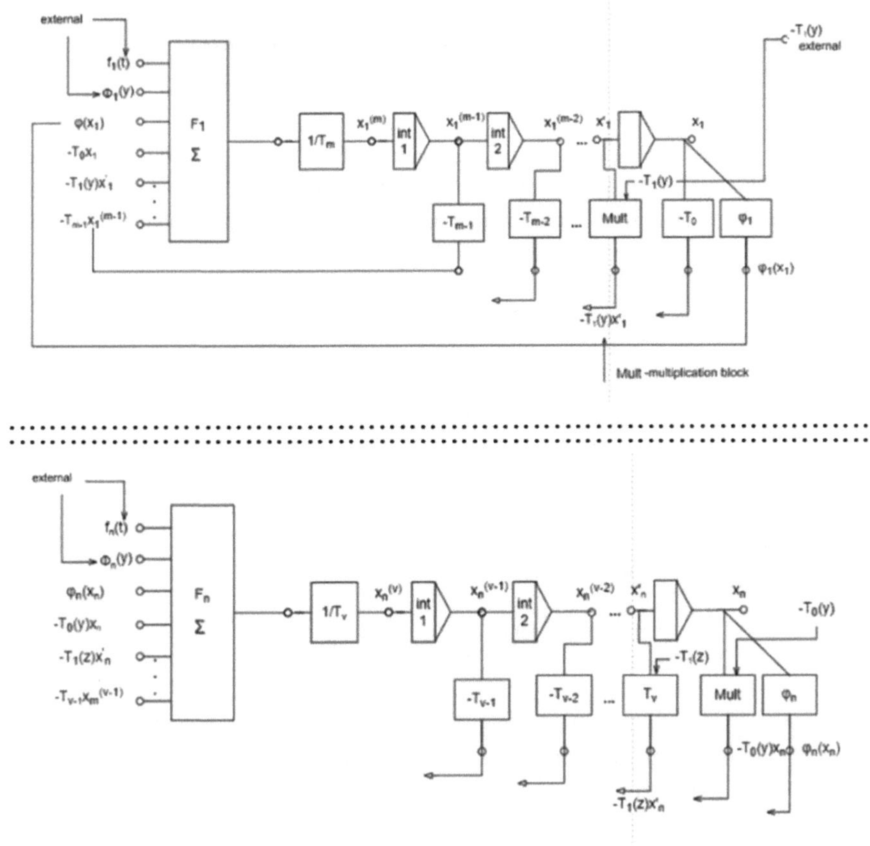

Fig. 2 SL-simulator based on a system of equations of different orders

Example: Simulation models realized by the scheme of Fig. 2.

$$\begin{cases} \dfrac{d^m x_1}{dt^m} = \left[f_1(x_1) + \varphi_1(y) + \varphi_1(x_1) - T_{m-1}\dfrac{d^{(m-1)}x_1}{dt^{(m-1)}} - ::::::: - \mathrm{T}_1(\mathrm{y})\dfrac{\mathrm{dx}_1}{dt} - \mathrm{T}_0\mathrm{x}_1 \right] / T_m \\ :: \\ \dfrac{d^y x_n}{dt^y} = \left[f_{n1}(x_n) + \varphi_n(y) + \varphi_n(x_n) - T_{y-1}\dfrac{d^{(y-1)}x_n}{dt^{(y-1)}} - ::::::: - \mathrm{T}_1(\mathrm{z})\dfrac{\mathrm{dx}_\mathrm{n}}{dt} - \mathrm{T}_0\mathrm{x}_\mathrm{n} \right] / T_y \end{cases} \tag{6}$$

The scheme of Fig. 2 is similar to Fig. 1, but in the calculation chains x_1, ..., x_n, there should be m and v of integrators, respectively. At the inputs F_1,..., F_n, there will be derivatives $\dot{x}_1, \ldots, x_1^{(m-1)}, \dot{x}_2, \ldots, x_2^{(m-1)}, \ldots, \dot{x}_n, \ldots, x_n^{(m-1)}$ of orders 1,..., $m - 1$ (after the first integrator, we have m – 1st order; m – the order of the equation, where $m > 1$).

2.3 DAE-Problem, in Which Differential Equations Contain Derivatives and Integrals of Different Variables (Unknowns)

This kind of description is attributed to mathematical models of network dynamic objects with concentrated parameters (NDOCP). Let us consider two simple objects – an electrical and an aerodynamic, which have the topology represented by the graph in Fig. 3.

The electrical NDOCP branch is an *R-L-C* scheme (Fig. 4).

The equation for the current *I*(*t*) is as follows:

$$L\frac{dI}{dt} + RI + \frac{1}{C}\int Idt = U_P - U_K \tag{7}$$

where U_P, U_K – the potentials at the initial and end nodes.

A branch of an aerodynamic NDOCP is an air duct (pipe, mine roadway, ventilation duct, etc.). If we do not take into account the ability of the airflow to compress, then the equation of air flow *Q* looks like this [2]:

$$K\frac{dQ}{dt} + RQ^2 = P_P - P_K \tag{8}$$

where $K = \frac{\rho l}{F}$; ρ – air density; l – length of the duct; F – cross-sectional area of the duct; R – aerodynamic resistance; and P_P, P_K – pressure at the initial and end nodes of the NDOCP branch.

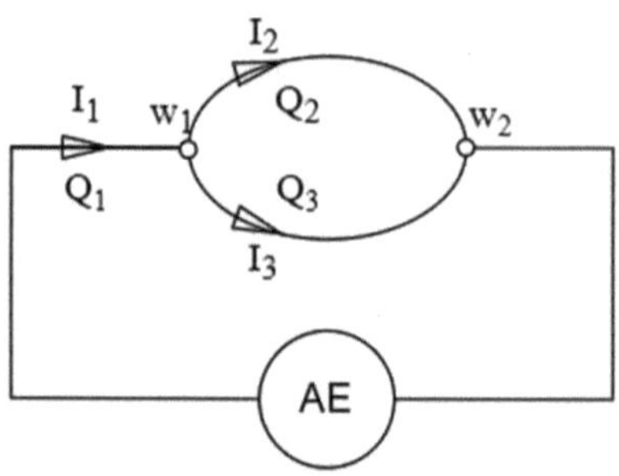

Fig. 3 NDOCP graph

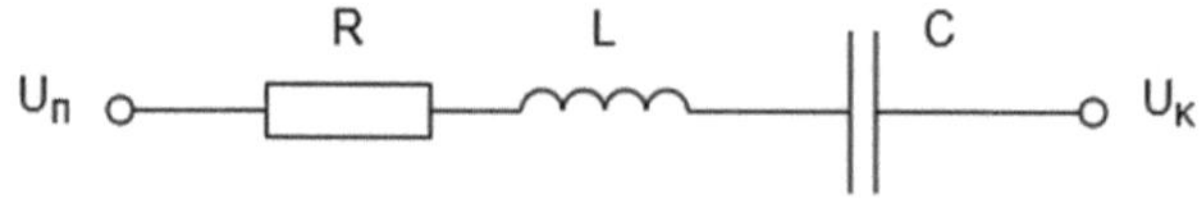

Fig. 4 Branch with *R*, *L*, and *C* elements

The DAE-problems are as follows:

$$\begin{cases} I_1 = I_2 + I_3 \\ L_1\frac{dI_1}{dt} + R_1 I_1 + \frac{1}{C_1}\int I_1 dt + L_2\frac{dI_2}{dt} + R_2 I_2 + \frac{1}{C_2}\int I_2 dt = E_{\mathrm{AE}} \\ L_1\frac{dI_1}{dt} + R_1 I_1 + \frac{1}{C_1}\int I_1 dt + L_3\frac{dI_3}{dt} + R_3 I_3 + \frac{1}{C_3}\int I_3 dt = E_{\mathrm{AE}} \end{cases} \tag{9}$$

where E_{AE} – voltage of the active element (generator, battery, etc.).

$$\begin{cases} Q_1 = Q_2 + Q_3 \\ K_1\frac{dQ_1}{dt} + R_1 Q_1^2 + K_2\frac{dQ_2}{dt} + R_1 Q_2^2 = H_{\mathrm{AE}} \\ K_1\frac{dQ_1}{dt} + R_1 Q_1^2 + K_3\frac{dQ_2}{dt} + R_3 Q_3^2 = H_{\mathrm{AE}} \end{cases} \tag{10}$$

where H_{AE} – the difference between atmospheric pressure and the pressure of the active element (fan, pump, compressor, etc.).

For the NDOCP graph, we introduce the incident matrix *A*, the contour matrix *S*, the vectors *I, Q, E, H*, and the diagonal matrices of parameter *L, R, 1/C, K*, and write the DAE-problem equation in the vector-matrix form:

$$\begin{cases} A\overline{I} = 0 \\ SL\frac{d\overline{I}}{dt} + SR\overline{I} + S\frac{1}{C}\int \overline{I}dt = SE \end{cases} \tag{11}$$

$$\begin{cases} A\overline{Q} = 0 \\ SK\frac{d\overline{Q}}{dt} + SRZ = SH \end{cases} \quad Z = \left(Q_1^2, \ldots, Q_n^2\right)^{\mathrm{T}} - \text{vector of squares } Q \tag{12}$$

The equations in the form of a simulation model can be obtained by observing the rule that one unknown variable *I j(t)*, *Q j(t)* must be found as a solution to one of the equations (*j* is the number of the branch). In NDOCP of the considered type, we have m branches and *n* nodes, systems (11, 12) have $n - 1$ nodal algebraic equations and $\gamma = m - (n - 1)$ contour equations. We agree to find $n - 1$ of unknown variables from the algebraic nodal equations, and the remaining γ unknowns from the differential (differential-integral) equations. We will write the simulation model through the following steps:

1. Define the variables to be calculated as solutions to algebraic equations. For NDOCP of real complexity, it is advisable to use the division of the graph into a tree and an anti-tree. The tree has $n - 1$ branch, and the currents (flows) in these branches can be found from $n - 1$ nodal equations. In our example,

$$\left.\begin{matrix} I_1 = I_2 + I_3 \\ Q_1 = Q_2 + Q_3 \end{matrix}\right\} \tag{13}$$

2. The rest of the currents (flows) can be found by solving the differential contour equations. However, the presence of a sum of derivatives in each equation makes it necessary to transform the equations in such a way as to avoid differentiation operations. Thus, if we simply solve the equations of the first contour with respect to $\frac{dI_2}{dt}\left(\frac{dQ_2}{dt}\right)$, then the SL-scheme is possible only if the operation of differentiation of the variable $I_1(Q_1)$ found by (13) is performed, that is, $\frac{dI_2}{dt}\left(\frac{dQ_2}{dt}\right)$. It is proposed to solve the contour equations with respect to the sum of derivatives by selecting the variable, which is to be found from the given equation (in our case, it is I_2 I_3 Q_2 Q_3):

$$\begin{aligned} \frac{d}{dt}\left(I_2+\frac{L_1}{L_2}I_1\right) &= \left[E_{AE}-R_1I_1-R_2I_2-\frac{1}{C_1}\int I_1dt-\frac{1}{C_2}\int I_2dt\right]/L_2 \\ \frac{d}{dt}\left(I_3+\frac{L_1}{L_3}I_1\right) &= \left[E_{AE}-R_1I_1-R_3I_3-\frac{1}{C_1}\int I_1dt-\frac{1}{C_3}\int I_3dt\right]/L_3 \end{aligned} \tag{14}$$

$$\begin{aligned} \frac{d}{dt}\left(Q_2+\frac{K_1}{K_2}Q_1\right) &= \left[H_{AE}-R_1Q_1^2-R_2Q_2^2\right]/K_2 \\ \frac{d}{dt}\left(Q_3+\frac{K_1}{K_3}Q_1\right) &= \left[H_{AE}-R_1Q_1^2-R_3Q_3^2\right]/K_3 \end{aligned} \tag{15}$$

According to these systems, the structures of SL-simulators are built "almost" according to the standard methodology, without using differentiator blocks (Fig. 5):

2.4 DAE-Problem: NDOCP in Vector-Matrix Form

For NDOCP of real complexity ($m > 100$, $n > 50$), writing the equations in the form considered is a time-consuming and error-prone task. It is necessary to develop model generators and simulation models with a subsequent generation of block

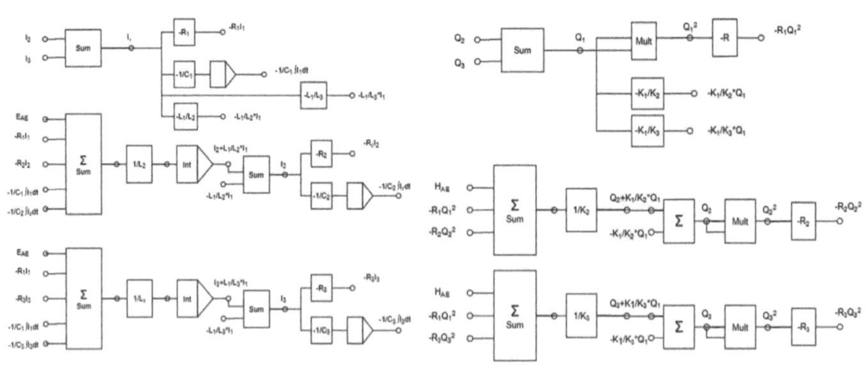

Fig. 5 SL-simulator for an electric and aerodynamic NDOCP

diagrams of SL-simulators as well. A high level of automation of generation can be provided by the vector-matrix form of the NDOCP model of the type (11, 12). If we distinguish a tree and an anti-tree in the NDOCP graph, the vector of currents can be divided into two subvectors:

$$\overline{I} = \left(\overline{X}, \overline{Y}\right)^{\mathrm{T}} \tag{16}$$

where X is the vector of currents in the branches of the tree and Y is the vector of currents in the branches of the anti-tree.

The matrices of incidents A and contours S are structured with respect to X, Y:

$$A = \left(A_x A_y\right), S = \left(S_x S_y\right) \tag{17}$$

The nodal equation $A_I = 0$ is transformed as follows:

$$\left(A_x A_y\right) * \begin{pmatrix} X \\ Y \end{pmatrix} = A_x X + A_y Y = 0 \tag{18}$$

and is solved with respect to X by multiplying by the inverse matrix A_x^{-1} on the left:

$$X = -WY \tag{19}$$

where

$$W = A_x^{-1} A_y \tag{20}$$

Taking into account (19) and (20), the contour equation can be solved with respect to the currents Y by the following operations:

1. Substitute the matrices S, L, R, $1/C$ structured with respect to X, Y and the vectors $\overline{I} = \left(\overline{X}, \overline{Y}\right)^{\mathrm{T}}$ $E = (E_X E_Y)^{\mathrm{T}}$:

$$(S_X S_Y)\begin{pmatrix} L_X & 0 \\ 0 & L_Y \end{pmatrix}\frac{d}{dt}\begin{pmatrix} X \\ Y \end{pmatrix} + (S_X S_Y)\begin{pmatrix} R_X & 0 \\ 0 & R_Y \end{pmatrix}\begin{pmatrix} X \\ Y \end{pmatrix} + (S_X S_Y) \times \begin{pmatrix} 1/C_X & 0 \\ 0 & 1/C_Y \end{pmatrix}\int\begin{pmatrix} X \\ Y \end{pmatrix}dt = (S_X S_Y)\begin{pmatrix} E_X \\ E_Y \end{pmatrix} \tag{21}$$

$$S_X L_X \frac{dx}{dt} + S_Y L_Y \frac{dy}{dt} + S_X R_X X + S_Y R_Y Y + S_X \frac{1}{C_X}\int Xdt + S_Y \frac{1}{C_Y}\int Ydt = S_X E_X + S_Y E_Y \tag{22}$$

2. Substitute (19) to (22):

$$-S_X L_X W\frac{dy}{dt} + S_Y L_Y\frac{dy}{dt} = S_X E_X + S_Y E_Y - (S_Y R_Y Y - S_X R_X W Y) - \left(S_Y\frac{1}{C_Y}\int Ydt - S_X\frac{1}{C_X}W\int Ydt\right) \tag{23}$$

3. In (23) factor out the variables $\frac{dY}{dt}, Y, \int Ydt$:

$$(S_Y L_Y - S_X L_X W)\frac{dy}{dt} = S_X E_X + S_Y E_Y - (S_Y R_Y - S_X R_X W)Y - \left(S_Y\frac{1}{C_Y} - S_X\frac{1}{C_X}W\right)\int Ydt \tag{24}$$

4. Find the inverse matrix $V = (S_Y L_Y - S_X L_X W)^{-1}$ and multiply Eq. (24) on the left by it:

$$\frac{dy}{dt} = \mathrm{VSE} - V(S_Y R_Y - S_X R_X W)Y - V\left(S_Y\frac{1}{C_Y} - S_X\frac{1}{C_X}W\right)\int Ydt \tag{25}$$

Thus, we obtained a simulation model in the vector-matrix form:

$$\begin{cases} X = -WY \\ \frac{dy}{dt} = \mathrm{VSE} - V(S_Y R_Y - S_X R_X W)Y - V\left(S_Y\frac{1}{C_Y} - S_X\frac{1}{C_X}W\right)\int Ydt - \mathrm{VSRI} - \mathrm{VS}\frac{1}{C}\int Idt \end{cases} \tag{26}$$

one can do this without substitutions

For the aerodynamic NDOCP, such a simulation model is obtained by similar vector-matrix operations.

3 Specifications of SL-Simulators Based on the Vector-Matrix Form of Equations of Network Dynamic Objects with Concentrated Parameters (NDOCP)

The following methodology for constructing SL-simulators using the schemes for solving Eq. (26) is proposed.

3.1 Development of a Topological Analyzer, in Which the Following Operations Are Performed

1. Encoding NDOCP graph by Table 1:

where Q_j is the branch number (I_j for an electric NDOCP); IN_i and EN_i are the numbers of the initial and end nodes of the branch, $i = 1, 2, ..., n$; K_j, R_j are the coefficients.

2. According to Table 1, a tree and an anti-tree are built, two tables are formed – for the tree and the anti-tree (BAUMTAB, ANTITAB, for example).
3. The vectors X, Y are formed, the diagonal matrices are reformatted to fit X, Y.
4. According to Table 1, a restructured coding table (TABXY) is formed taking into account X, Y (Table 2):

5. The incident matrix A is formed (Table 3):

6. The matrix of contours S is formed: the basic branches of the contours are the flows/branches Y (Table 4)

Table 1 (TABUR)

IN_i	EN_i	Q_j	K_j	R_j	H_j	Commentary
$w5$	$w1$	Q_j	K_j	R_j	H_j	
		...	...	...	...	
		Q_m	K_m	R_m	H_m	

Table 2 (TABXY)

IN_i	EN_i	X/Y	K_X/K_Y	R_X/R_Y	H	Commentary
$w5$	$w1$	x_1	K_{x1}	R_{x1}	H_{x1}	
...	...	...	...	...	...	
			K_{xn-1}	R_{xn-1}	H_{xn-1}	
$w2$	$w3$	y_1	K_{Y1}	R_{Y1}	H_{Y1}	
...	...	...	...	...	...	
		y_1	K_{yy}	R_{YY}	H_{yy}	

Table 3 Incident matrix

	Branches							
Nodes	X_1	X_2	X_{N-2}	X_{N-1}	Y_1	Y_2	Y_{Y-1}	Y_Y
$w1$	1	−1	0		0	0	−1	−1
$w2$	0	1	0		−1	−1	0	0
$w3$	0	0	−1		1	1	0	0
$w4$	0	0	1	1	0	0	1	1
...								
w_{N-1}								

Table 4 Matrix of contours

	Branch							
Contour	X	X_2	X_3	X_4	Y_1	Y_2	Y_3	Y_4
k_1	1	1	1	1	1	0	0	0
k_2	1	1	1	1	0	1	0	0
k_3	1	0	0	1	0	0	1	0
k_{4Y}	1	0	0	1	0	0	0	1
			S_X				S_Y	

For the first "maximum" contour, which includes all tangent branches of the tree and anti-tree, we take the base Y1.

We have obtained the topological information about graph of NDOCP – matrices *A*, *S*, vectors *X*, *Y*, *H*, *Z*, diagonal matrices *K*, *R* (similarly for the electric NDOCP). This information is provided by Top Analyzer, a program that implements operations 1–6.

3.2 Development of the Equation Generator of the Simulation Model in the Following Order

1. The inverse matrix A_X^{-1} is found (A_X – a part of the matrix *A* of size $(n - 1) * (n - 1)$).
2. The following matrix is calculated:

$$W = A_X^{-1} A_Y \tag{27}$$

3. The following matrix is calculated:

$$V = (S_Y L_Y - S_X L_X W)^{-1} \tag{28}$$

4. Matrix differences are calculated:

$$\left.\begin{aligned} & S_Y L_Y - S_X L_X W \\ & S_Y \frac{1}{C_Y} - S_X \frac{1}{C_X} W \end{aligned}\right\} \text{electrical object} \tag{29}$$

5. In an electrical object, you can write the side on the right in (26) without substituting $X = -WY$ and take $I = (X, Y)^\gamma$

$$\frac{dY}{dt} = \mathrm{VSE} - \mathrm{VSRI} - \mathrm{VS}\frac{1}{c}\int I dt \tag{30}$$

Here, one needs to calculate:

$$\mathrm{VSE} = \mathrm{VS} * \left(E_{X1} \ldots E_{Xn-1} E_{Y1} \ldots E_{Y\gamma}\right)^{\mathrm{T}} \tag{31}$$

$$\mathrm{VSRI} = \mathrm{VS} * \left[\begin{pmatrix} R_X & 0 \\ 0 & R_Y \end{pmatrix} * \left(X_1 \ldots X_{n-1} Y_1 \ldots Y_\gamma\right)^{\mathrm{T}}\right] \tag{32}$$

$$\mathrm{VS}\frac{1}{C}\int Idt = \mathrm{VS} * \left[\begin{pmatrix} 1/C_X & 0 \\ 0 & 1/C_Y \end{pmatrix} * \int \left(X_1 \ldots X_{n-1} Y_1 \ldots Y_\gamma\right)^{\mathrm{T}} dt\right] \tag{33}$$

Subsequently, we denote VS = SN as the transformed matrix S.

6. As a result of these calculations, we obtain a simulation model concretized by coefficients, that is, a system of equations divided with respect to *X* and *dY/dt*.

For the electrical object:

$$\left.\begin{array}{l} \dfrac{dy_1}{dt} = \left(S_P E - S_P RI - S_P \dfrac{1}{C}\displaystyle\int Idt\right)_{Y1} \\ \cdots\cdots\cdots\cdots\cdots\cdots\cdots\cdots \\ \dfrac{dy_\gamma}{dt} = \left(S_P E - S_P RI - S_P \dfrac{1}{C}\displaystyle\int Idt\right)_{Y\gamma} \\ x_1 = -(WY)_{X1} \\ \cdots\cdots\cdots\cdots\cdots\cdots\cdots\cdots \\ x_{n-1} = -(WY)_{Xn-1} \end{array}\right\} \tag{34}$$

The initial conditions for (27) are assigned as follows:

$$\overline{Y}(0) = \overline{Y}_0, \overline{X}(0) = -W\overline{Y}(0) \tag{35}$$

System (27–35) is the result of the equation generator of the simulation model.

3.3 Construction of the SL-Simulator Specification Based on Generated Equations

The system (27) obtained by the equation generator is distinguished by the fact that it achieves resolution with respect to the components of *Y* vector, that is, this is the form of equations required by the modeling language. The SL-simulator for an aerodynamic object has γ integrators, the input values for which form the block diagrams of the right sides. The components of vector *X* are calculated by $n - 1$ adders.

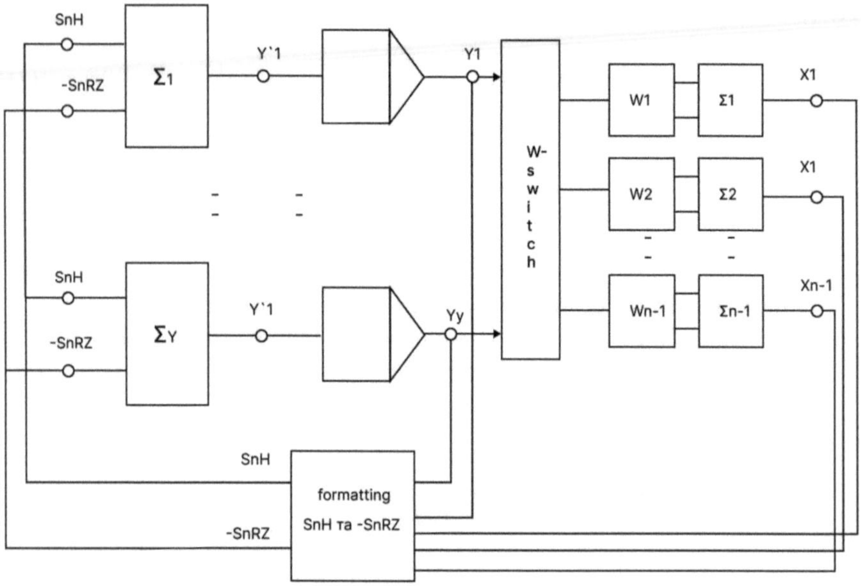

Fig. 6 Generalized block diagram/specification of the SL-simulator based on a vector-matrix system of equations for an aerodynamic object

The matrix W plays the role of a switch, each of its rows "connects" to the adder those variables Y_j that are incident to the i-node ($i = 1, 2, \ldots, n - 1$) (Fig. 6).

For an electrical NDOCP, the SL-simulator has γ "main" integrators and theoretically m integrators according to the number of branches with capacitors (Fig. 7).

4 CDS with Distributed Parameters (CDSDP)

SL-simulators of these complex dynamic systems are developed in the following sequence:

1. Formulation of equations with partial derivatives, of boundary and initial conditions (statement of the problem stage).
2. Approximation of partial differential equations (PDEs) by spatial coordinates (1D-, 2D-, and 3D-problems).
3. The approximated PDEs are reduced to the form of simulation models:

   3.1. 1D problem: a one-dimensional object is divided into $M = l/\Delta x$ elements with a step of Δx at length l. For each element, the usual differential equations are written, and a "chain" system of ODE is obtained [3].

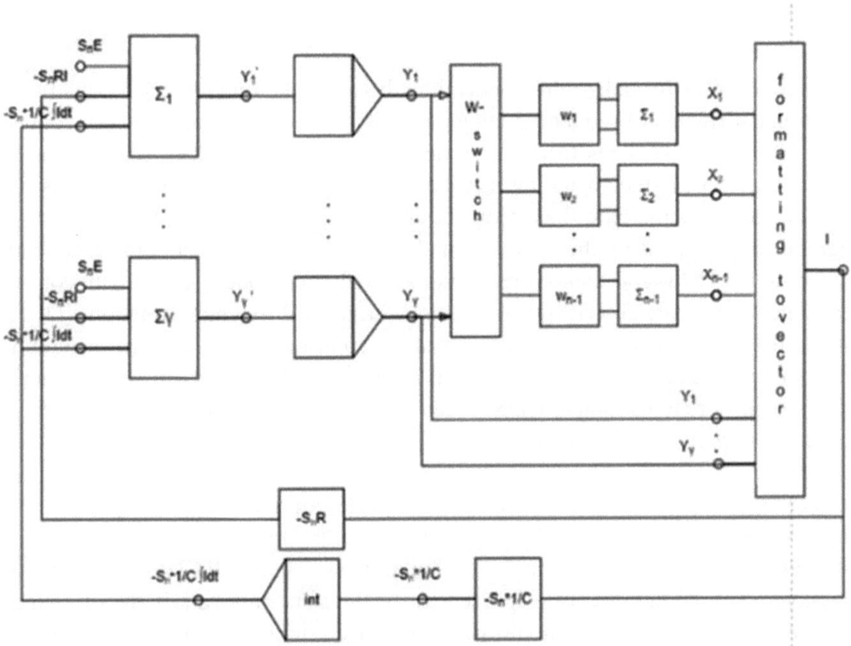

Fig. 7 Generalized block diagram/specification of the SL-simulator based on the vector-matrix system of Eq. (27)

3.2. 2D problem: a two-dimensional object is approximated by a lattice with steps Δx, Δy; ODEs are written for the lattice nodes, and a difference in spatial coordinates simulation model is formed.
3.3. 3D problem: similar to the 2D problem.

4. SL-simulators are considered similarly to those discussed in Parts II and III. CDSDPs are distinguished by various mathematical descriptions, especially for technological schemes with physical processes of different nature (chemical reactions, heat, electricity, hydraulics, gas dynamics, hydromechanics, etc.) As a result of the approximation with steps $\Delta\Delta$, Δy, and Δz, large-dimensional systems of ordinary differential equations appear. SL-simulators contain corresponding sets of functional elements [4].

It is expedient to consider this type of problem in the process of developing problem-oriented parallel modeling environments.

5 Conclusions

It is proposed to formalize the connections between the outputs and inputs of the functional elements of sequential languages (SL) of modeling using sets of connections (1). Their use leads to a formulaic description of the block diagram of the SL-simulator using the switching matrices (2) and state matrices (3). The construction of parallel simulators based on the analogy "SL-element – MIMD-process" significantly depends on the forms of mathematical description of complex dynamical systems (CDS) and their corresponding simulation models. CDSs with concentrated parameters are described by systems of ordinary differential equations (ODEs) of the first order, systems of equations of different orders, systems of differential algebraic equations (DAE-Problem), and in network CDS there are sums of derivatives of different variables. CDSs with distributed parameters are described by systems of partial differential equations; hence, the development of simulation models is connected with approximation of equations by spatial coordinates and formation of ODE systems. For these forms of description, we consider simulation models and corresponding SL-specifications in the form of block diagrams of SL-elements, which are the basis for building virtual MIMD simulators of two possible levels of granularity of processes. At the same time, a scheme of connections between SL-elements should be formed according to the form of simulation models, taking into account the specifics of functional SL-elements in terms of the "input-output" indicator. In further works on virtual switching and devitalization, we are going to solve the following tasks:

1. Comparison of the workload of processes similar to SL-elements.
2. Development of structures of MIMD simulators similar to SL-simulators of CDS, described by the considered systems of equations.
3. MIMD switches, aspects of architectural relevance.
4. Approaches to generalization of switch synthesis. Implementation in Infiniband.
5. A priori analysis of virtual MIMD simulators.

References

1. Svjatnij, V.A., Liubymov, A.S., Miroshkin, O.M., Kushnarenko, V.H.: MIMD-Simulators based on parallel simulation language. Inform. Math. Methods Simul. **8**(3), 189–199 (2018)
2. Svjatnij, V.A: Modeling of aerogasdynamic processes and development of control systems of coal mines ventilation. Thesis of Doctor in Technical Sciences, Donetsk, 407p. (1985)
3. Braunl, T.: Parallel programming. Textbook (translated from German by V. A. Svyatnyi). Kyiv, VSh, 407p. (1997)
4. Svyatnyy, V., Kushnarenko, V.: Ein Ansatz zur gleichmäßigen Lastverteilung zwischen Prozessoren des MIMD-Simulators der dynamischen Netzobjekte mit verteilten Parametern. Scientific works of DonNTU, series "Problems of Simulation and Design Autimatization". **1**(15), 5–14 (2020)

A Smart Autonomous E-Bike Fail-Safe System

Haneen Mahmoud, Hassan Soubra, and Ahmed Mazhr

1 Introduction

When it comes to mobility vehicles, bikes are considered one of the oldest, since they started out in the 1800s. Until now, they have been favored by many due to the fact that they are a relatively cheaper vehicle than cars and motorcycles, easy to maintain, and that cycling has a lot of health benefits.

Over the years, there have been a lot of modifications to bicycles to increase their efficiency and add comfort to the cyclists. A particularly interesting evolution of bicycles, like many other vehicles, was electrifying them. E-bikes have gained a lot of popularity. They are ahead of other EVs, according to Deloitte's Sector Briefing study report on e-mobility in Germany [3], as shown in Fig. 1. The effects of electric bikes compared to different modes of transport, for example, cars and motorbikes, in terms of emission rates, are studied in [2].

Next came connecting E-bikes wirelessly to the infrastructure and other road vehicles and adding partial and full autonomous modes to them. Autonomous e-bicycles comprise a lot of modules. They are electrified; hence, they are powered by batteries and have power management systems [7]. They have modules for communication [1, 5, 17], path planning [10, 19], obstacle detection and avoidance, and many features to assist the cyclist [9]. Moreover, renting an autonomous bike from a bicycle rental can be easier as the cyclist can use the bike to reach their desired destination and the bicycle can go to the nearest specified parking

H. Mahmoud (✉) · A. Mazhr
German University in Cairo, New Cairo, Egypt
e-mail: haneen.hassan@student.guc.edu.eg; ahmed.amgad@guc.edu.eg

H. Soubra
ECE- Ecole Centrale d'Electronique, Lyon, France
e-mail: hsoubra@ece.fr

M. N. Seyman (ed.), *2nd International Congress of Electrical and Computer Engineering*, EAI/Springer Innovations in Communication and Computing,
https://doi.org/10.1007/978-3-031-52760-9_19

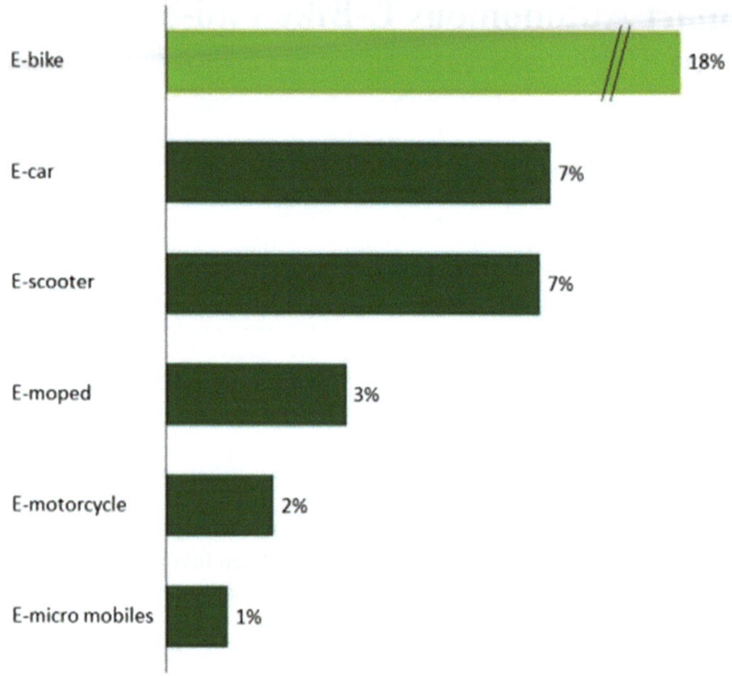

Fig. 1 Use of selected means of electric transportation

autonomously, which will save a lot of time and energy for the cyclists. Nevertheless, with autonomous E-bikes comes the risk of software or hardware failures. A fail-safe system that handles these failures by taking the appropriate action to guarantee the safety of the cyclist, the bike, surrounding vehicles and pedestrians becomes a necessity. While there is a lot of research done in the field of failure detection and fail-safe systems on automotive and aerial vehicles, the literature review showed that research in the context of autonomous E-bikes is scarce, and there is in fact only one work, to our best knowledge, on Smart Autonomous Bike Hardware Safety Metrics [6].

This paper first introduces the fail-safe paradigm for smart autonomous E-bikes by identifying the possible failures and their severity. In addition, it proposes the implementation of a fail-safe system on a smart autonomous E-bike prototype as a proof of concept (POC).

The paper is organized as follows: Section 2 presents the literature review on related works. Section 3 presents the fail-safe paradigm in the context of smart autonomous E-bikes. Section 4 focuses on the implementation of the fail-safe system on an autonomous E-bike. Section 5 discusses system testing and analysis. Finally, conclusions follow in Sect. 6.

2 Literature Review

2.1 Architectures for Fail-Safe Systems

There are many architectures for fail-safe systems for autonomous vehicles. The choice of the architecture or method of implementation of the fail-safe system depends on the vehicle, its Electronic Control Units (ECUs), sensors, actuators, and features. Another factor that affects the choice of architecture is the cost and effectiveness of the implementation of that architecture. For example, [14] offers a comparison between fail-safe architectures. There is the 1-out-of-1 with Diagnostic (1oo1D) architecture, which has 1 channel and 1 safe state for all failures. Another quite popular architecture is the Dual-CPU (lock-step) architecture, where there is a master CPU, a checker CPU, and a comparing unit. Both CPUs compute all instructions, and the comparing unit compares the output of both CPUs. Should an inconsistency arise, the system reverts to a safe state. A third architecture is called the Challenge-Response architecture, where there is a computing element and a monitoring element that execute a sequence control of the Microcontroller Unit (MCU), and depending on the output, it decides whether there are failures or not and whether it should revert to a safe state.

Another method for a fail-safe system, clarified in [4, 8, 16], is to add a redundant CPU and Error Correction Codes to mask faulty outputs. However, these architectures can result in high power consumption, software overhead, and hardware overhead, especially because they deal with failures by adding hardware redundancies like redundant CPUs and redundant sensors that are used instead of faulty components. In order to limit these problems, an architecture has been introduced called FaultRobust [14, 15]. In this architecture, multiple Fault-Robust Intellectual Properties, which are supervising modules, are connected to each sub-module in the MCU to monitor their functions, detect faults, and tolerate them. These are all architectures that rely on adding hardware components to have a fail-safe system.

Some approaches to designing a fail-safe system are more software-based. For example, [12] offers an approach for a fail-safe system through sliding mode approach control and emergency control, in which both algorithms use inputs from sensors to control the motion of the vehicle and bring (slide) it to a safe state. Another technique offered by [13] is to control the maneuver of the vehicle in case of failures and to have a safe trajectory to limit or eliminate any collisions with other vehicles or objects through the Maneuver Recognition Module (MRM). Both of these techniques prove to be efficient, especially in vehicles where space and power are very limited, which will be the case in this research, as the vehicle we are working on is an E-bike.

2.2 Failure Mode Effective Analysis

In order to design a fail-safe system, a lot of aspects should be analyzed first. For example, what are the functions or tasks that run on the bike and what could go wrong with each of them in order to take the necessary measures if any failure occurs? A detailed explanation of all the aspects that need to be analyzed is offered in [11]. The paper describes a Failure Mode Effective Analysis for ECUs, taking the Functional Safety Standards of ISO 26262 into account. The steps can be summed up into the following points:

1. Determine what systems and/or subsystems or components considered
2. Determine possible functions for systems and subsystems
3. Determine all possible list of failures from functions listed in point 2, their severity (4: Very High; 1: Very Low), occurrence, and detection
4. Determine effects from failures listed in point 3
5. Determine causes of failures identified from point 4
6. List current actions or controls for failures
7. Determine recommended action from list of failures noted
8. Determine any other relevant actions or necessary modification in design

3 Smart Autonomous E-Bike Life Situations and Risk Analysis

The fail-safe system should, depending on the failure, either perform an emergency stop, park itself in a safe zone, or regulate the speed of the E-bike. Prior to implementing the fail-safe system, the possible failures and their severity need to be identified. Next, for each failure, a risk management technique has to be followed. For this purpose, a Failure Mode Effective Analysis has been done.

3.1 Failure Mode Effective Analysis

First, three life situations of the smart autonomous E-Bike are identified, as presented in Table 1.

Second, 12 main functions are identified, as presented in Table 2. Third, for each function, we identify the possible failure scenario and the effect on the rider and the bike. This can be seen in Table 3. Finally, for each failure, we identify its severity and the failure management action, as seen in Table 4.

Table 1 Life situations

	Risk analysis – life situations	
N	Life situation	Detail
S.1	Parked	With or without rider
S.2	Autonomous riding	The bike rides at a maximum speed of 25 km/h in forward gear (no reverse gear)
S.3	Presence of an obstacle in the E-bike's lane	A dynamic or static obstacle is in the path of the bike

Table 2 Risk analysis – main functions

	Risk analysis – main functions	
N	Life situation	Description of life situation
F.1	Initialize autonomous systems when starting the E-Bike	Initialize autonomous systems when starting the E-Bike
F.2	Change from *manual* to *autonomous*	Switch to stand-alone mode if all requirements are met
F.3	Change from *autonomous* to *manual*	Switch to manual mode if all the required conditions are met
F.4	Shut down the bike and stand-alone systems	Switch off the electric motor and autonomous systems
F.5	Follow the road with its markings on the ground	E-Bike riding on a widening / narrowing / turning lane (side marking)
F.6	Exceed	Perform an overtaking maneuver
F.7	Pass roundabouts	Cross a roundabout (brake, engage, accelerate)
F.8	Go through crossroads with traffic lights	Crossing an intersection governed by traffic lights
F.9	Ride with signage	Ride respecting the signs (including crossroads without traffic lights)
F.10	Follow the E-Bike in front	Adapt your speed to the E-Bike in front which is moving
F.11	Ride according to the map	Follow planned path
F.12	Avoiding a crossing pedestrian or bicycle	Adapt your speed to let pedestrians or cyclists pass

3.2 *Failure Detection Triggered Actions*

Upon the detection of a failure, two elements need to be identified:

1. If the bike is in manual mode or autonomous mode
2. The nature of the detected failure

On the one hand, if the bike is in manual mode, then there should not be any autonomous maneuvers because that might cause an accident. Hence, the only action that would be taken is to warn the cyclist about the failure, and the cyclist can take the appropriate action. On the other hand, if the bike is in autonomous mode, then depending on the failure and its severity, the bike will either perform an emergency

Table 3 Risk analysis – possible failures

	Risk analysis possible failures		
N	Bike dreaded event	Scenario	Rider Dreaded event
F.1	No Autonomous System Initialization	The bike does not have the autonomous system	Stand-alone mode not available
F.2	No activation of autonomous mode (No change from *manual* to *autonomous*)	Bike remains in manual mode	Stand-alone mode not available
F.3	No transition from *autonomous* to *manual*	The bike remains in autonomous mode	Collision, or accident
F.4	Bike shutdown not performed	The bike remains powered, battery discharge	Breakdown
F.5	No longer follows the road with its markings	The bike can go off the road	Collision, or accident
F.6	No overtaking (over-taking not possible)	If no obstacle detection, collision with the obstacle in front	Collision, or pedestrian impact
F.7	No roundabout passage	The bike does not spot that there is a round- about coming. With the give way sign and obstacle detection OK, the bike enters the roundabout but will lose the markings	Collision, or pedestrian impact
F.8	Passage of the untimely crossroads	The bike goes through a red light	Collision, or impact
F.9	Non-application of panels	The bike no longer has panel information	Collision, or pedestrian impact
F.10	Accidental E-Bike tracking	The E-Bike in front exceeds 25 km/h and the bike continues to follow it	Highway code violation
F.11	No longer drives (or drives badly) according to the map	The bike does not follow planned path	Collision, or accident
F.12	Don't slow down or stop	Obstacle detection and avoidance module not working	Pedestrian impact

stop or an emergency park (on the side of the road). The decision whether to stop or park will be based on the Failure Mode Effective Analysis done in the previous section. Before doing any maneuvers, the bike first has to handle the already-running tasks. The easiest way to do this is to terminate tasks that will not be needed during the emergency maneuver. This will be useful for removing tasks that take a lot of computational power and could affect the performance of the fail-safe system.

After that, the planned path of the journey should be saved in memory in case there is a possibility of continuing the trip.

Emergency Park: If the failure requires an emergency park, then the bike needs to check its surroundings for surrounding obstacles and nearby sidewalks, which it could park next to.

The easiest way to implement this autonomously is to make the bike behave like a human who decided to park at the side of the road. The way it is done is by checking where the side of the road is and then cycling toward it and stopping when it is reached.

Table 4 Risk analysis – failure management actions

	Risk analysis – failure management actions	
N	Severity	Failure management action
F.1	3	Stay manual mode but warn the cyclist
F.2	3	Stay manual mode but warn the cyclist
F.3	4	Emergency park
F.4	3	Not covered in our study
F.5	4	Emergency park
F.6	4	Emergency stop
F.7	4	Emergency stop
F.8	4	Emergency park
F.9	4	Emergency park
F.10	3	Limit bike's speed to 25 km/h
F.11	4	Emergency park
F.12	4	Emergency stop

Our system will mimic this behavior by using the camera to check where the sidewalk is and will plan a path to reach it using Dubins path planning [18] algorithm. Based on the planned path, a kinematic bicycle model using Stanley controller [18] is used to steer the bike to the sidewalk. The speed of the bike will be kept constant at 15 km/h, and when the sidewalk is reached, the bike will decelerate by cycling next to it until 0 km/h is reached, and then brakes will be applied.

Emergency Stop: If the failure requires an emergency stop, then the bike will simply decelerate until 0 km/h is reached. While an emergency stop might seem not very appropriate, it is sometimes the only solution if the failure is in the obstacle detection module or in the camera or Lidar.

Speed Regulation: In some scenarios, the bike only needs speed regulation. For example, during platooning, if the bike follows the speed of another bike that is exceeding the speed limit. Once that failure is detected, the required speed will be sent to the PID controller to get the required throttle to adjust the speed of the bike. Details on how that is done are explained in the Implementation section.

4 Smart Autonomous E-Bike Fail-Safe System

4.1 Warning the Cyclist

If the bike is in manual mode, then the only needed action is to warn the cyclist of the detected failure, and it would be up to them to take the action they deem necessary. Warning the cyclist is done by picking an audio file stored on the Jetson Nano board based on the detected failure and playing it to the cyclist through a speaker connected to the board.

4.2 Handling Running Tasks

To make sure that our system will always be running and not preempted, most non-critical tasks will be terminated, especially the tasks that lead to failure and the tasks that have a lot of CPU consumption. Some of the tasks that we cannot terminate are those responsible for handling the battery that powers up the bike. Moreover, if the obstacle detection and avoidance tasks were not the source of the failure, then these tasks should also not be terminated because obstacle detection and avoidance will be needed during the maneuver from the current coordinates to a safe place. Another task we should not terminate unless it was the source of the failure is the path planning task. If there was no problem with the path planning task, then we save the planned path in memory before terminating the path planning task.

4.3 Emergency Park

Road Segmentation: In case the failure requires the park to park on the side of the road, the main file calls the "park()" function. The first step in moving the bike safely to the side of the road is to check where the side of the road is. For that purpose, we use YOLOP [20]. YOLOP, like any other version of YOLO, can be used for object detection. It can take input an image or a video and output an image or a video with the detected objects. In our system, the input live video will be captured using a monocular camera, which will be mounted on the bike. YOLOP has added features like lane detection and road segmentation, which are useful for our purpose.

As seen in the image below, the road is segmented into drivable area segments, highlighted in green, and non-drivable area segments. In addition to that, we can see that lanes and road edges are highlighted in red. We can simply choose the most-right edge in the image to be where we park the bike before the failure can cause any further damage to it or its surroundings.

Choosing Destination: After we have an output image from YOLOP, this image will be processed in the "capture-view()" function, which will be explained in this section. Since the right-most part of the road will be highlighted in red because of the road edge, our destination coordinates will be on that highlighted line.

Capture-view(): To get the X coordinates of the destination of the bike, the pixel that will be selected will be the red pixel with the largest X coordinates because that means that it is on the right-most side of the drivable area. Multiple pixels could have the same X coordinates, so one of them will be selected based on the Y coordinates.

To get the Y coordinates, we will select the point that is farthest from the initial Y coordinates of the bike to make the trajectory as smooth as possible. This is done using OpenCV, which is a library that can be used for image processing in Python. After getting the coordinates in pixels, we need to get them in world coordinates to be able to apply the path-planning algorithm. For that purpose, a library called CameraTransform available in Python will be used, where there is a function that converts from image to space coordinates: cam.spaceFromImage([xCoordsPixels, yCoordsPixels]).

In order for the function to work as accurately as possible, we first need to create an instance of the camera with the correct projection and orientation parameters. These parameters are given as follows:

1. focallengthpx: focal length in pixels, which we got from camera matrix after camera calibration
2. center: center of the camera, which we also got from the camera matrix
3. image: image resolution
4. headingdeg = 0: our camera is facing north
5. elevationm: distance of camera from the ground in meters
6. tiltdeg: tilt of the camera (90 if parallel to the ground)

After creating our camera, we can use the image to use the space transform function to get destination coordinates in world coordinates.

Path Planning: For the path planning, we will use the Dubins path planner algorithm [18], which is implemented in Python. The algorithm will take the initial coordinates and destination coordinates of the bike as input and plan the path accordingly. The output is a list of coordinates that the bike will follow, in addition to the yaw angle at each point, which is essential for the steering control. The path planning function calls the steering control function after the path has been computed and gives it the parameters: x coordinates array, y coordinates array, and yaw angles array.

Steering to Destination: In order to steer the bike, we need to control the steering angle and the velocity.

The velocity will be constant at 15 km/h while the bike is following the planned path until the destination is reached. Next, the bike will stop by making the throttle equal zero and then applying the brakes.

The steering angle will change based on the coordinates of the planned path and the yaw angles. To calculate it, we will use Stanley Controller [18], which is a steering controller for vehicles and is widely used for kinematic bicycle models, also implemented in Python. The controller has used PID control to make the speed as stable as possible because, while we will keep it constant at 15 km/h, there is still a transition from the speed the bike was going at before entering fail-safe mode. The controller also takes into account the wheelbase of the bike (the distance between the center of the front wheel and the center of the back wheel) and the maximum steering angle. The maximum steering angle will be kept at 28° to avoid sudden trajectories as much as possible. The controller then outputs the steering angle and the speed until the destination is reached. When the destination is reached, the throttle will be set to 0, and brakes will be applied.

4.4 Emergency Stop and Limit Speed

When the decision is made to either limit the speed of the bike or stop it, based on the failure, the desired speed is sent to a PID controller, implemented in Arduino, to output an optimal throttle for the bike. This is done by connecting and reading the speed of the bike, sending it to the PID controller along with the desired speed, and then sending the output throttle to the bike along with the steering angle in case we choose emergency parking.

5 Testing and Results

To test our system, a 4×4 matrix keypad was used to test different scenarios the bike could be in. A and C keys represent whether the bike is autonomous or manual, respectively, and keys 1, 2, and 3 were to simulate if the bike needed to stop, park, or regulate its speed, respectively.

5.1 Warning the Cyclist

When C is pressed, that means the bike is in manual mode, so an audio file stored on a Jetson Nano board mounted on the bike is played through a speaker, warning the cyclist of the detection of a failure.

5.2 Emergency Park

After A and 2 are pressed, YOLOP analyzes input from the monocular camera that is mounted on the bike, and the first frame of the output will be further processed to steer the bike to the sidewalk, as depicted in Fig. 2. Next, the initial coordinates and destination coordinates are specified and sent to the Dubins path planner, where we get a planned course, as shown in Fig. 3, (array of X coordinates, an array of respective Y coordinates, and an array of respective yaw angles). The Stanley controller function is called at the end of the path planner, where the planned path is passed to it. The Stanley controller outputs steering angles to perform a trajectory, as can be seen in Fig. 4, that are sent to the bike along with the throttle that will be specified using the PID controller, where we specify the setpoint to be 15 (speed 15 km/h).

Fig. 2 YOLOP output

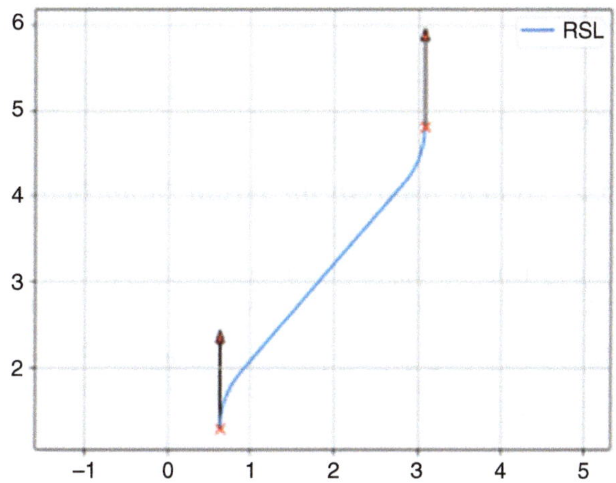

Fig. 3 Planned path

5.3 Emergency Stop and Limit Speed

After A and 1 are pressed, speed 0 will be sent to the PID controller as a setpoint along with the value of the current speed of the bike. The output throttle values are limited between 50 and 100 because a value less than 50 does not move the bike, and a value greater than 100 exceeds the speed limit. These steps are repeated until speed 0 is reached.

If A and 3 are pressed, then the same steps for the emergency stop will be taken, but instead of speed 0, we pass speed 25 to the PID controller.

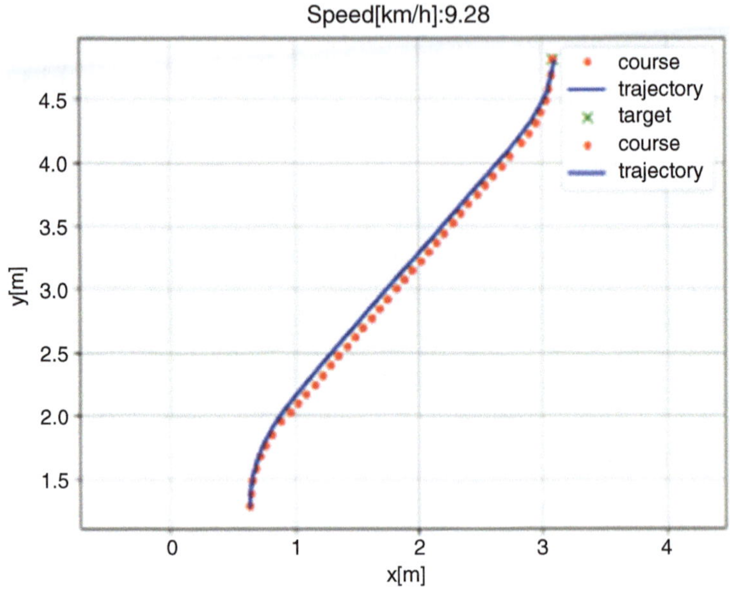

Fig. 4 Stanley controller output

5.4 System Analysis on Jetson Nano Board

The system was implemented on a quad-core Jetson Nano board mounted on the bike, where we were able to connect the camera and the Arduino boards to it.

1. When running YOLOP on Jetson Nano, %CPU = 144 and %MEM = 13.8
2. When running the rest of the system on Jetson Nano, %CPU = 104 and % MEM = 4.5

6 Conclusion

This paper introduced the fail-safe paradigm for smart autonomous E-bikes by identifying the possible failures, their severity and the actions required to guarantee safety. In addition, it proposes the implementation of a fail-safe system on a smart autonomous E-bike prototype as a proof of concept (POC).

The system implemented is designed to take action based on the detected failure. The possible actions are: stop the bike, regulate its speed, or park next to a sidewalk. The latter is implemented using YOLOP to detect the sidewalk and pick a destination; path planning using the Dubins path planner; and a Stanley controller for steering the bike according to the planned path. For stopping the bike or limiting its speed to 25 km/h, a PID controller is used to throttle accordingly.

The results show that YOLOP is quite effective in identifying sidewalks through road segmentation. Also, the use of the Dubins path planner proved effective, since it takes the start and end yaw angel into account during path planning. Additionally, the Stanley controller proved to be a good choice since the resulting trajectory follows the planned course.

Despite the promising results, some limitations need to be acknowledged. First, no camera input was taken outside the university campus in which the system was tested due to campus policies. Furthermore, different approaches to emergency parking or maneuvers in general could be explored, like the approach introduced in [13]. The development of a fail-safe system for autonomous electric bikes opens up several areas for future research and improvement. For instance, failure management actions for failures labeled to be studied in Table 4 should be defined.

References

1. Abdelrahman, A., Youssef, R., ElHayani, M., Soubra, H.: B2x communication system for smart autonomous bikes. In: 2021 16th International Conference on Computer Engineering and Systems (ICCES), pp. 1–6, Cairo (2021)
2. Cherry, C., Weinert, J., Xinmiao, Y.: Comparative environmental impacts of electric bikes in China. Transp. Res. Part D: Transp. Environ. **14**, 281–290 (2009)
3. Deloitte: Consumer sector briefing e-bikes are charged up (2023), https://s3.eu-north-1.amazonaws.com/vmn-bike-eu.com/2022/06/deloitte-e-bike-sector-briefing-1.pdf
4. Dörflinger, A., Kleinbeck, B., Albers, M., Michalik, H., Moya, M.: A framework for fault tolerance in RISC-V. In: 2022 IEEE Intl Conf on Dependable, Autonomic and Secure Computing, Intl Conf on Pervasive Intelligence and Computing, Intl Conf on Cloud and Big Data Computing, Intl Conf on Cyber Science and Technology Congress (DASC/PiCom/CBDCom/CyberSciTech), pp. 1–8, Falerna (2022)
5. Elhusseiny, N., Sabry, M., Soubra, H.: B2x multiprotocol secure communication system for smart autonomous bikes. In: 2023 IEEE Conference on Power Electronics and Renewable Energy (CPERE), pp. 1–6, Luxor (2023)
6. Elnemr, M., Soubra, H., Sabry, M.: Smart autonomous bike hardware safety metrics. In: García Márquez, F.P., Jamil, A., Eken, S., Hameed, A.A. (eds.) Computational Intelligence, Data Analytics and Applications, pp. 132–146. Springer International Publishing, Cham (2023)
7. Elwatidy, M.A., Sabry, M., Soubra, H.: Energy needs estimation for smart autonomous bikes. In: 2023 IEEE Conference on Power Electronics and Renewable Energy (CPERE), pp. 1–6, Luxor (2023)
8. Fetzer, C., Cristian, F.: Fail-awareness: an approach to construct fail-safe systems. Real-Time Syst. **24**, 203–238 (2003)
9. Halim, C.E., Sabry, M., Soubra, H.: Smart bike automatic autonomy adaptation for rider assistance. In: 2023 IEEE Conference on Power Electronics and Renewable Energy (CPERE), pp. 1–6, Luxor (2023)
10. Khalifa, H.H., Sabry, M., Soubra, H.: Visual path odometry for smart autonomous e-bikes. In: 2023 IEEE Conference on Power Electronics and Renewable Energy (CPERE), pp. 1–6, Luxor (2023)
11. Kosuru, V.S.R., Venkitaraman, A.K.: Conceptual design phase of fmea process for automotive electronic control units. Int. Res. J. Mod. Eng. Technol. Sci. **4**(9), 1474–1480 (2022)

12. Lee, J., Oh, K., Yoon, Y., Song, T., Lee, T., Yi, K.: Adaptive fault detection and emergency control of autonomous vehicles for fail-safe systems using a sliding mode approach. IEEE Access. **10**, 27863–27880 (2022)
13. Magdici, S., Althoff, M.: Fail-safe motion planning of autonomous vehicles. In: 2016 IEEE 19th International Conference on Intelligent Transportation Systems (ITSC), pp. 452–458. IEEE, Rio de Janeiro (2016)
14. Mariani, R., Fuhrmann, P.: Comparing fail-safe microcontroller architectures in light of iec 61508. In: 22nd IEEE International Symposium on Defect and Fault-Tolerance in VLSI Systems (DFT 2007), pp. 123–131. IEEE, Rome (2007)
15. Mariani, R., Kuschel, T., Shigehara, H.: A flexible microcontroller architecture for fail-safe and fail-operational systems. In: Proc. of the HiPEAC Workshop on Design for Reliability (2010)
16. Pasagadugula, S., Verma, G., Harmalkar, J.: Techniques and procedure to be followed for developing fail safe system in complaince with ISO26262. Asian J. Converg. Technol. (AJCT). **5**(3), 67–71. ISSN-2350-1146 (2019)
17. Sabry, N., Abobkr, M., ElHayani, M., Soubra, H.: A cyber-security proto type module for smart bikes. In: 2021 16th International Conference on Computer Engineering and Systems (ICCES), pp. 1–5, Cairo (2021)
18. Sakai, A., Ingram, D., Dinius, J., Chawla, K., Raffin, A., Paques, A.: Python-robotics: a python code collection of robotics algorithms. arXiv preprint arXiv:1808.10703 (2018)
19. Seoudi, M.S., Mesabah, I., Sabry, M., Soubra, H.: Virtual bike lanes for smart, safe, and green navigation. In: 2023 IEEE Conference on Power Electronics and Renewable Energy (CPERE), pp. 1–6, Luxor (2023)
20. Wu, D., Liao, M.W., Zhang, W.T., Wang, X.G., Bai, X., Cheng, W.Q., Liu, W.Y.: Yolop: you only look once for panoptic driving perception. Mach. Intell. Res. **19**(6), 550–562 (2022)

Parallel Implementation of Discrete Cosine Transform (DCT) Methods on GPU for HEVC

Mücahit Kaplan and Ali Akman

1 Background

Streaming services and the popularity of video traffic in the world's internet traffic have significantly increased. However, there is an ongoing struggle to deliver high-definition videos with limited bandwidth. In this scenario, encoding techniques that enable more efficient compression without compromising video quality have become necessary.

The High-Efficiency Video Coding (HEVC), also known as H.265, was proposed by the ISO/IEC MPEG and ITU-T VCEG joint collaboration team (JCT-VC) in January 2010 and reviewed during the JCT-VC meeting in April 2010 [2, 7]. This coding standard, which emerged in 2013 and is also referred to as ×265, has significantly improved compression compared to the Advanced Video Coding (AVC) standard, commonly known as ×264, achieving more than a 50% reduction in bit rate without compromising quality.

In recent years, due to the rapid development of Graphics Processing Units (GPUs), the use of General-Purpose GPU (GPGPU) for parallel acceleration in video coding has become a prominent trend. Simultaneously, "Compute Unified Device Architecture" (CUDA), released by NVIDIA, has made GPU parallelization more programming-friendly [10]. Therefore, it is possible to apply parallel intra-prediction for large-scale coding blocks in video sequences. However, parallelizing intra-prediction using graphics hardware is a challenging task. Particularly, there is a high restructuring dependency between the intra-prediction unit (prediction block)

M. Kaplan (✉)
Maltepe University, Istanbul, Turkey

A. Akman
Istanbul Ticaret University, Istanbul, Turkey
e-mail: aakman@ticaret.edu.tr

M. N. Seyman (ed.), *2nd International Congress of Electrical and Computer Engineering*, EAI/Springer Innovations in Communication and Computing,
https://doi.org/10.1007/978-3-031-52760-9_20

and neighboring blocks, often leading to synchronization issues when processing simultaneously with current and reference samples.

The Discrete Cosine Transform (DCT) is an important step used in many image and video compression standards due to its high energy compaction property [13]. The High-Efficiency Video Coding (HEVC) standard, like other compression standards, also employs DCT. However, the computational complexity of real-valued DCT is quite high, which is why HEVC uses integer DCT [3]. Nevertheless, in HEVC, the computational complexity of DCT remains high due to the numerous multiplication operations involved. In the reference software developed for HEVC, a quarter of the total encoding time is spent on the transform and quantization processes, including rate-distortion analysis [1, 13]. Due to this complexity, various techniques have been proposed in the literature to reduce energy consumption.

In application design, to cope with increasing computational complexity, multiplication operations are executed in parallel. Parallel execution can be achieved using homogeneous systems (such as GPUs) or heterogeneous systems (combining CPUs and GPUs) [15]. In this context, transform blocks for all supported transform sizes in the HEVC standard are efficiently distributed into thread blocks for parallel processing, and results obtained efficiently using shared memory addressing on the GPU have been demonstrated through studies [8, 14].

In this study, we present a highly parallel High-Efficiency Video Coding (HEVC) Discrete Cosine Transform (DCT) designed to run on a GPU. Memory bandwidth and memory access have been balanced for high resource utilization and efficient execution. Optimization techniques to mitigate additional costs due to overlaps during data transfer have been included to enhance performance. Experimental results leveraging the Compute Unified Device Architecture (CUDA) programming model [12] on the Nvidia Jetson Xavier Nx developer device have been evaluated using timing benchmarks.

2 Methods

HEVC (High-Efficiency Video Coding) is a block-based video coding standard that employs a quad-tree structure. It divides the image into flexible-sized blocks called Coding Tree Units (CTUs) to provide better coding in areas with varying details. These blocks are square-shaped and have a size of 64×64 pixels. Each CTU can be further subdivided into Coding Units (CUs) of sizes 4×4, 8×8, 16×16, and 32×32, forming the leaves of the quad-tree structure. Each CU is further divided into Prediction Units (PUs). PUs are then fragmented into Transform Units (TUs). Discrete Cosine Transform (DCT) is applied to TUs of sizes 32×32, 16×16, 8×8, and 4×4 pixels.

2.1 Discrete Cosine Transform (DCT)

The 2D DCT operation can be performed as a matrix multiplication (1). In this section, a 32 × 32 DCT operation is applied and compared on the GPU. In HEVC video coding, transform units (TU) can be partitioned into blocks of sizes 4 × 4, 8 × 8, 16 × 16, and 32 × 32. In HEVC, integer coefficient matrices are used for DCT.

The transform size in HEVC, denoted as $N \times N$, where N belongs to {4, 8, 16, 32}. In HEVC, the 2D DCT operation is performed as follows:

$$[C] = [D]\,[X]\,[D]^{\mathrm{T}} \quad (1)$$

In this context, it appears that the 2D DCT operation is performed by first applying a 1D DCT to the columns of the matrix $[X]$ and then applying a 1D DCT to the rows of the resulting intermediate matrix. After the transform process is completed, the resulting output undergoes a sequence of quantization, inverse quantization, and inverse transform steps within the HEVC structure. In this study, the original image was reconstructed by applying only the transform and inverse transform steps to the original image, instead of using the residual block.

The 2D inverse transform equation is as follows:

$$[C] = [D]^{\mathrm{T}}[X][D] \quad (2)$$

The HEVC matrices are scaled by a factor of $2^{\left(6+\frac{M}{2}\right)}$, relative to an orthonormal DCT transform, and additional scaling factors S_{T1}, S_{T2}, S_{IT1}, and S_{IT2} need to be applied through 2D forward and inverse transform to preserve the norm of the residual block. These factors are applied as shown in Fig. 1.

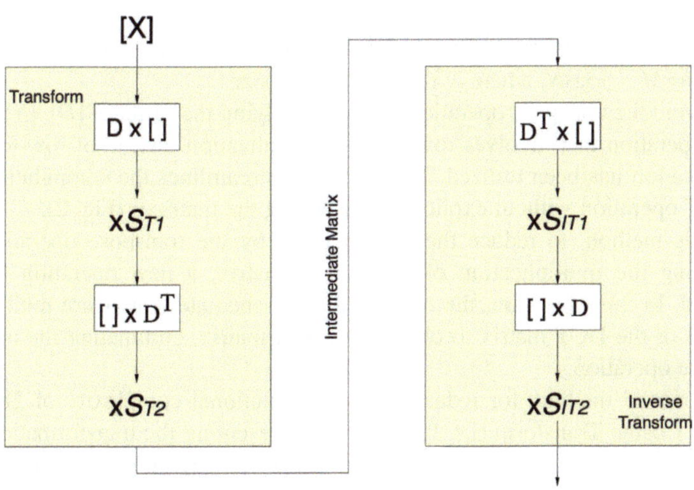

Fig. 1 The application of additional scaling factors to the transform and inverse transform steps

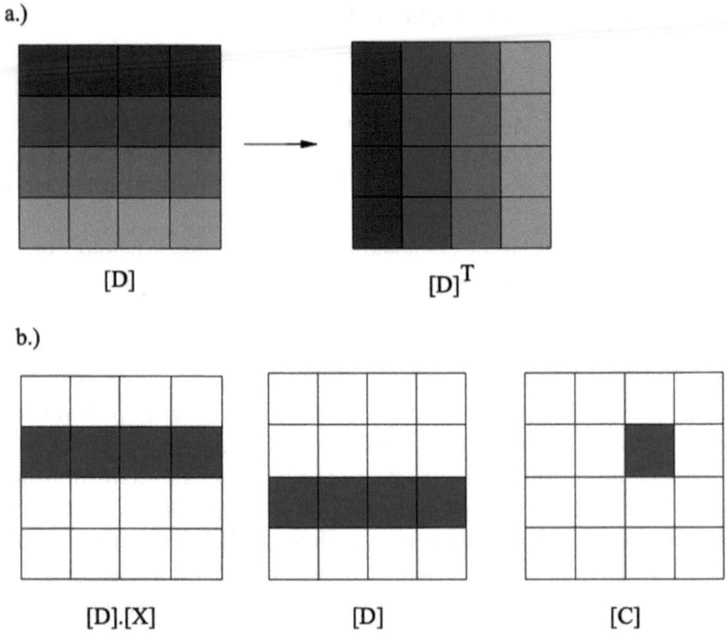

Fig. 2 (**a**) Transpose operation and (**b**) proposed operation for multiplying the rows of two matrices

The scaling factors obtained after different transform stages are as follows:

- After the first forward transform stage: $S_{T1} = 2^{(-B + M - 9)}$
- After the second forward transform stage: $S_{T2} = 2^{-(M + 6)}$
- After the first inverse transform stage: $S_{IT1} = 2^{(-6)}$
- After the second inverse transform stage: $S_{IT2} = 2^{-(21 - B)}$

Here, *B* represents the bit depth of the input/output signal (e.g., 8 bits), and *M* is defined as $M = \log_2 N$, where *N* is the transform size.

To avoid the transpose operation when multiplying the result of [*D*]. [*X*] by $[D]^T$, a new operation that involves row-to-row multiplication instead of row-to-column multiplication has been utilized. This approach streamlines the computation of the 2D DCT operation without explicitly transposing the matrices (Fig. 2).

In this method, to reduce the cost incurred by the transpose operation when performing the multiplication of the second matrix, a new operation has been proposed. In this operation, the rows of the intermediate matrix are multiplied by the rows of the DCT matrix to obtain the result matrix, eliminating the need for a transpose operation.

The second method, for reducing the computational complexity of 2D integer Discrete Cosine Transform (DCT), involves representing the transformations using smaller degrees. Integer DCT matrices possess symmetrical, anti-symmetrical, and butterfly properties, which can be leveraged to reduce the matrix size. In the case of the integer DCT matrix used in High-Efficiency Video Coding (HEVC), even-numbered rows exhibit symmetrical properties, while odd-numbered rows display anti-symmetrical properties. Consequently, by combining even-numbered and odd-numbered rows and utilizing their symmetry properties, an N-point 1D DCT is as follows:

$$\begin{bmatrix} C_c \\ C_t \end{bmatrix} = \begin{bmatrix} D_{N/2XN/2} & 0 \\ 0 & M_{N/2XN/2} \end{bmatrix} \cdot \begin{bmatrix} a_{n/2} \\ b_{n/2} \end{bmatrix} \tag{3}$$

Here, $a_{n/2} = x_n + x_{N-i-1}$, $b_{n/2} = x_n - x_{N-i-1}$, $a_{N/2} = x_i + x_{N-i-1}$, $b_{N/2} = x_i - x_{N-i-1}$, Cc ve Ct are vectors representing the even and odd coefficients of C. $D_{N/2xN/2}$ is an $N/2$-point DCT matrix, $M_{N/2xN/2}$ is an $N/2xN/2$ sized diagonal symmetric matrix [1].

Hence, in this equation, since half of the elements in the $N \times N$ matrix are zero, multiplying by zero will not affect the resulting element. Therefore, zero multiplications will not be performed. This will lead to a reduced number of multiplication operations and an increase in processing speed.

2.2 CUDA Structure and Parallelization in GPUs

Parallelization is a processing technique in which multiple processors or cores work simultaneously to accelerate data operations. In the CUDA architecture, parallel execution occurs primarily through the principle of breaking tasks into smaller pieces. CUDA employs elements such as blocks and threads to enable the parallel execution of tasks or jobs. Within CUDA, the aggregation of many threads gives rise to blocks, and the aggregation of blocks forms structures referred to as "grids." A task is executed in parallel by the blocks within the grid contained within a core (Fig. 3).

The portion of code executed on a GPU in CUDA is referred to as a kernel. The kernel code is called by the CPU (host) and executed on the GPU (device). To define a kernel function, "__global__" is used [5].

In the CUDA architecture, the smallest programming units are referred to as threads. These threads can be found within blocks in one-dimensional, two-dimensional, or three-dimensional configurations. Within the same block, all threads execute the same code segment concurrently. Each thread has its own set of registers, a state register, and a program counter. Within a block, there is also a unique identifier to facilitate the identification of threads. These indices are named "threadIdx.x," "threadIdx.y," and "threadIdx.z" based on their dimensions [6].

Block structures consist of concurrently executed threads. They can be one-dimensional, two-dimensional, or three-dimensional within a grid. Blocks are organized in groups within the grid, and each block is identified by an index. These indices are represented as "blockIdx.x," "blockIdx.y," and "blockIdx.z" [16]. The number of threads in a block is determined based on the rows and columns of the

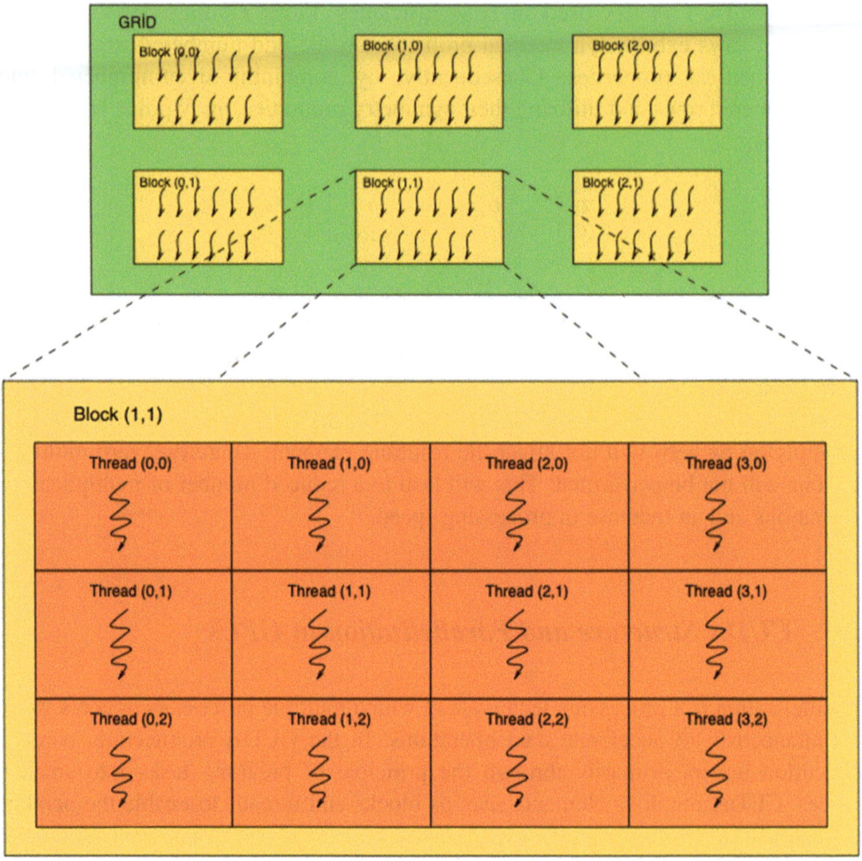

Fig. 3 Grid, block, and thread hierarchy [9]

block and is denoted as "blockDim.x," "blockDim.y," and "blockDim.z." These expressions also indicate the size of the block.

Grids are structures formed by assembled blocks. With each invocation of a kernel, a grid is created. This means that the code segment executed in parallel, while running on the GPU, has a copy of this code segment executed in each grid. Like other structures, grids can also be one-dimensional, two-dimensional, or three-dimensional. The dimensions of grids are denoted as "gridDim.x," "gridDim.y," and "gridDim.z" depending on their size.

CUDA code is executed in a heterogeneous manner on both the CPU and GPU. The GPU hardware contains different types of memory. Local memory is a type of memory found in each thread within the GPU. Additionally, there is shared memory that belongs to a block, and all threads within that block can access this memory. In the GPU architecture, global memory is accessible by all units (Fig. 4) [11].

Due to their multitude of processing units, GPUs provide a significant advantage in handling large-scale data operations. In the approach used in this study, image

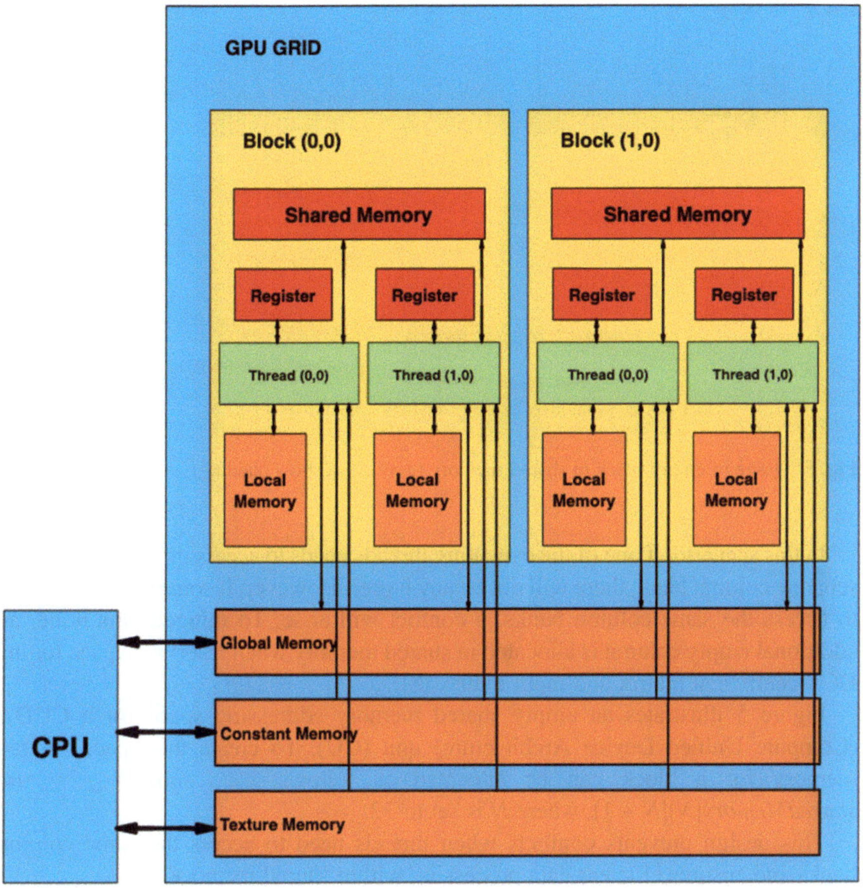

Fig. 4 CUDA memory model [4]

blocks are transferred from the main memory to the GPU memory. Within the GPU, these blocks are then moved to shared memory. All pixels within a block are computed in parallel by a thread.

Shared memory is organized into banks to achieve higher bandwidth. Each bank can serve one address per thread. The Volta GPU architecture, found in the Jetson Xavier Nx developer kit, has 32 banks, each 4 bytes wide. When an array is stored in shared memory, adjacent 4-byte words go to consecutive rows.

In this study, the data that threads need to access has been transferred to shared memory. The "__shared__" function is used to allocate space in shared memory. In the transform and inverse transform steps, since all threads need to access the image block (*X*), the HEVC DCT integer matrix, and the matrix resulting from the first stage of the DCT, shared memory space has been allocated for all these blocks, and the necessary portions have been copied to shared memory.

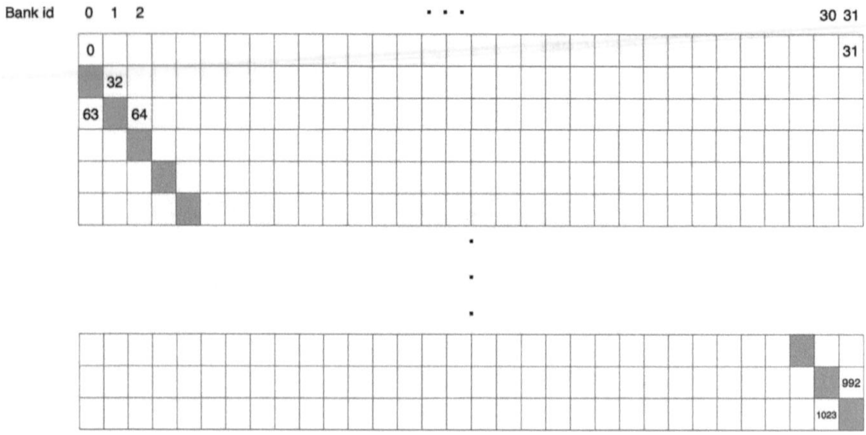

Fig. 5 Usage of empty space in shared memory (gray areas show padding)

In this scenario, if one of the executing threads needs to access information from a separate column bank, there will not be any issues. However, if some threads attempt to access the same column banks, a conflict will arise. To mitigate this issue, an additional empty column is allocated in shared memory when reserving space for the DCT coefficient matrix and input matrix [8].

Figure 5 illustrates an empty shared memory addressing space for a CUDA (Compute Unified Device Architecture) unit (CU). To create this space, shared memory for a block can be allocated as follows: _ _ *shared* _ _ *int sharedMemory*[*N*][*N* + 1], where *N* is set to 32.

This design prevents conflicts when threads need to access the same column banks and ensures efficient data processing within shared memory.

2.3 Implementation of DCT on GPU

In the first proposed method (DCT-1), a new operation is introduced instead of the transpose operation. This operation is applied to the intermediate matrix obtained in the first stage of 2D DCT and the DCT coefficient matrix, resulting in a 1D DCT. In this study, the actual image block is directly used instead of the residue matrix mentioned in Eq. (1) since the other steps of HEVC are not considered.

An example image of size 512 × 512, initially located in the CPU main memory, is divided into blocks of size 32 × 32. These blocks are then transferred to the GPU memory for execution on the GPU core. Within the GPU core, the coefficient matrix and the transferred block matrix are optimally placed in shared memory with gaps for optimization purposes [8]. Each pixel of the blocks is calculated in parallel, with 1024 threads, each of which computes each element of the intermediate matrix. Threads can access the shared memory allocated for each block. Subsequently, the

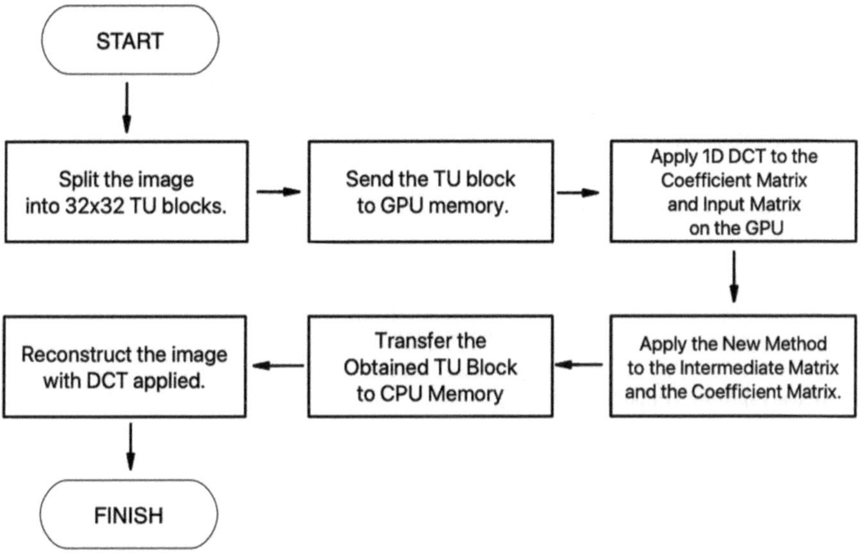

Fig. 6 DCT-1 flowchart

proposed new operation is applied in parallel between the obtained intermediate matrix and the DCT coefficient matrix. The resulting result matrix is then sent from the GPU to the CPU main memory. The blocks are combined to obtain the DCT-transformed image. The flowchart of this process is illustrated in Fig. 6.

In the proposed second method (DCT-2), DCT coefficient matrix and input matrix are subjected to 1D DCT using Eq. (2). When the coefficient matrix is manipulated using its characteristics, it is observed that half of the coefficient matrix consists of zeros, resulting in a reduction in multiplication operations by half. To accommodate this, certain modifications have been made to the input matrix. The elements in the columns of the input matrix have been modified in a way that the elements at positions (0 to $N-1$, 1 to $N-2$, ...) are summed and written to the upper half of the column, while the same elements are subtracted and written to the lower half of the column (3).

During the implementation of DCT-2, when transferring the residue matrix to shared memory, the sum of column elements of the matrix is copied to the upper half of the matrix, and the difference of elements is copied to the lower half of the matrix. Subsequently, if the row value of the pixel to be processed is even, 1D DCT is applied to the matrix consisting of even rows of the coefficient matrix and the upper half of the residue matrix. If the row value of the pixel is odd, 1D DCT is applied to the matrix consisting of odd rows of the coefficient matrix and the lower half of the residue matrix, resulting in an intermediate matrix. The intermediate matrix is stored in shared memory with rows and columns swapped. Then, 1D DCT is applied to the coefficient matrix and the obtained intermediate matrix, and the results are stored again with rows and columns reversed to obtain the final result matrix. Reversing rows and columns is necessary because, in the second stage, transposing the matrix is required due to the separate multiplication of the upper and lower halves of the DCT matrix. To eliminate the need for transpose operation, Eq. (1) has been modified as follows.

In the DCT Eq. (1), it is as follows:

$$[A] = [D].[X] \tag{4}$$

This leads to:

$$[C] = [A].[D]^{\mathrm{T}} \tag{5}$$

If we take the transpose of both sides:

$$[C]^{\mathrm{T}} = [D].[A]^{\mathrm{T}} \tag{6}$$

In other words, if the result $[A]$ is stored in memory with rows and columns reversed, it becomes $[A]^{\mathrm{T}}$. Therefore, by multiplying this result with the $[D]$ matrix again and storing the result with rows and columns reversed, we obtain the $[C]$ matrix. Consequently, the issue that could arise from the transpose operation of the $[D]$ matrix is eliminated. The flowchart of the DCT-2 method is illustrated in Fig. 7.

3 Experimental Results

In experimental studies aimed at parallel processing of DCT on the GPU during the HEVC transform step and determining its efficiency, the Lena reference image (512 × 512) was selected. The tests were conducted on an Nvidia Xavier Nx developer kit using the Nvidia Nsight Eclipse Editor development environment and the CUDA programming model. The test image, the two proposed methods, and the serial execution of DCT were evaluated for shared memory utilization and non-utilization scenarios.

DCT processing was performed on a 512 × 512 image using 32 × 32 transform blocks. The proposed parallel DCT methods and the serial execution of DCT on the GPU were separately assessed for shared memory utilization and non-utilization scenarios.

All tests were examined in terms of GPU kernel execution time and overall procesing time. In this study, kernel time and overall time are considered as test criteria. Kernel time corresponds to the time when all blocks perform the prediction process, starting from the first block. This duration does not include memory transfers and the division of the image into blocks. Total time, on the other hand, represents the time it takes from the very beginning of the HEVC DCT application until the final reconstructed image is obtained. This duration includes image

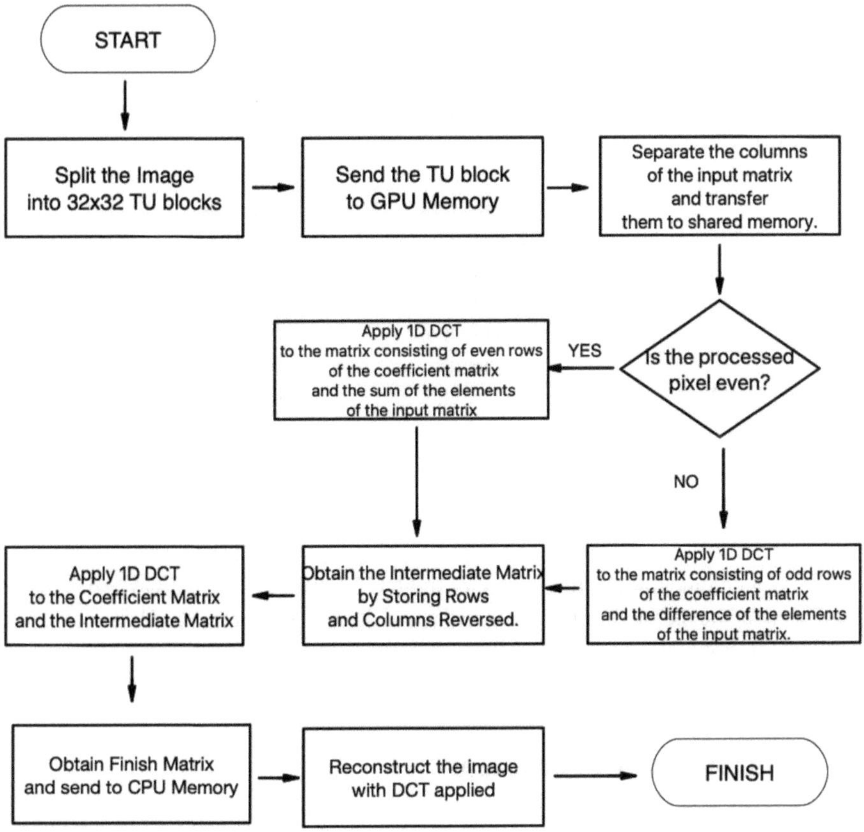

Fig. 7 DCT-2 flowchart

partitioning and memory transfers for the GPU.Time data were recorded by analyzing them using the Nvidia Nsight Profiler tool. As seen in Table 1, leaving padding in shared memory and omitting the transpose operation significantly improved the execution time.

When examining the table data, it can be observed that leaving gaps in shared memory for both DCT-1 and DCT-2 scenarios has resulted in acceleration in both overall time and kernel execution time. Shared memory padding utilization for DCT-1 and DCT-2 has led to an approximately twofold increase in kernel time. In comparison to the serial implementation of DCT, the best kernel execution time for DCT-1 has achieved approximately a 4.9 times acceleration, while DCT-2 has achieved a 6.62 times acceleration. When comparing the overall time, the fastest DCT-1 result achieved a speedup of 1.95 times, while the fastest DCT-2 result provided a speedup of 2.44 times.

In the case of DCT-1, the best kernel execution time has increased by approximately 1.34 times compared to DCT-2. The DCT-1 method has led to an acceleration of approximately 1.95 times for both overall time and kernel time compared to

Table 1 DCT processing test data

Methods	Experimental results			
	Using padding in shared memory	Duration measurements	Execution time	Acceleration
DCT serial	Yes	Overall time	0.3720 sn	1
		Kernel time	0.2219 sn	1
	No	Overall time	0.4399 sn	1
		Kernel time	0.2181 sn	1
DCT-1	Yes	Overall time	0.1903 sn	1.95
		Kernel time	0.0452 sn	4.9
	No	Overall time	0.4121 sn	1.06
		Kernel time	0.0518 sn	4.21
DCT-2	Yes	Overall time	0.1623 sn	2.29
		Kernel time	0.0335 sn	6.62
	No	Overall time	0.1801 sn	2.44
		Kernel time	0.0472 sn	4.62

the serial implementation. In the case of the DCT-2 method, 2.29 times acceleration has been achieved in both overall time and kernel time compared to the serial implementation.

4 Discussion and Conclusions

When we examine the literature, it is observed that in DCT (Discrete Cosine Transform) applications, parallelization has provided an average speedup of approximately 1.22 times compared to serial [8]. Accordingly, in this study, an average speedup of 1.9 times has been achieved. Therefore, in comparison to the existing literature, it yields a speedup of 1.55 times.

In this study, the acceleration of the Discrete Cosine Transform (DCT) used in the High-Efficiency Video Coding (HEVC) transform step has been investigated. It has been observed that the execution and parallel processing of DCT on the GPU have a positive impact on efficiency. Two different DCT methods were proposed for this process, and these methods were separately tested for efficient utilization of shared memory. When examining the test results, it was observed that the most efficient DCT processing was achieved when the DCT-2 method was used, and a column padding was left in shared memory.

Considering the implementation times for all cases, for 32×32 transform blocks, both proper addressing in shared memory and the application of new methods have resulted in approximately a 94% acceleration compared to the serial implementation. It has been demonstrated in this study that the proposed method can be effectively used in obtaining DCT in the HEVC transform step.

References

1. Akman, A., Cekli, S.: Design of approximate discrete cosine transform architecture for image compression with HEVC intra prediction. In: 12th International Conference on Electrical and Electronics Engineering (ELECO), pp. 155–158, Bursa, Turkey (2020)
2. Bross, B., Han, W.-J., Sullivian, G.J., Ohm, J.-R., Wiegand, T.: High Efficiency Video Coding (HEVC) Text Specification Draft 9. Available: http://phenix.it-sudparis.eu/jct/doc_end_user/current_document.php?id=6803. Last accessed 15 June 2023
3. Budagavi, M., Fuldseth, A., Bjontegaard, G., Sze, V., Sadafale, M.: Core transform design in the high efficiency video coding (HEVC) standard. IEEE J. Sel. Top. Signal Process. **7**(6), 1029–1041 (2013)
4. Corana, A.: Architectural Evolution of Nvidia Gpus for High-Performance Computing. IEIIT-CNR, Tech. Rep, Genova (2015)
5. CUDA Refresher: The CUDA Programming Model. (2019). https://developer.nvidia.com/blog/cuda-refresher-cuda-programming-model/. Last accessed 18 July 2023
6. Ghorpade, J., Parande, J., Kulkarni, M., Bawaskar, A.: GPGPU processing in CUDA architecture. arXiv preprint arXiv:1202.4347 (2012)
7. JTC1/SC29/WG11, ITU-TSG16Q6 and ISO/IEC, Joint call for proposals on video compression technology. ITU-T SG16 Q6 document VCEG-AM91 and ISO/IEC TC1/SC29/WG11 document N11113, Kyoto (2010)
8. Mate, Č., Alen, D., Leon, D., Igor, P., Mario, K.: Performance engineering for HEVC transform and quantization kernel on GPUs. Automatika. **61**(3), 325–333 (2020)
9. Mawson, M., Leaver, G., Revell, A.: Real-time flow computations using an image based depth sensor and GPU Acceleration. In: Proceedings of the NAFEMS World Congress (2013)
10. NVIDIA: Nvidia Jetson Xavier Nx Developer Kit. https://openzeka.com/wp-content/uploads/2020/03/Jetson_Xavier_NX_Developer_Kit-One-Pager.pdf. Last accessed 23 July 2023
11. Ozkurt, C.: Yerçekimsel N-cisim simülasyonundaki karşılıklı kuvvetler optimizasyonunun gpu üzerindeki analizi. Doctoral dissertation, Karadeniz Teknik Üniversitesi (2018)
12. Ryoo, S., Rodrigues, C., Baghsorkhi, S.S., et al.: Optimization principles and application performance evaluation of a multithreaded GPU using CUDA. In: Proceedings of the 13th ACM SIGPLAN Symposium on Principles and Practice of Parallel Programming (PPoPP), pp. 73–82. ACM, New York (2008)
13. Singhadia, A., Bante, P., Chakrabarti, I.: A novel algorithmic approach for efficient realization of 2-D-DCT architecture for HEVC. IEEE Trans. Consum. Electron. **65**(3), 264–273 (2019)
14. Souza, D.F.D., Ilic, A., Roma, N., et al.: GHEVC: an efficient HEVC decoder for graphics processing units. IEEE Trans. Multimedia. **19**(3), 459–474 (2017)
15. Souza, D.F.D., Roma, N., Sousa, L.: Opencl parallelization of the HEVC de-quantization and inverse transform for heterogeneous platforms. In: 2014 22nd European Signal Processing Conference (EUSIPCO), pp. 755–759, Lisbon (2014)
16. Thurley, M.J., Danell, V.: Fast morphological image processing open-source extensions for GPU processing with CUDA. IEEE J. Sel. Top. Signal Process. **6**(7), 849–855 (2012)

Mathematical Models and Methods of Observation and High-Precision Assessment of the Trajectories Parameters of Aircraft Movement in the Infocommunication Network of Optoelectronic Stations

Andriy Tevjashev, Oleg Zemlyaniy, Igor Shostko, and Anton Paramonov

1 Introduction

Modern optoelectronic stations (OES) with video cameras in the visible and infrared frequency range and laser range finders (LRF) have significantly increased the efficiency and expanded the capabilities of an airspace continuous monitoring, detection, recognition, and high-precision tracking of small-sized and highly dynamic aircraft, including those manufactured using stealth technology. Moreover, OES, compared to radar stations (RS), provides a significant increase in the accuracy of determining the parameters of aircraft trajectories. OES, like radars, during open observation provide direct measurements of three parameters characterizing the aircraft location in airspace in a spherical coordinate system: azimuth, elevation angle, and slant range. When using the infocommunication network (ICN) of OES, it becomes possible to covertly monitor the aircraft trajectory, in which the aircraft is not irradiated by LRF and is not able to establish the fact of its detection and tracking. This is possible when the detected aircraft is accompanied by ICN with at least two OES. In this case, each OES measures only two parameters: azimuth and elevation.

A. Tevjashev (✉) · I. Shostko · A. Paramonov
Kharkiv National University of Radio Electronics, Kharkiv, Ukraine
e-mail: andrew.teviashev@nure.ua

O. Zemlyaniy
O. Ya. Usikov Institute for Radio Physics and Electronics of the National Academy of Sciences of Ukraine, Kharkiv, Ukraine

M. N. Seyman (ed.), *2nd International Congress of Electrical and Computer Engineering*, EAI/Springer Innovations in Communication and Computing,
https://doi.org/10.1007/978-3-031-52760-9_21

2 Problem Analysis

The problem of geolocation of different types of objects in air-land and sea space attracts wide attention due to its importance for many applications such as radar, sonar, video surveillance, trajectory measurements, wireless communications, sensor, and infocommunication networks of radar and optoelectronic stations (OES) [1–12].

In [1, 2], mathematical models for geolocation of radio sources by time-difference-of-arrival (TDOA), frequency-difference-of-arrival (FDOA), and angle-of-arrival (AOA) methods, which are based on a group of sensor nodes with known locations, are presented. In [3–5], the estimates of the positioning accuracy limits of radio sources using TDOA and AOA methods, such as the Cramer-Rao lower bound and the circular probability deviation, are discussed. In [6], the detailed review and main tactical-technical characteristics of the modern ground-based radio-technical and radio-telemetric means of high-precision external trajectory measurements are provided. A large number of works are devoted to the problems of using OES for airspace monitoring, external trajectory measurements, and metrological certification of OES [7–12]. In [7], the description and some characteristics of ground-based optical, optoelectronic, quantum-optical, laser-television means, and systems of high-precision trajectory measurements, which are used at research test sites, are presented. However, the problem of high-precision trajectory measurements and methods of their processing, providing maximum efficiency and minimum bias in the estimation of aircraft trajectory parameters, remains extremely actual.

Solving this problem requires a detailed analysis of the causes of occurrence, the degree of influence, and methods of suppressing (neutralizing) all types of errors in measuring the OES of aircraft trajectory parameters [9, 13]. Modern OES for trajectory measurements are implemented in a modular design and include the following modules [10–12]: a rotary support with a two-axis platform with rotation in azimuth and elevation, on which an optical-electronic module with a global positioning and platform leveling system, is placed; an optical-electronic module with a television camera, thermal imager, and laser rangefinder. OES provides all-round visibility of the airspace in the optical and infrared frequency ranges, detection, identification, recognition, and automatic aircraft tracking. When building an effective airspace video surveillance system over a significant area with the help of one OES, it is practically impossible due to the limitation of its viewing area, so it is necessary to use ICN of OES. The systematic solution to the problem of building an ICN OES includes solving an ordered set of interrelated problems; one of the primary tasks is to determine the minimum number and optimal placement of OES in such a way as to minimize the total volume of "dead zones" in the airspace video control zone [14]. In addition, the OES ICN must be spatially distributed, scalable, highly reliable, and survivable throughout the period of its operation.

The aim of the work is to develop and investigate mathematical models of video surveillance of aircrafts by infocommunication network of OES, methods of high-precision estimation of parameters of their motion trajectories, and statistical properties of the obtained estimations.

3 Mathematical Models of Aircraft Observation

3.1 Coordinate Systems

Instrumental coordinate systems of OES To use OES as an instrument for conducting external trajectory measurements, the video camera is installed on a support-swivel platform (SSP), which provides the ability to rotate the video camera in two planes – a horizontal plane (in azimuth) by an angle of $n \times 360$ and a vertical plane (by an elevation angle) by the angle 0–90°. An SSP with a video camera installed on it are the main elements of any OES. Two OES instrument coordinate systems are used: the OES Instrument Spherical Coordinate System (ISCS) and the Instrument Cartesian Coordinate System (ICCS). It is assumed that the beginning of the instrumental systems coincides with the O_C point – the beginning of the video camera coordinate system, as well as with the intersection point of the rotation axes along the azimuth and the SSP angle.

The *Y*-axis points vertically upwards; the *X*-axis points north; the direction of the *Z*-axis makes the coordinate system right. Instrumental OES coordinate systems are used for direct measurements of the aircraft spatial position relative to the OES location [15–18].

The use of ISCS and ICCS allows you to determine the aircraft location estimates relative to the location of a specific OES. To carry out external trajectory measurements along the entire aircraft trajectory, it is necessary to use world coordinate systems.

World geodetic coordinate system WGS-84 It is intended for the quantitative description of the position and movement of objects located on the surface of the Earth and in space around the Earth. The WGS-84 coordinate system is based on the following provisions: the coordinate system origin is located at the mass center of the Earth; the *Z*-axis is directed to the IERS Reference Pole (IRP); the *X*-axis is directed to the intersection point of the plane (IRM) with the plane that passes through the origin of the WGS-84 coordinate system and is perpendicular to the *Z*-axis; and the *Y*-axis complements the system to the right coordinate system.

The points position in the WGS-84 system can be obtained in the form of spatial rectangular Greenwich (GCS) or geodetic coordinate systems (GDCS). Geodetic coordinates depend on the global ellipsoid, the dimensions and shape of which are determined by the values of the semi-major axis and compression. The ellipsoid center coincides with the origin of the WGS-84 coordinate system, the ellipsoid rotation axis coincides with the *Z*-axis, and the plane of the initial meridian coincides with the XOZ plane.

In GCS, the object location is determined by its coordinates $(X, Y, Z)^{\mathrm{T}}$, where X, Y, Z are the object coordinates in km. In GDCS, the object location is determined by its coordinates $(B, L, H)^{\mathrm{T}}$, where B – geodetic latitude – the angle between the normal and the plane of the equator [rad]; L – geodetic longitude – the angle between the meridian plane of a given point and the initial meridian plane [rad]; H – geodesic height – segment of the normal from the point to the surface of the ellipsoid [m].

Notations We will use the following notations for random variables, their parameters, and estimates: (Ω, B, P) Cartesian product of probability spaces (Ω_i, B_i, P_i), $i = 1, 2, \ldots, n; \Omega = \Omega_1 \times \Omega_2 \times \ldots \times \Omega_n; B = B_1 \times B_2 \times \ldots \times B_n, P = P_1 \times P_2 \times \ldots \times P_n$; Ω_i is the space of elementary events; B_i is σ-algebras of events with Ω_i; P_i is probability measures on B_i; C is a deterministic value; $C(\omega)$ is a random value, $\omega \in \Omega$; $\overline{c} = \underset{\omega}{\mathrm{M}}\{C(\omega)\}$ is the mathematical expectation of a random value; $\sigma_c^2 = \underset{\omega}{\mathrm{M}}\{C(\omega) - \overline{c}\}^2$ is the variance of a random value; $C(\tilde{\omega})$ is an implementation of a random value for a fixed $\tilde{\omega} \in \Omega$; $\widehat{C}(\omega)$ is an estimate of a random value; $C(\omega) \cong N(\overline{c}, \sigma_c^2)$ is the normal distribution of the random value $C(\omega)$ with mathematical expectation $\overline{c}$ and variance σ_c^2.

Let $T = \{t_0, t_1, t_2, \ldots, t_k, \ldots, t_K\}$ denote the set of moments of an aircraft trajectory observation in the airspace monitoring area of OES ICN. All measurements are carried out with time redundancy, which allows for their optimal filtering and associating the received estimates to the moment of time $\forall t \in T$ synchronized with all OES. Every i-th OES in the ICN (where $i = 1, 2, \ldots, N$) for each time moment $t \in T$ provides: airspace video control in the observation area; an aircraft detection at the time moment $t \in T$; detection, recognition, automatic an escort point formation on an aircraft surface, high-precision an aircraft escort, recapture of a lost aircraft; calculations by filtering direct measurements of mathematical expectation estimates of the aircraft location coordinates in ISCS (mandatory $\widehat{\overline{\alpha}}_{it}(\omega)$ – azimuth, $\widehat{\overline{\beta}}_{it}(\omega)$ – declension and, if possible, the distance to an object $\widehat{\overline{D}}_{it}(\omega)$) for each moment of time $t \in T$ the aircraft spatial location coordinates are estimated; transmission of video data and estimates of the aircraft location coordinates $\widehat{\overline{\alpha}}_{it}(\omega)$, $\widehat{\overline{\beta}}_{it}(\omega)$, $\widehat{\overline{D}}_{it}(\omega)$ to the information and analytical center of the ICN, which provides reception, accumulation, and storage of data coming from all OES for each moment of time $t \in T$; the formation of overdetermined systems of algebraic equations, the solution of which allows obtaining estimates of the aircraft location coordinates in the GCS at the time moment $t \in T$; displaying the current position and trajectory of the aircraft in the GCS for each time moment $t \in T$ according to the data received from the OES network. Figure 1 schematically shows the observation of a flying object by three OES.

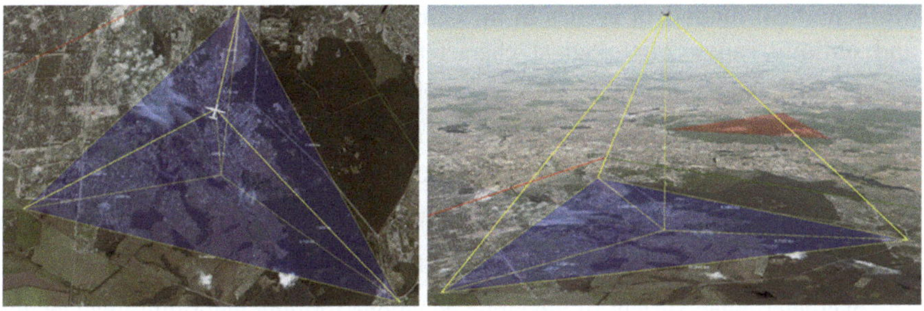

Fig. 1 Observation of a flying object by three OES

3.2 Mathematical Models of Observation Errors

Mathematical model of OES geolocation errors The results of OES ICN aircraft trajectory observations are divided into two classes: static and dynamic. Static data include estimates of location coordinates mathematical expectation and variance of each *i*-th OES ICN in GDCS and GCS. A GPS receiver is installed on each OES for its geolocation for automatic, continuous acquisition of current coordinates regardless of weather conditions, which ensures the reception and consideration of differential corrections in the differential DGPS mode and ensures the accuracy of determining of the coordinate vector components estimates $\left(\widehat{X}_{0\text{ir}}(\omega), \widehat{Y}_{0\text{ir}}(\omega), \widehat{Z}_{0\text{ir}}(\omega)\right)^{\text{T}}$ of the *i*-th OES position in GCS with $\sigma = 0.5$ m. Thus, the mathematical model of geolocation errors of each *i*-th OES is accepted as a system of three uncorrelated normally distributed random values with known variances.

Mathematical model of OES alignment errors After estimating the coordinate vector components $\left(\widehat{X}_{0\text{ir}}(\omega), \widehat{Y}_{0\text{ir}}(\omega), \widehat{Z}_{0\text{ir}}(\omega)\right)^{\text{T}}$ of the *i*-th OES position, its adjustment is performed (leveling the OES rotary support platform and linking the direction of the *X*-axis to the local meridian). As shown by studies [14], it is almost impossible to completely eliminate alignment errors, it is only possible to minimize them using various technologies and equipment, including laser, optical, and navigation. In [14], the results of the analysis of the dependence of the instrumental (systematic) errors of the aircraft location estimates associated with random errors of adjustment are presented. As the most adequate model of OES adjustment, a model with random errors is used with nonzero mathematical expectation values of the platform axes rotation angles α_x, α_z, α_y of the form of a system of three uncorrelated normally distributed random values with known variances of the OES platform adjustment of no more than ± 0.1 mrad on each of the axes.

In this case, the ISCS and ICCS coordinate systems have a common origin, but, as a result of alignment errors, differ in direction by the corresponding angles along each coordinate axis.

Mathematical models of errors in measurements of azimuth, elevation angle, and oblique range Metrological characteristics (variances) estimates of direct measurements of azimuth – $\widehat{\sigma}^2_{\alpha_i}(\omega)$, elevation angle – $\widehat{\sigma}^2_{\beta_i}(\omega)$, and oblique range – $\widehat{\sigma}^2_{D_i}(\omega)$ of each *i*-th OES sensor network are assumed to be known a priori.

Estimates of the mathematical expectation of the aircraft location coordinates coming from each *j*-th OES $(\widehat{\overline{\alpha}}_{jt_k}(\omega), \widehat{\overline{\beta}}_{jt_k}(\omega), \widehat{\overline{D}}_{jt_k}(\omega))$ for each time moment t_k in ISCS are entered into the database (DB) for post-session processing.

The overdetermined system of algebraic equations forming algorithm includes the sequential execution of the following steps.

STEP 1. Formation of the OES numbers list, which at the time t_k were monitoring (surveillance) the aircraft and conducted direct measurements of the coordinates of its location in the OES ISCS. The result is a OES numbers list $I = \{i, j, \ldots, l\}$, that, at the time moment t_k performed direct measurements of the aircraft location coordinates in the OES ISCS, while the $|I| = n$, $n = \{1,2,3,\ldots\}$.

STEP 2. Calculation of line-of-sight directional coefficients estimates from OES to the aircraft at the time moment t_k.

The calculation of line-of-sight direction coefficients estimates from OES to the aircraft at time t_k is performed sequentially for each OES whose number is in the list $\mathrm{I} = \{i, j, \ldots, l\}$. Mathematical expectations estimates of the directional coefficients $\widehat{\overline{\boldsymbol{l}}}_{it_k}(\boldsymbol{\omega}), \widehat{\overline{\boldsymbol{m}}}_{it_k}(\boldsymbol{\omega}), \widehat{\overline{\boldsymbol{n}}}_{it_k}(\boldsymbol{\omega})$ of the sight line from the *i*-th OES to the aircraft at the time moment t_k in the ICCS are calculated as follows:

$$\widehat{\overline{\boldsymbol{l}}}_{it_k}(\boldsymbol{\omega}) = \cos\widehat{\overline{\beta}}_{it_k}(\omega)\cos\widehat{\overline{\alpha}}_{it_k}(\omega), \tag{1}$$

$$\widehat{\overline{\boldsymbol{m}}}_{it_k}(\boldsymbol{\omega}) = \sin\widehat{\overline{\beta}}_{it_k}(\omega), \tag{2}$$

$$\widehat{\overline{\boldsymbol{n}}}_{it_k}(\boldsymbol{\omega}) = \cos\widehat{\overline{\beta}}_{it_k}(\omega)\sin\widehat{\overline{\alpha}}_{it_k}(\omega), \tag{3}$$

where $\widehat{\overline{\alpha}}_{it_k}(\omega), \widehat{\overline{\beta}}_{it_k}(\omega)$ – mathematical expectations estimates of azimuth and elevation angle from the *i*-th OES to the aircraft in ISCS, the numerical values of which are taken from the database.

The mathematical expectations estimates of line-of-sight directional coefficients from each OES from the list $I = \{i, j, \ldots, l\}$ are calculated in a similar way, replacing the number i with the next number j, and so on until the end, including the last number l of the list $I = \{i, j, \ldots, l\}$. The mathematical expectations estimates of line-of-sight directional coefficients from each OES are entered in the database.

STEP 3. Formation of line-of-sight equations from OES to an aircraft in GCS.

Let, as before, the number of the first OES in the list $I = \{i, j, \ldots, l\}$ equals i. Then, the equations of the lines of sight from the *i*-th OES to an aircraft at the time moment t_k in the GCS have the form:

$$\frac{x_{t_k}(\omega)-\widehat{\overline{x}}_{0ir}(\omega)}{\widehat{\overline{l}}_{it_k}(\omega)}=\frac{y_{t_k}(\omega)-\widehat{\overline{y}}_{0ir}(\omega)}{\widehat{\overline{m}}_{it_k}(\omega)}=\frac{z_{t_k}(\omega)-\widehat{\overline{z}}_{0ir}(\omega)}{\widehat{\overline{n}}_{it_k}(\omega)}, \tag{4}$$

where $\left(\widehat{\overline{x}}_{0ir}(\omega),\widehat{\overline{y}}_{0ir}(\omega),\widehat{\overline{z}}_{0ir}(\omega)\right)^{\mathrm{T}}$ – the vector of the mathematical expectations estimates of the location coordinates of the of the i-th OES polygon in the GCS; $\left(\widehat{\overline{l}}_{it_k}(\omega),\widehat{\overline{m}}_{it_k}(\omega),\widehat{\overline{n}}_{it_k}(\omega)\right)^{\mathrm{T}}$ – mathematical expectations estimates of the directional vector of sighting lines from the i-th OES to an aircraft in the GCS; vector $\left(x_{t_k}(\omega),y_{t_k}(\omega),z_{t_k}(\omega)\right)^{\mathrm{T}}$ of unknown (to be estimated) aircraft location coordinates at the time moment t_k.

The vector $\left(\widehat{\overline{l}}_{it_k}(\omega),\widehat{\overline{m}}_{it_k}(\omega),\widehat{\overline{n}}_{it_k}(\omega)\right)^{\mathrm{T}}$ of mathematical expectations of the directional coefficients of the line of sight from the i-th OES to an aircraft at the time moment t_k in the GCS is equal to:

$$\left\|\begin{matrix}\widehat{\overline{l}}_{it_k}(\omega)\\ \widehat{\overline{m}}_{it_k}(\omega)\\ \widehat{\overline{n}}_{it_k}(\omega)\end{matrix}\right\|=\mathbf{A}_{ir}\cdot\left\|\begin{matrix}\widehat{\overline{\boldsymbol{l}}}_{\boldsymbol{it_k}}(\boldsymbol{\omega})\\ \widehat{\overline{\boldsymbol{m}}}_{\boldsymbol{it_k}}(\boldsymbol{\omega})\\ \widehat{\overline{\boldsymbol{n}}}_{\boldsymbol{it_k}}(\boldsymbol{\omega})\end{matrix}\right\|,$$

where $\left(\widehat{\overline{\boldsymbol{l}}}_{\boldsymbol{it_k}}(\boldsymbol{\omega}),\widehat{\overline{\boldsymbol{m}}}_{\boldsymbol{it_k}}(\boldsymbol{\omega}),\widehat{\overline{\boldsymbol{n}}}_{\boldsymbol{it_k}}(\boldsymbol{\omega})\right)^{\mathrm{T}}$ – the vector of the directional coefficients mathematical expectations estimates of the line of sight from the i-th OES an aircraft at the time moment t_k in ICCS; $\mathbf{A}_{ir}$ – transformation matrix for the i-th OES from ICCS to GCS having the form:

$$\mathbf{A}_{ir}=\left\|\begin{matrix}-\cos L_i\cdot\sin B_i & \cos L_i\cdot\cos B_i & -\sin L_i\\ -\sin L_i\cdot\sin B_i & \sin L_i\cdot\cos B_i & \cos L_i\\ \cos B_i & \sin B_i & 0\end{matrix}\right\|,$$

where B_i, L_i, H_i – geodetic latitude, longitude, and altitude of the i-th OES ICCS origin.

From Eq. (4), we have a system of two linear equations, which specifies the position of the lines of sight from the i-th OES to an aircraft in the GCS at the time moment t_k:

$$\begin{cases}\widehat{\overline{m}}_{it_k}(\omega)\left(x_{t_k}(\omega)-\widehat{\overline{x}}_{0ir}(\omega)\right)=\widehat{\overline{l}}_{it_k}(\omega)\left(y_{t_k}(\omega)-\widehat{\overline{y}}_{0ir}(\omega)\right);\\ \widehat{\overline{n}}_{it_k}(\omega)\left(y_{t_k}(\omega)-\widehat{\overline{y}}_{0ir}(\omega)\right)=\widehat{\overline{m}}_{it_k}(\omega)\left(z_{t_k}(\omega)-\widehat{\overline{z}}_{0ir}(\omega)\right),\end{cases}. \tag{5}$$

The system of line of sight Eq. (5) is underdetermined because it contains two equations and three unknown variables. To unambiguously define the system, it is supplemented with an equation that defines the slope range (distance) from the location point of the i-th OES in the GCS to the location point of the aircraft in the GCS at the time moment t_k. We present the slope distance equation in the form:

$$\widehat{\overline{D}}_{it_k}(\omega) = \sqrt{\left(x_{t_k}(\omega) - \widehat{\overline{x}}_{0ir}(\omega)\right)^2 + \left(y_{t_k}(\omega) - \widehat{\overline{y}}_{0ir}(\omega)\right)^2 + \left(z_{t_k}(\omega) - \widehat{\overline{z}}_{0ir}(\omega)\right)^2}, \quad (6)$$

where, as before, $\left(\widehat{\overline{x}}_{0ir}(\omega), \widehat{\overline{y}}_{0ir}(\omega), \widehat{\overline{z}}_{0ir}(\omega)\right)^{\mathrm{T}}$ is the mathematical expectations estimates vector of the i-th OES location coordinates in the GCS; $\widehat{\overline{D}}_{it_k}(\omega)$ – the mathematical expectation estimate of the sloped range from the i-th OES location point to the aircraft location point at the time moment t_k; $\left(x_{t_k}(\omega), y_{t_k}(\omega), z_{t_k}(\omega)\right)^{\mathrm{T}}$ – the unknown vector of the aircraft location coordinates in the GCS at the time moment t_k.

In this case, the system of line-of-sight and slant-range equations for the i-th OES contains three equations and three unknown variables:

$$\widehat{\overline{m}}_{it_k}(\omega)\left(x_{t_k}(\omega) - \widehat{\overline{x}}_{0ir}(\omega)\right) = \widehat{\overline{l}}_{it_k}(\omega)\left(y_{t_k}(\omega) - \widehat{\overline{y}}_{0ir}(\omega)\right);$$

$$\widehat{\overline{n}}_{it_k}(\omega)\left(y_{t_k}(\omega) - \widehat{\overline{y}}_{0ir}(\omega)\right) = \widehat{\overline{m}}_{it_k}(\omega)\left(z_{t_k}(\omega) - \widehat{\overline{z}}_{0ir}(\omega)\right);$$

$$\widehat{\overline{D}}_{it_k}(\omega) = \sqrt{\left(x_{t_k}(\omega) - \widehat{\overline{x}}_{0ir}(\omega)\right)^2 + \left(y_{t_k}(\omega) - \widehat{\overline{y}}_{0ir}(\omega)\right)^2 + \left(z_{t_k}(\omega) - \widehat{\overline{z}}_{0ir}(\omega)\right)^2}. \quad (7)$$

STEP 4. Formation of the overdetermined system of algebraic equations of sighting lines and oblique range from OES to the aircraft at the time moment t_k.

The formation of the overdetermined system of algebraic equations of sighting lines and oblique range from OES to the aircraft at the time moment t_k is carried out by forming three equations of the form (7) for each OES from the OES numbers list $\mathrm{I} = \{i, j, \ldots, l\}$, which at the moment of time carried out direct measurements of the aircraft location coordinates in the OES ISCS.

For this, data from the database is used in accordance with the algorithm discussed in STEP 3. The system of equations formed by this method has the form for the i-th OES:

$$\widehat{\overline{m}}_{it_k}(\omega)\left(x_{t_k}(\omega) - \widehat{\overline{x}}_{0ir}(\omega)\right) - \widehat{\overline{l}}_{it_k}(\omega)\left(y_{t_k}(\omega) - \widehat{\overline{y}}_{0ir}(\omega)\right) = 0;$$

$$\widehat{\overline{n}}_{it_k}(\omega)\left(y_{t_k}(\omega) - \widehat{\overline{y}}_{0ir}(\omega)\right) - \widehat{\overline{m}}_{it_k}(\omega)\left(z_{t_k}(\omega) - \widehat{\overline{z}}_{0ir}(\omega)\right) = 0,$$

$$\widehat{\overline{D}}_{it_k}(\omega)$$

$$-\sqrt{\left(x_{t_k}(\omega)-\widehat{\overline{x}}_{0ir}(\omega)\right)^2+\left(y_{t_k}(\omega)-\widehat{\overline{y}}_{0ir}(\omega)\right)^2+\left(z_{t_k}(\omega)-\widehat{\overline{z}}_{0ir}(\omega)\right)^2}=0; \quad (8)$$

The system of Eq. (8) is the most complete for the autonomous processing of the aircraft location observations results by the OES network at the time moment t_k and significantly overdetermined.

3.3 *Methods of Solving the Overdetermined System of Algebraic Equations of the Results of Observations of the Trajectory of the Aircraft Movement by the OES Sensor Network*

If $|I| = n$, then the system of Eq. (8) contains $3n$ algebraic equations and three unknown variables, that is, it is significantly overdetermined, and its solution $\left(x^*_{t_k}(\omega), y^*_{t_k}(\omega), z^*_{t_k}(\omega)\right)^{\mathrm{T}}$ should be considered as unbiased and most effective mathematical expectations estimates of the aircraft location coordinates $\left(\widehat{\overline{x}}_{t_k}(\omega), \widehat{\overline{y}}_{t_k}(\omega), \widehat{\overline{z}}_{t_k}(\omega)\right)^{\mathrm{T}}$ in the GCS at the time moment t_k. In real-time systems, the most effective methods for solving a system of algebraic equations are exact analytical methods compared to numerical methods. Numerical methods of solving a system of algebraic equations are used in those cases when the systems do not have an exact analytical solution, and the conditions of the problem allow the use of approximate numerical solutions. Therefore, in the following sections, we will consider both exact analytical methods for some subsystems of the system of Eq. (8), and approximate numerical methods for its solution.

Analytical methods for solving a determined system of algebraic equations of the observations results of the aircraft trajectory Analytical methods of solving a system of algebraic equations are used for certain systems in which the number of equations coincides with the number of independent variables. The simplest case is the case when, at the time moment t_k only one, for example, the i-th OES of the network received estimates of the results of measurements: azimuth – $\widehat{\overline{\alpha}}_{it}(\omega)$; elevation angle – $\widehat{\overline{\beta}}_{it}(\omega)$ and slant distance – $\widehat{\overline{D}}_{it}(\omega)$. In this case, the equations of the i-th OES lines of sight to the aircraft at the time moment t_k and slant distance in the ICCS have the form:

$$\frac{x_{t_k}(\omega)}{\hat{\bar{\boldsymbol{l}}}_{it_k}(\boldsymbol{\omega})}=\frac{y_{t_k}(\omega)}{\hat{\bar{\boldsymbol{m}}}_{it_k}(\boldsymbol{\omega})}=\frac{z_{t_k}(\omega)}{\hat{\bar{\boldsymbol{n}}}_{it_k}(\boldsymbol{\omega})}; \tag{9}$$

$$\hat{\bar{D}}_{it_k}(\omega)=\sqrt{x_{t_k}^2(\omega)+y_{t_k}^2(\omega)+z_{t_k}^2(\omega)}$$

and the system of Eq. (9) in the ICCS takes the form:

$$\hat{\bar{\boldsymbol{m}}}_{it_k}(\boldsymbol{\omega})x_{t_k}(\omega)-\hat{\bar{\boldsymbol{l}}}_{it_k}(\boldsymbol{\omega})y_{t_k}(\omega)=0;$$

$$\hat{\bar{\boldsymbol{n}}}_{it_k}(\boldsymbol{\omega})y_{t_k}(\omega)-\hat{\bar{\boldsymbol{m}}}_{it_k}(\boldsymbol{\omega})z_{t_k}(\omega)=0;$$

$$\hat{\bar{D}}_{it_k}(\omega)-\sqrt{x_{t_k}^2(\omega)+y_{t_k}^2(\omega)+z_{t_k}^2(\omega)}=0. \tag{10}$$

It can be proven that the system of Eq. (10) has such an analytical solution:

$$x_{it_k}^*(\omega)=\hat{\bar{D}}_{it_k}(\omega)\cos\hat{\bar{\beta}}_{it_k}(\omega)\cos\hat{\bar{\alpha}}_{it_k}(\omega), \tag{11}$$

$$y_{it_k}^*(\omega)=\hat{\bar{D}}_{it_k}(\omega)\sin\hat{\bar{\beta}}_{it_k}(\omega), \tag{12}$$

$$z_{it_k}^*(\omega)=\hat{\bar{D}}_{it_k}(\omega)\cos\hat{\bar{\beta}}_{it_k}(\omega)\sin\hat{\bar{\alpha}}_{it_k}(\omega). \tag{13}$$

The solution vector of the system of equations $\left(x_{t_k}^*(\omega), y_{t_k}^*(\omega), z_{t_k}^*(\omega)\right)^{\mathrm{T}}$ should be considered as the unbiased and most effective mathematical expectations estimates of the aircraft location coordinates $\left(\hat{\bar{x}}_{t_k}(\omega), \hat{\bar{y}}_{t_k}(\omega), \hat{\bar{z}}_{t_k}(\omega)\right)^{\mathrm{T}}$ in the ICCS at the time moment t_k. Let us now consider the complete system of the equation of lines of sight and the oblique distance of the i-th OES to the aircraft at the time moment t_k в GCS:

$$\hat{\bar{m}}_{it_k}(\omega)\left(x_{t_k}(\omega)-\hat{\bar{x}}_{0ir}(\omega)\right)-\hat{\bar{l}}_{it_k}(\omega)\left(y_{t_k}(\omega)-\hat{\bar{y}}_{0ir}(\omega)\right)=0;$$

$$\hat{\bar{n}}_{it_k}(\omega)\left(y_{t_k}(\omega)-\hat{\bar{y}}_{0ir}(\omega)\right)-\hat{\bar{m}}_{it_k}(\omega)\left(z_{t_k}(\omega)-\hat{\bar{z}}_{0ir}(\omega)\right)=0,$$

$$\begin{aligned}&\hat{\bar{D}}_{it_k}(\omega)\\&-\sqrt{\left(x_{t_k}(\omega)-\hat{\bar{x}}_{0ir}(\omega)\right)^2+\left(y_{t_k}(\omega)-\hat{\bar{y}}_{0ir}(\omega)\right)^2+\left(z_{t_k}(\omega)-\hat{\bar{z}}_{0ir}(\omega)\right)^2}\\&=0.\end{aligned} \tag{14}$$

It can be proved that the system of Eq. (14) has such an analytical solution:

$$\begin{Vmatrix} x^*_{it_k}(\omega) \\ y^*_{it_k}(\omega) \\ z^*_{it_k}(\omega) \end{Vmatrix} = \mathbf{A}_{ir}(\boldsymbol{\omega}) \cdot \begin{Vmatrix} \widehat{\overline{x}}_{it_k}(\omega) \\ \widehat{\overline{y}}_{it_k}(\omega) \\ \widehat{\overline{z}}_{it_k}(\omega) \end{Vmatrix} + \begin{Vmatrix} \widehat{\overline{x}}_{0ir}(\omega) \\ \widehat{\overline{y}}_{0ir}(\omega) \\ \widehat{\overline{z}}_{0ir}(\omega) \end{Vmatrix}, \tag{15}$$

where $\left(\widehat{\overline{x}}_{it_k}(\omega), \widehat{\overline{y}}_{it_k}(\omega), \widehat{\overline{z}}_{it_k}(\omega)\right)^{\mathrm{T}}$ is a vector of the aircraft location coordinates in the ICCS at the time moment t_k:

$$\widehat{\overline{x}}_{it_k}(\omega) = \widehat{\overline{D}}_{it_k}(\omega) \cos \widehat{\overline{\beta}}_{it_k}(\omega) \cos \widehat{\overline{\alpha}}_{it_k}(\omega), \tag{16}$$

$$\widehat{\overline{y}}_{it_k}(\omega) = \widehat{\overline{D}}_{it_k}(\omega) \sin \widehat{\overline{\beta}}_{it_k}(\omega), \tag{17}$$

$$\widehat{\overline{z}}_{it_k}(\omega) = \widehat{\overline{D}}_{it_k}(\omega) \cos \widehat{\overline{\beta}}_{it_k}(\omega) \sin \widehat{\overline{\alpha}}_{it_k}(\omega). \tag{18}$$

Expression (15) is an affine transformation of the vector of mathematical expectations estimates of the aircraft location coordinates in the ICCS of the i-th OES into the vector of mathematical expectations estimates of the aircraft location coordinates $\left(x^*_{t_k}(\omega), y^*_{t_k}(\omega), z^*_{t_k}(\omega)\right)^{\mathrm{T}}$ in the GCS, given by the 3×3 transformation (rotations) matrix $\mathbf{A}_{ir}$ with a nonzero determinant and a transfer vector $\left(\widehat{\overline{x}}_{0ir}(\omega), \widehat{\overline{y}}_{0ir}(\omega), \widehat{\overline{z}}_{0ir}(\omega)\right)^{\mathrm{T}}$.

Analytical methods for solving the system of Eq. (15) are used for autonomous results processing of the aircraft trajectory observations by the OES sensor network in real time. Since the aircraft at the time moment t_k is located at different distances from each i-th OES, the estimates $\left(x^*_{it_k}(\omega), y^*_{it_k}(\omega), z^*_{it_k}(\omega)\right)^{\mathrm{T}}$, $i = 1,2,\ldots, n$ must be considered as the results of indirect not equal precision measurements. In this case, a weighted average should be used as an estimate of the mathematical expectation of the coordinates of the aircraft location in the GCS at a point in time. This value is also used as estimation of mathematical expectations in the GCS aircraft location coordinates. Taking into account the spatial redundancy of measurements of aircraft location coordinates by the OES sensor network, it becomes possible to get the estimation which are less biased and more efficient compared to the estimates obtained from each i-th OES separately.

4 Research Results

To confirm the model operation and analyze the accuracy of determining the aircraft position depending on the distance to the object and the number of OES participating in the observation, a network of OES was formed. OESs are located in groups of 8 stations around the aircraft at equal distances. The groups are located at distances

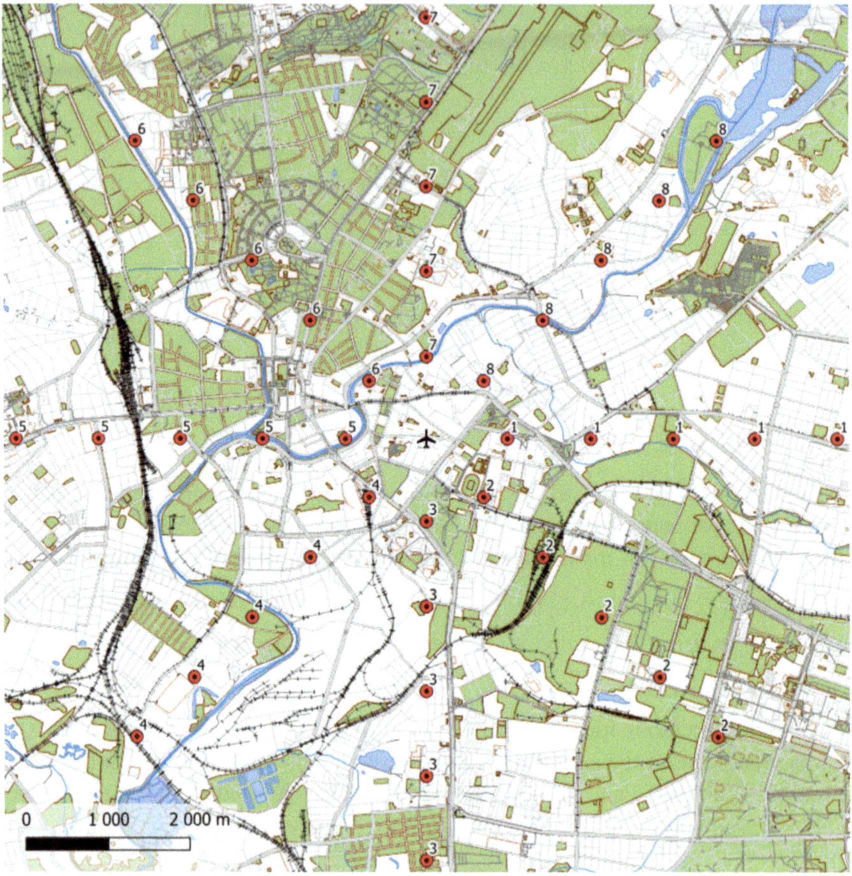

Fig. 2 Positions of OES equidistant from the aircraft

from 500 to 5000 m in increments of 500 m (Fig. 2). The station numbers indicate the order in which they are added to the aircraft's position calculations.

The calculation of the aircraft's position was carried out by two methods: hidden (Method α, β) and explicit (Method α, β, D) video surveillance of the aircraft by solving overdetermined systems of algebraic equations. The instrumental error in measuring the location coordinates of each *i*-th OES in WGS-84, the error in measuring azimuth, elevation angle, and slant distance from the OES to the aircraft were modeled as normally distributed random variables. The standard deviation of the OES location coordinates and the slant distance from the OES to the aircraft was taken equal to 0.5 m, for α, β the standard deviation was equal to 0.002 rad. Figures 3, 4, and 5 show the values of mathematical expectations and the standard deviation of errors in estimating the coordinates of the aircraft and the standard deviation of the methods of hidden (Method α, β) and explicit (Method α, β, D) video surveillance of aircraft by the OES infocommunication network.

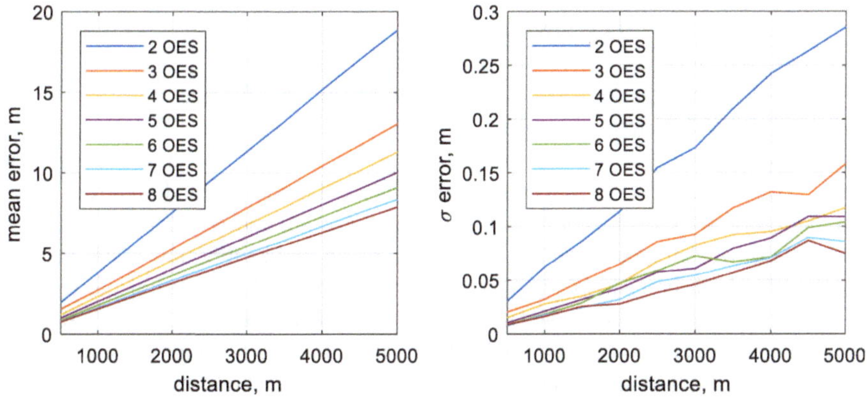

Fig. 3 Method α, β

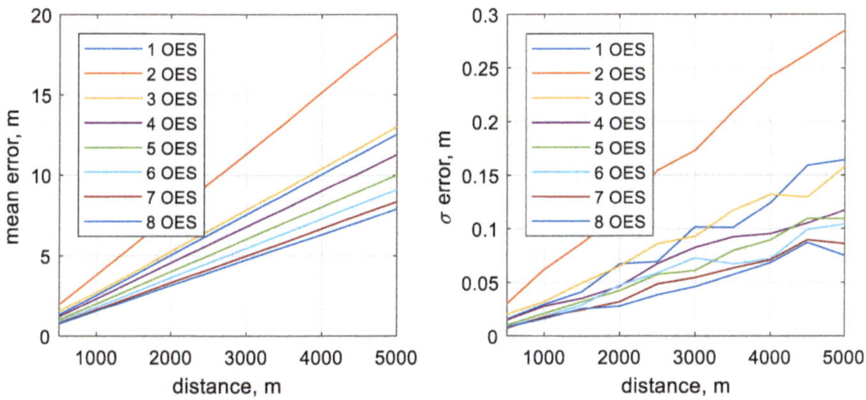

Fig. 4 Method α, β, D

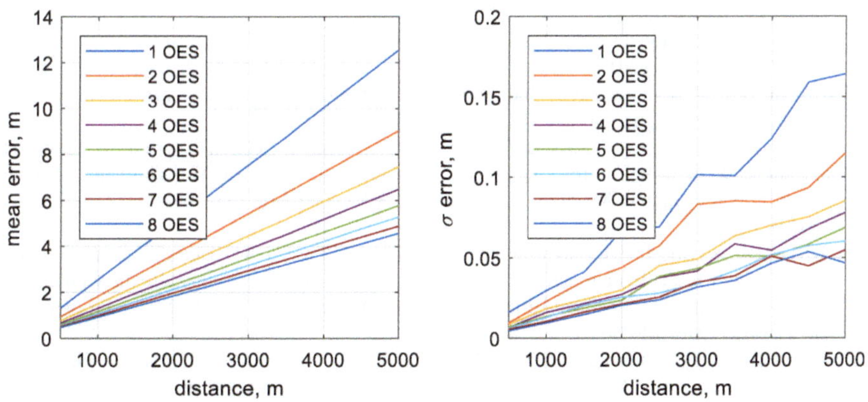

Fig. 5 Method average α, β, D

5 Conclusions

Mathematical models of open and covert video surveillance of aircrafts by infocommunication network of optoelectronic stations are given. The models take into account systematic errors in the leveling of the OES platform and the alignment of the axes of the OES coordinate system to the local meridian, random errors in the estimation of the OES location coordinates in WGS-84, and errors in the measurement of the azimuth, elevation angle, and slant range of the OES view to the aircraft. Methods of high-precision estimation of the parameters of aircraft trajectories as a result of solving overdetermined systems of linear and nonlinear systems of algebraic equations are considered. It is shown that the use of temporal and spatial redundancy allows to significantly increase the efficiency and reduce the bias of the estimations of the aircraft trajectories parameters.

References

1. Høye, G.: Analyses of the geolocation accuracy that can be obtained from shipborne sensors by use of time difference of arrival (TDOA), scanphase, and angle of arrival (AOA) measurements. Forsvarets forskningsinstitutt. **178** (2010). https://www.ffi.no/en/publications-archive/analyses-of-the-geolocation-accuracy-that-can-be-obtained-from-shipborne-sensors-by-use-of-time-difference-of-arrival-tdoa-scanphase-and-angle-of-arrival-aoa-measurements
2. Zekavat, R.: Buehrer Handbook of Position Location: Theory, Practice and Advances, p. 1376. John Wiley and Sons, Hoboken (2019)
3. Kireev, A., Fedorenko, R., Fokin, G.: Accuracy evaluation of positioning by Cramer-Rao bound. Proc. Educ. Inst. Commun. **3**(2), 77–83 (2017)
4. Lazarev, V., Fokin, G.: Positioning accuracy evaluation of radio emission sources using time difference of arrival and angle of arrival methods. Part 1. Proc. Telecommun. Univ. **5**(2), 88–100 (2019)
5. Fokin, G., Lazarev, V.: Positioning accuracy evaluation of radio emission sources using time difference of arrival and angle of arrival methods. Part 2. 2D-simulation. Proc. Telecommun. Univ. **5**(4), 65–78 (2019)
6. Dodonov, A.G., Putyatin, V.G.: Radio-technical means of external vector measurements. Math. Mach. Syst. **1**, 3–30 (2018)
7. Putjatin, V.G., Dadonov, A.G.: One problem of high-precision trajectory measurements with optical facilities. Regist. Storage Data Process. **19**(2), 36–54 (2017)
8. Shostko, I., Tevyashev, A., Kulia, Y., Koliadin, A.: Optical-electronic system of automatic detection and high-precise tracking of aerial objects in real-time. In: The Third International Workshop on Computer Modeling and Intelligent Systems, CMIS, pp. 784–803, Zaporizhzhia (2020)
9. Tevjashev, A., Shostko, I., Neofitnyi, M., Koliadin, A.: Laser opto-electronic airspace monitoring system in the visible and infrared ranges. In: 2019 IEEE 5th International Conference Actual Problems of Unmanned Aerial Vehicles Developments, APUAVD 2019 – Proceedings, pp. 170–173, Kiev (2019)
10. Tevjashev, A.D., Shostko, I.S., Neofitnyi, M.V., Kolomiyets, S.V., Kyrychenko, I.Y., Pryimachov Yu, D.: Mathematical model and method of optimal placement of optical-electronic systems for trajectory measurements of air objects at test. Odessa Astron. Publ. **32**, 171–175 (2019)

11. Shostko, I., Tevyashev, A., Neofitnyi, M., Kulia, Y.: Information-measuring system of polygon based on wireless sensor infocommunication network. In: Chapter in the Book Lecture Notes on Data Engineering and Communications Technologies, vol. 48, pp. 649–674. Springer Nature, Cham (2021)
12. Shostko, I., Tevyashev, A., Neofitnyi, M., Ageyev, D., Gulak, S.: Information and measurement system based on wireless sensory infocommunication network for polygon testing of guided and unguided rockets and missiles. In: 2018 International Scientific-Practical Conference on Problems of Infocommunications Science and Technology, PIC S and T 2018 – Proceedings, pp. 705–710, Kharkiv (2019)
13. Lysenko, I.V.: A method for automated operational leveling control of radio engineering facilities of trajectory measurements. Transp. Bus. Russia. **6**, 45–50 (2016)
14. Stoyan, Y., Romanova, T., Pankratov, O., Tevyashev, A.: Lattice coverage of cuboid with minimum number of hemispheres. Cybern. Syst. Anal. **58**(4), 542–551 (2022)
15. Puk, J., Habarov, E.: Ground-based optical-electronic and quantum-optical facilities of the European Space Control System. Foreign Mil. Rev. **2**(9), 69–74 (2016)
16. Chupahin, A.P., Savin, M.L.: Features of the construction of optoelectronic recording systems for trajectory measurements. Izvestia TulGU. Ser. Tech. Sci. **11**(2), 273–279 (2014)
17. Zhdanyuk V.F.: Fundamentals of Statistical Processing of Trajectory Measurements, 384 p. Soviet Radio, Moscow (1978)
18. Shapiro L., Stockman D.: Computer Vision, p. 580. Publishing house "BINOM. Knowledge Laboratory" (2015)

German-Ukrainian Research and Training Center for Parallel Simulation Technology

Artem Liubymov, Volodymyr Svyatnyy, and Oleksandr Miroshkin

1 Introduction

Computing systems have emerged as indispensable universal tools in various fields, including mechanics, electrical engineering, mining, electronics, chemistry, biology, molecular dynamics, medicine, biomechanics, and astronomy. These advancements have been made possible through the development of mathematics and computer science.

The evolution of computing systems has brought us to a point where recent generations of computer processors are nearing physical limits in microcircuit production. Further progress is contingent upon substantial changes in the material foundation used and the principles governing information storage and processing. Scientists have contemplated alternative computing systems like quantum and biological computing, but these are not yet ready for widespread implementation.

However, the evolution of widely used computer systems has its unique challenges. In the progression of computer technology, there is a notable issue of imbalance between the development of computing power and network resources. This imbalance is particularly prominent in distributed (network) computing systems, where information exchange processes consume a significant portion of time. In the context of network systems, the term "computer" has evolved to encompass new roles in the era of Industry 4.0, including functions such as monitoring, data synchronization, storage, and security. Since data exchange speed among these

A. Liubymov (✉) · V. Svyatnyy
Donetsk National Technical University, Lutsk, Volyn, Ukraine
e-mail: artemliubymov@gmail.com; vsvjatnyj@gmail.com

O. Miroshkin
Ulm University, Ulm, Germany
e-mail: oleksandr.miroshkin@uni-ulm.de

M. N. Seyman (ed.), *2nd International Congress of Electrical and Computer Engineering*, EAI/Springer Innovations in Communication and Computing,
https://doi.org/10.1007/978-3-031-52760-9_22

network agents is considerably lower than the processing speed of their computing resources, it results in a reduction in the potential performance of both computing devices and the entire distributed system. This aspect is crucial and considered in the design of high-performance parallel computer systems and the corresponding software algorithms.

Despite these complexities, some computing systems theoretically possess the resources required to address intricate scientific and engineering challenges. However, realizing this potential hinges on meeting other essential conditions, including a sufficient level of user expertise.

For routine office tasks or web browsing, acquiring the necessary skills is relatively straightforward, often taking only a few hours. In contrast, mastering the use of complex high-performance computing resources for tasks like setting up and managing a parallel computer cluster for scientific and technical computations or simulating complex dynamic processes with various programming languages and mathematical libraries is a multi-year learning process, even with suitable training materials.

Addressing the dissemination of specialized knowledge becomes crucial in fields where dedicated organizations are lacking. An example is the simulation of processes in complex dynamic systems. Proficiency in this domain can typically be obtained only through access to data centers dealing with similar problems. Practical seminars and workshops offer an alternative path to acquiring knowledge but are limited in scope, providing overviews and solutions to a limited range of problems.

To tackle the challenge of disseminating specialized knowledge and skills in high-performance computing, particularly in the simulation of complex dynamic systems, the Research and Training Center for Parallel Simulation Technology has been established. The Center possesses specialized hardware resources, enabling the installation of high-performance computer clusters for resource-intensive scientific and technical tasks. Additionally, there are plans to implement a distributed parallel simulation environment, with theoretical groundwork already completed over an extended period.

Within the context of international collaboration, the Center is set to facilitate extensive educational programs comprising extended training courses, as well as brief seminars and workshops, with the aim of fostering the exchange of knowledge between foreign experts and the Center's users. Furthermore, there is a concerted effort to facilitate the transfer of educational resources between the staff of the Training Center and HLRS, with appropriate permissions granted for subsequent utilization. Through the dedicated endeavors of the Center's personnel, their accrued expertise will be systematically structured into educational modules, instructional materials, and training manuals.

2 Research and Development Activity

2.1 Problem-Oriented Distributed Parallel Simulation Environment

General simulation approach. Innovative projects of modern subject areas are related to the development, research, and implementation of complex dynamic systems (CDS), which despite the different nature of physical processes have the same formal description structure: equations of dynamic processes and means of topologies representation [1–3].

For simulation reasons, a distinction between dynamic systems with lumped (DSLP) and spatially distributed (DSDP) parameters is made. Research and development practices show that the DSLP and DSDP are often investigated as the same objects: usually an object is first analyzed as an object with distributed parameters, then further investigations are carried out through an approximation to a certain number of objects with lumped parameters and compliance with the appropriate approximation conditions.

An analysis shows that despite the wide physical diversity of dynamic systems in various subject areas, their topologies are described with the relatively limited number of representation means: process flow diagrams (PFD), structure diagrams of automation systems (SDAS), dynamic network objects (DNO) represented by graphs, and the secondary topologies (ST) that arise during the approximation of DSDP (Figs. 1 and 2).

A Distributed Parallel Simulation Environment (DPSE) is proposed, developed, and implemented for processing complex DSLP and DSDP simulation tasks. A functionality of the environment is determined by the development stages of models, Simulation models and parallel simulators of DSLP, DSDP, and their research using the developed simulators (Simulation). These stages are shown in Figs. 3 and 4 with a focus on complex dynamical systems of three main topologies. Functionality is provided by the HW/SW-resources of the DPSE structure. Experience shows that a development of problem-oriented DPSE remains a current challenge in various subject areas.

The following solutions have scientific and practical significance: specification languages of DSLP and DSDP description on a technological level; a formal description of topologies and physical processes, development of algorithms for topology analysis and equation generation; the approaches to parallelization; development of virtual parallel simulation models based on the block numerical methods (BNM) and their theoretical analysis; devirtualization and process allocation on parallel target computer systems (TCS); adaptation of numerical procedures and construction of parallel equation solvers; development of effective subsystems; analysis and optimization of the parallel simulators efficiency; integration with subject area-specific CASE tools; model-driven development, project planning, and process management; friendly user interface organization of parallel resources usage.

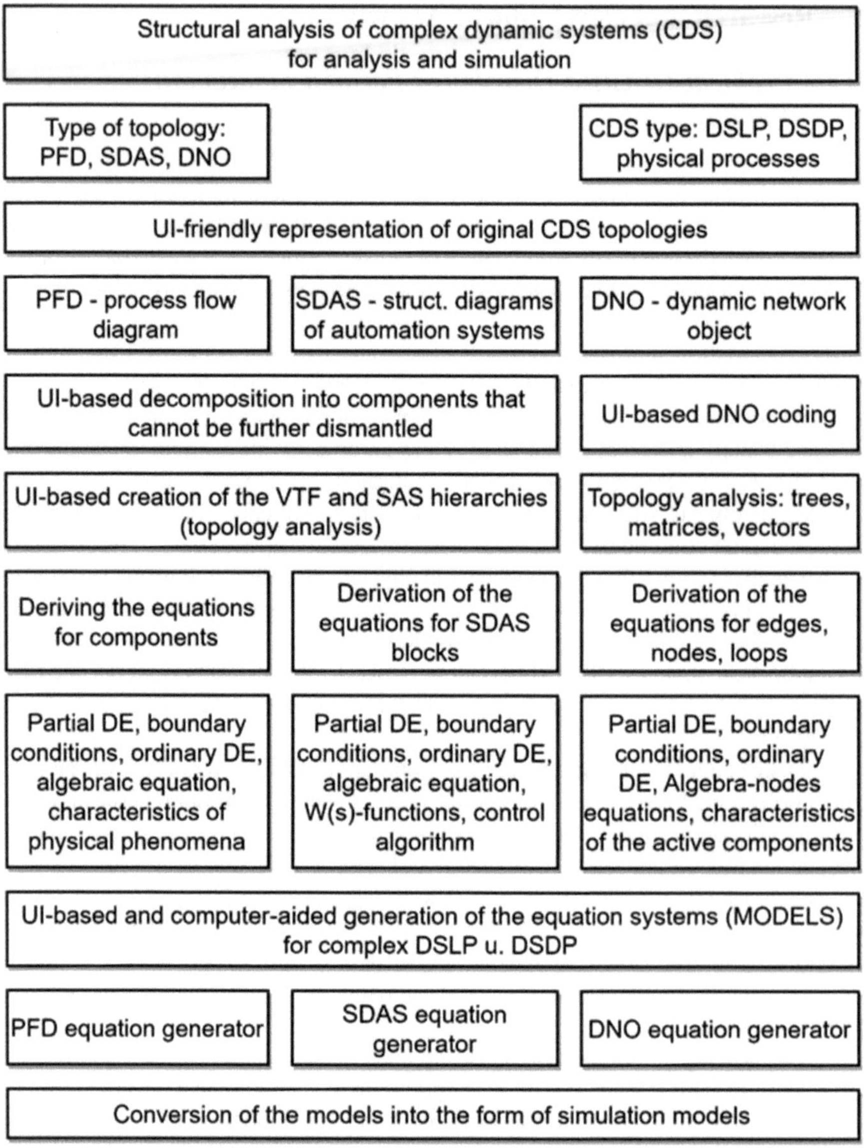

Fig. 1 Stages of the computer-aided development of DSLP and DSDP simulation models

Simulation environment for dynamic network objects (DNO topology). A dynamic technical and procedural networks are objects of investigation, project planning, automation, monitoring, quality assurance, optimal process management, safety analysis and forecasting, avoidance of safety-critical operating conditions and liquidation of accidents in several subject areas. Network objects belong to a class of complex (often safety-critical) dynamic systems.

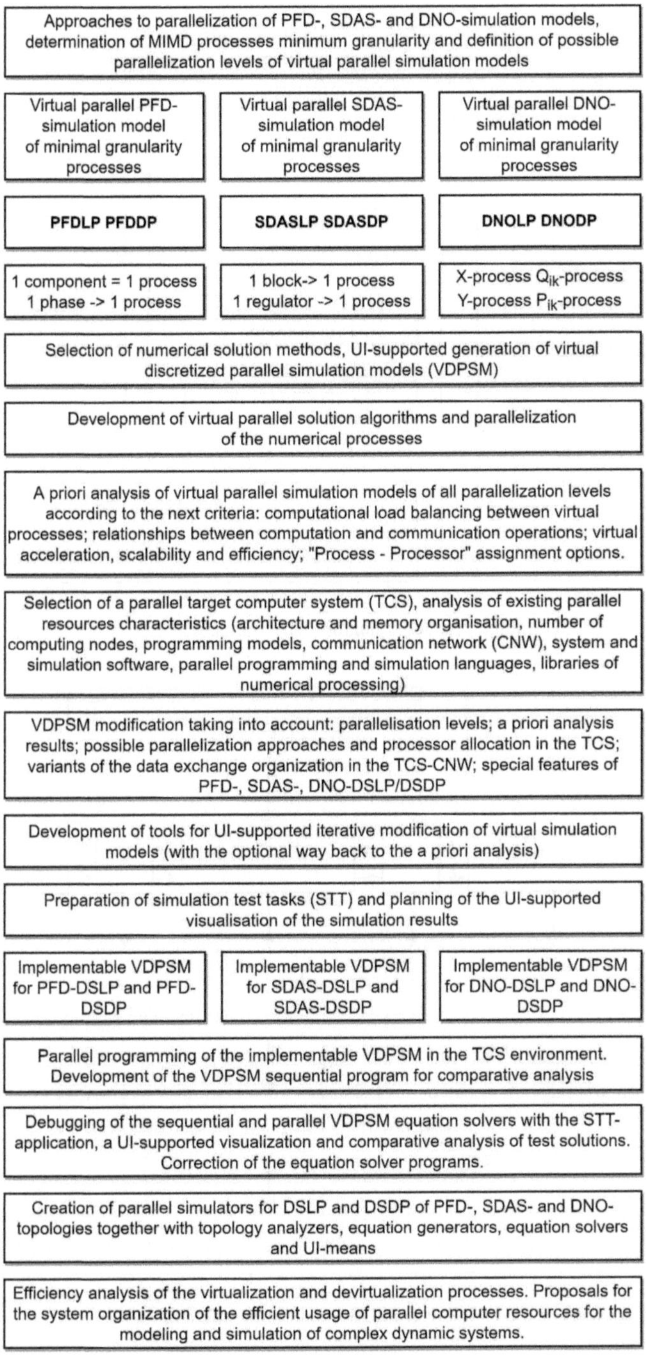

Fig. 2 Stages of parallel CDS modeling and computer-aided development of parallel simulators

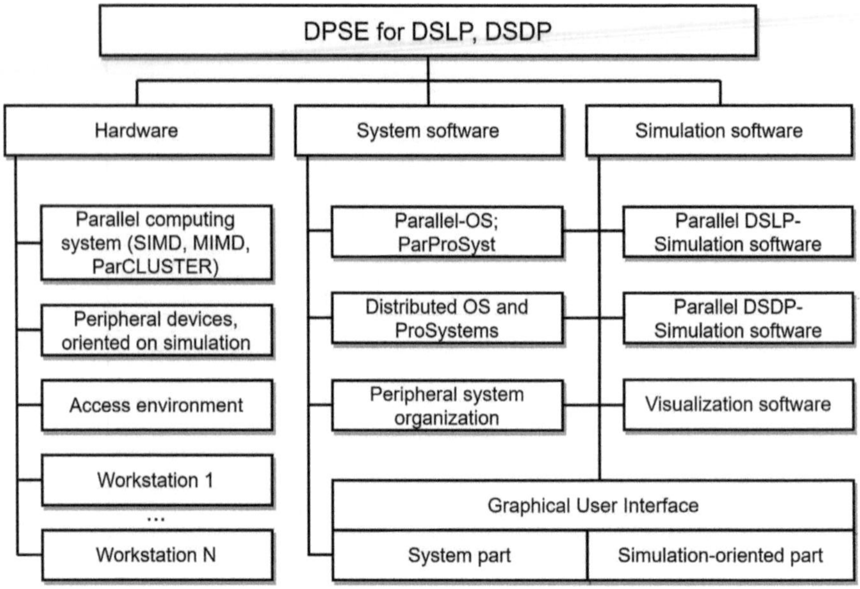

Fig. 3 Structure of distributed parallel simulation environment

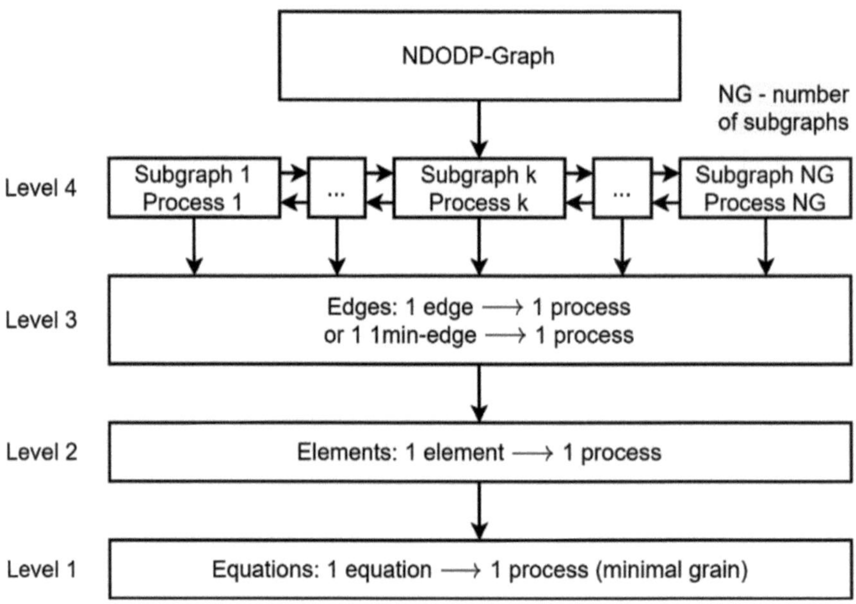

Fig. 4 Virtual levels of the DNODP simulation model parallelization

Non-linearity of the process-describing functions, spatial distribution of process parameters, large network dimensions (number of edges $m > 100$, number of nodes $n > 50$), as well as several active elements with non-linear current-dependent characteristics, significant multiple and hierarchical interaction of the controllable process parameters as well as simultaneous influences of deterministic and stochastic disturbances are the main features of this complexity.

However, only a few, very simplified tasks of dynamic network objects can be solved analytically. That is why the methods and means of modeling and simulation of this object class are of increasing theoretical and practical importance, both during project planning and during operation. As a DNODP example, we consider a mine ventilation network. The object is described topologically as a graph $G(m, n)$ and encoded with m rows and $s + 5$ columns, where s is the number of object parameters.

Model of the dynamic processes in the j-th edge is described by the next equations:

$$\begin{cases} -\dfrac{\partial P_j}{\partial \xi} = r_j Q_j^2 + \dfrac{\rho}{F_j}\dfrac{\partial Q_j}{\partial t} + r_j(\xi_r, t) Q_j^2; \\ \dfrac{\partial P_j}{\partial t} = \dfrac{\rho a^2}{F_j}\dfrac{\partial Q_j}{\partial \xi}, \end{cases} \tag{1}$$

where P_j and Q_j are, respectively, an air pressure and an air flow along the ξ coordinate, that is calculated from the AK_j toward the EK_j node; r_j is a specific aerodynamic resistance; F_j is the cross-sectional area of the j-th edge (air path); ρ is the air density; a is the speed of a sound in the air; $r_j(\xi_r, t)$ is an adjustable aerodynamic resistance; ξ_r is the location coordinate of the adjustable resistor (e.g., a slide).

Boundary conditions for Eq. (1) are the pressure functions $P(AK_j)$, $P(EK_j)$ at the ends of the j-th edge. There are three types of edges and nodes according to boundary conditions in DNODP:

1. The edges, that are adjacent to the inner n_1 DNO nodes; here, the pressure values P_{wi} during the solving of the DNODP equation system correspond to the dynamic node conditions:

$$-\frac{\partial P_{wi}}{\partial t} = \frac{\rho a^2}{F_{wi}} \frac{\partial Q_{wi}}{\partial \xi} \tag{2}$$

2. The edges that are adjacent to the n_2 nodes of the fan connections; here, the pressure P_{wi} is specified as the fan (edge's active element) characteristic:

$$P_{wi} = P_{AE_j}(Q_j) \tag{3}$$

3. The edges that are adjacent to the n_3 nodes that are connected to the atmosphere:

$$P_{wi} = P_{\text{ATM}} = \text{const.} \tag{4}$$

The DNODP has a total of

$$n = n_1 + n_2 + n_3 \tag{5}$$

nodes and accordingly n boundary conditions. The initial conditions are as follows:

$$P_j(\xi, 0), Q_j(\xi, 0)\ (j = 1, 2, \ldots, M) \tag{6}$$

Problem Definition For DNODP, whose graph is encoded with the topological Table 1 and each edge of which is described with the equation systems (1), boundary conditions (2), (3), (4), the initial condition (6), the parallel software simulation tools are developed and implemented that adequately reflect the dynamic processes $P_j(\xi, t)$, $Q_j(\xi, t)$, $(j = 1, 2, \ldots, M)$ with the consideration of defined working conditions, disturbances and regulations of air flows.

By approximating the Eq. (1) using the linear method with the local increment $\Delta\xi$, we get the system of equations for the k-th element of the j-th edge:

$$\begin{cases} \dot{Q}_{j,k} = \alpha_j (P_{j,k} - P_{j,k+1}) - \beta_j Q_{j,k} |Q_{j,k}| - \beta_{r,j} Q_{j,k} |Q_{j,k}|; \\ \dot{P}_{j,k+1} = \gamma_j (Q_{j,k} - Q_{j,k+1}). \end{cases} \tag{7}$$

The α_j, β_j, $\beta_{r,\,j}$, γ_j are the coefficients dependent on aerodynamic parameters of the j-th edge.

The inner boundary conditions of type (2) are represented with the next equation:

$$-\frac{dP_{wi}}{dt} = \frac{\rho a^2}{F_{wi}} \frac{Q_{j.k} - \sum\limits_{jwi} (Q_{jwi1} - Q_{jwiMj})}{\Delta\xi_{j,k}} \tag{8}$$

Equation (8) uses the following designations: $P_{wi} = P_{jMj\,+\,1}$ is the air pressure at the end node of the last element $Q_{jk} = Q_{jMj}$ is the air flow in the j-th edge, that is directed into the wi-th node; $jwiMj$ – the edge indices from set $j = 1, 2, \ldots, M$, the wi-th node is adjacent to; wherein $jwi1$ is the first element of j-th edge with the wi-th as the initial node (outflow) and $jwiMj$ is the last element with the wi-th as end node (inflow). Each edge gives Mj elements $Q_{j,\,1}, \ldots, Q_{j,\,Mj}$ after the spatial approximation with the step $\Delta\xi$. The numbering $P_{j,\,1}$, $P_{j,\,2}, \ldots, P_{j,\,Mj\,+\,1}$ of pressure values is used. It is important to mention that j-th edge is the initial node wi with the pressure $P_{wi} = P_{j,\,1}$ and the end node $wi + b$ ($b = \text{const}$) with the pressure $P_{wi\,+\,b} = P_{j,\,Mj\,+\,1}$. For dynamic network objects with distributed parameters, each edge is represented by two vectors according to the above approximation Q_j, P_j, $(j = 1, 2\ldots, M)$:

$$Q_j = \left(Q_{j,1} \;\; Q_{j,2} \;\; \cdots \;\; Q_{j,Mj} \right)^T \tag{9}$$

is the air flow in j-th edge and

$$P_j = \left(P_{j,1} \quad P_{j,2} \quad \ldots \quad P_{j,Mj+1} \right)^T \tag{10}$$

is the air pressure in the j-th edge. Thereby, the value M_j is calculated as the amount of elements in the j-th edge depending on the edge lengths l_j with the same local spatial coordinate increment $\Delta\xi$ for whole DNODP as $M_j = l_j/\Delta\xi$.

When developing the spatially discretized simulation model of the entire network object, M equation systems of type (7) for all edges is required, d. h. ($j = 1, 2..., M$):

$$\begin{cases} \begin{cases} \dot{Q}_{1,k} = \alpha_1 (P_{1,k} - P_{1,k+1}) - \beta_1 Q_{1,k} |Q_{1,k}| - \beta_{r,1} Q_{1,k} |Q_{1,k}|; \\ \dot{P}_{1,k+1} = \gamma_1 (Q_{1,k} - Q_{1,k+1}); \\ k = 1, 2, \ldots, M_1; \end{cases} \\ \vdots \\ \begin{cases} \dot{Q}_{m,k} = \alpha_m (P_{m,k} - P_{m,k+1}) - \beta_m Q_{m,k} |Q_{m,k}| - \beta_{r,m} Q_{m,k} |Q_{m,k}|; \\ \dot{P}_{m,k+1} = \gamma_m (Q_{m,k} - Q_{m,k+1}); \\ k = 1, 2, \ldots, M_m. \end{cases} \end{cases} \tag{11}$$

According to the Eq. (8), we formulate n_1 boundary conditions for internal nodes of the mesh object ($wi = 1, 2, ..., n_1$):

$$\begin{aligned} &-\frac{dP_{w1}}{dt} = \frac{\rho a^2}{F_{w1}} \frac{Q_{j.k} - \sum_{jw1} \left(Q_{jw11} - Q_{jw1Mj} \right)}{\Delta\xi_{j,k}}; \\ &\vdots \\ &-\frac{dP_{wn1}}{dt} = \frac{\rho a^2}{F_{wn1}} \frac{Q_{j.k} - \sum_{jw1} \left(Q_{jwn11} - Q_{jwn1Mj} \right)}{\Delta\xi_{j,k}}. \end{aligned} \tag{12}$$

According to the Eq. (3), the fan characteristics form further n_2 boundary conditions:

$$\begin{aligned} &P_{wi} = P^1_{AEJ}(QJ); \\ &\vdots \\ &P_{wi} = P^{n2}_{AEJ}(QJ). \end{aligned} \tag{13}$$

Active elements (fans) are in (13) from 1 to n_2 additionally numbered. For the n_3 nodes of the atmosphere connections, the boundary conditions according to the Eq. (4) look like:

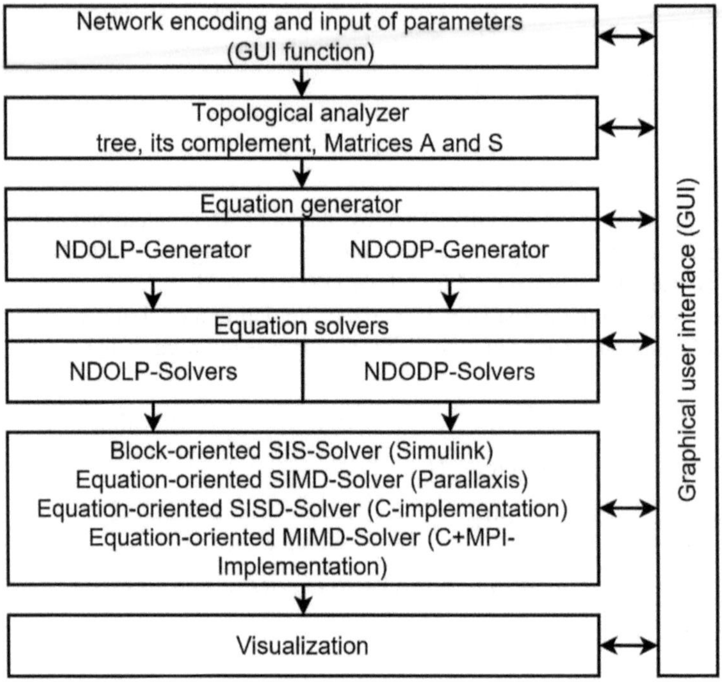

Fig. 5 Structure of a parallel DNOLP/DNODP simulator

$$
\begin{aligned}
&P_{wi} = P_{(n1+n2)+1} = P^{1}_{\mathrm{ATM}} = \mathrm{const};\\
&\vdots\\
&P_{wi} = P_{n} = P^{n3}_{\mathrm{ATM}} = \mathrm{const}.
\end{aligned}
\tag{14}
$$

Equation (11) together with the boundary conditions (12), (13) and (14) form the spatially discretized DNODP simulation model. For the industry-related DNODP ($m \geq 1000$, $n \geq 300$, $M\_j \geq 50$), the computer-aided creation of the DNODP simulation models is currently realized with the help of the topology analyzer and the equation generator. The parallelization approaches give four possible parallelism levels of the virtual parallel simulation model (Fig. 6). A parallel DNOLP/DNODP simulator (Fig. 5) is developed and implemented. Simulator tests are performed on test network object models (Fig. 6).

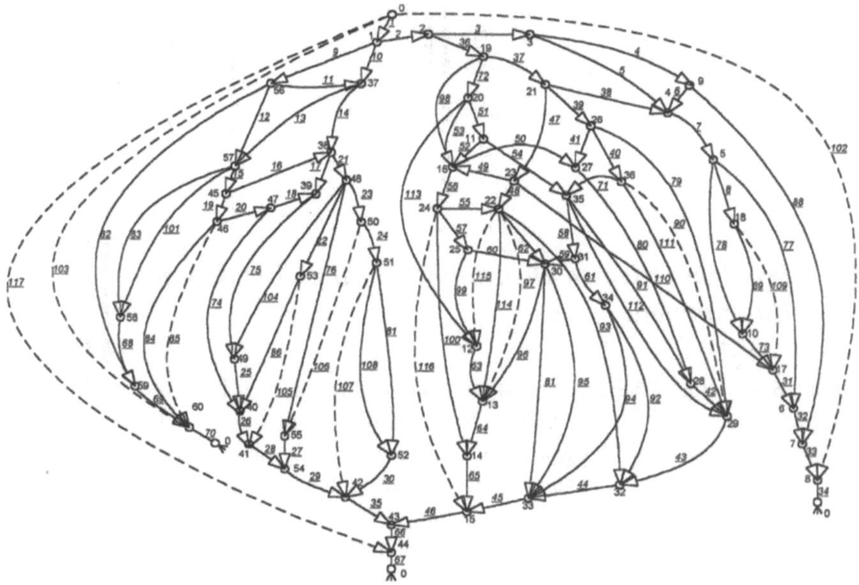

Fig. 6 Test network object ($m = 117, n = 61$)

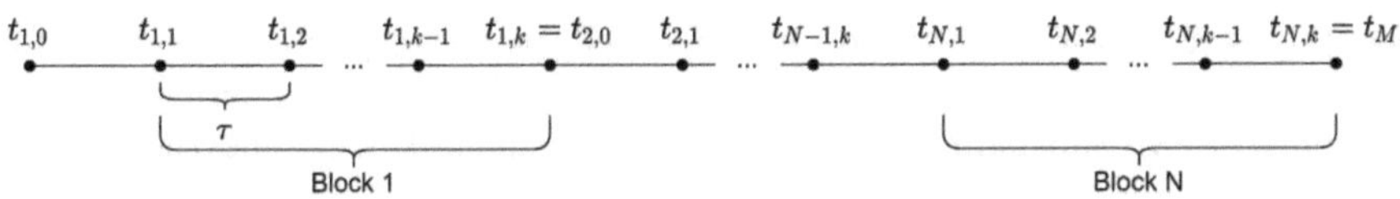

Fig. 7 Scheme of disassembly into blocks

2.2 *Parallel Equations Solvers Based on Block Difference Numerical Methods (BDM)*

For the ordinary and partial differential equation systems, the equation solvers are developed on the basis of the block numerical methods [2]. Let us consider solving the Cauchy problem:

$$\begin{cases} x' = f(t, x); \\ x(t_0) = x_0 \end{cases} \tag{15}$$

with the help of the *k-point* block method (Fig. 7). We decompose the set of M nodes of the homogeneous lattice $\{t_m\}$, $m = 1, 2, \ldots, M$ with the step τ on blocks so that every block includes k points, $k \cdot N > M$. Let us number the points in each block $i = 0, 1, \ldots, k$ and refer to $t_{n,\ i}$ as to the i-th point of n-th block. The points $t_{n,\ 0}$ and $t_{n,\ k}$ are, respectively, the beginning and the end points of the n-th block. It follows that $t_{n,\ k} = t_n + t_{n,\ 0}$, the starting point will not belong to the n-th block.

The general calculation formula for the definition of the new k values according to [5] is:

$$u_{n,i} = u_{n,0} + i \cdot \tau \left[b_i F_{n-1,j} + \sum\nolimits_{j=1}^{k} a_{i,j} F_{n,j} \right]. \tag{16}$$

2.3 Problem-Oriented Distributed Parallel Simulation Environment

Figure 8 illustrates the primary stages and methodologies involved in modeling complex dynamic systems (CDS). A mathematical model of a CDS comprises equations representing dynamic processes under study and a formal depiction of the system's structure (such as technological schemes, graphs, and automation system configurations). This also includes secondary structures resulting from approximations of systems with distributed parameters, among other factors, as shown in Fig. 2. The model, once adapted to the requirements of computational methods and equipped with the necessary software and hardware to solve equation systems, is referred to as the simulation model of CDS [3].

Given the intricacy of CDS models, characterized by high-dimensional equation systems, spatial distribution, and interconnected parameters, various methods for approximating models concerning spatial coordinates, constructing Virtual Testbed Systems (VTS) Simulation models is a challenging task that relies heavily on computer support. The selection of a specific computational approach dictates the discrete nature of the CDS Simulation model, which, during the process of hardware and software implementation, transforms into a CDS simulator. Initially, sequential CDS simulators were developed using programming languages and have evolved to encompass block- (BO), equation- (EO), and object-oriented (OO) modeling languages. Additionally, parallel Multiple Instruction, Multiple Data (MIMD) simulators are developed using programming languages with the assistance of MPI and OpenMP libraries for data exchange and synchronization of MIMD processes [4, 5].

However, it is worth noting that subject matter experts working on parallel simulators are often compelled to utilize tools from conventional second- and third-generation modeling systems, which are less user-friendly and offer a lower level of service compared to sequential modeling languages. In the context of Parallel Simulation Technology (Par-SimTech), one of the central challenges lies in creating a Distributed Parallel Simulation Environment (DPSE) equipped with comprehensive software for the development, debugging, and operation of parallel CDS simulators.

To approach the level of service provided by fifth-generation modeling tools, DPSE should feature parallel modeling languages that can convert CDS model specifications into executable software modules for parallel simulators. An examination of the developed and experimentally tested "topological analyzers -

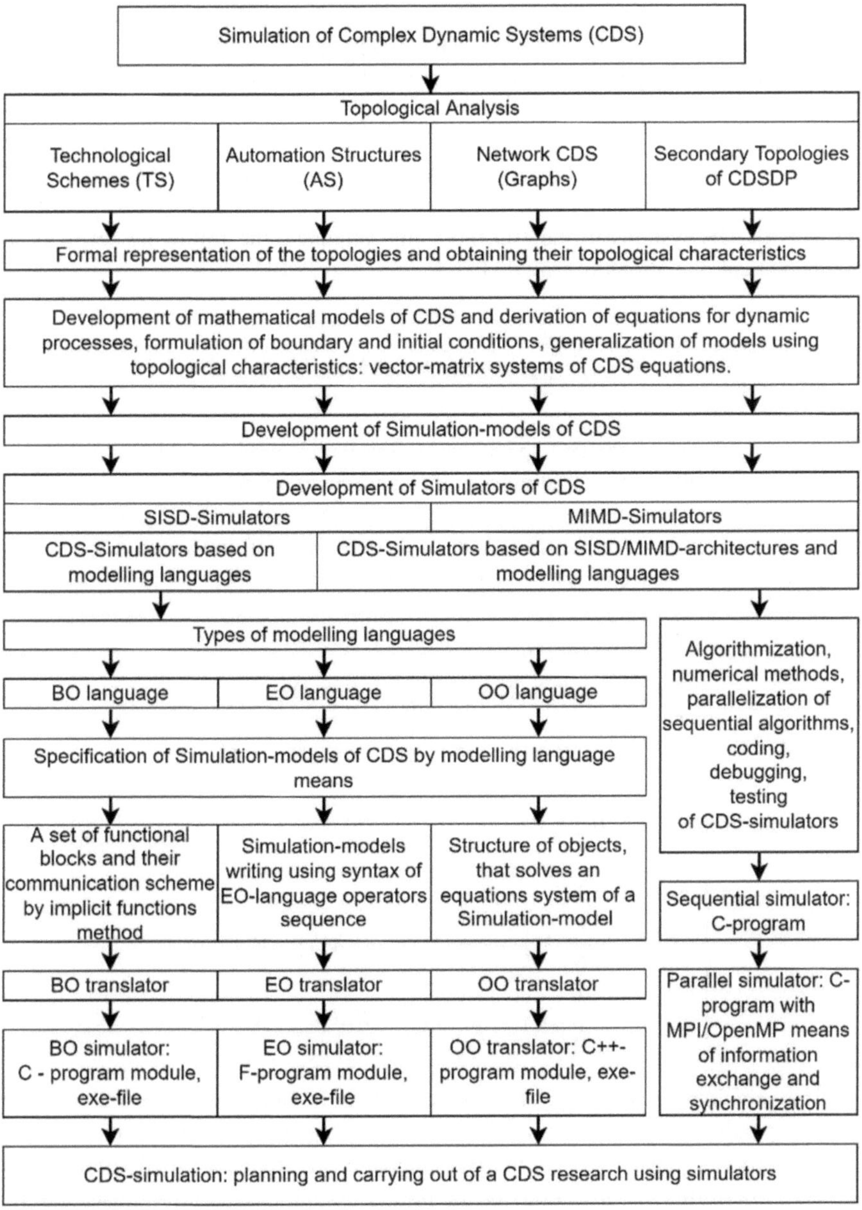

Fig. 8 Stages and means of mathematical modeling of complex dynamical systems

generators of equations" pairs within Simulation models reveals their capability to directly apply principles from block-oriented, object-oriented, and equation-oriented simulation languages to address equation systems. This underscores the importance of exploring the concept of developing parallel modeling languages.

The main component of the BO modeling language [6–8] is the functional block, which is implemented in software. The block has n inputs and 1 output; the coefficients of the input variables and the initial conditions, which is the result of a certain operation, can be set in it:

$$Y = F(X_1, X_2, \ldots, X_n, a_1, a_2, \ldots, a_n, t).. \tag{17}$$

To solve ODE systems in the BO modeling language, the following set of basic mathematical operations is provided:

$$F \in \left\{ a_i x_i, \sum_{i=1}^{n} a_i x_i; \int \sum_{i=1}^{n} a_i x_i dt; f(x_i), \varphi(x_i, x_k); f(t), x_i \cdot x_k; x_i / x_k \right\}.. \tag{18}$$

The analogy between the BO-specification Simulation model of CDS and MIMD-principle of parallelization is shown on the example of a model of a simple network dynamic object with concentrated parameters (DNOLP), described by the system of Eq. (20) with air flows in branches X, Y_1, Y_2, coefficients of flows inertia K_x, K_1, K_2, aerodynamic resistances R_x, R_1, R_2 and fan characteristics $f(X)$:

$$\begin{cases} X = Y_1 + Y_2; \\ K_x \dfrac{dX}{dt} + R_x X|X| + K_1 \dfrac{dY_1}{dt} + R_1 Y_1 |Y_1| = f(X); \\ K_x \dfrac{dX}{dt} + R_x X|X| + K_2 \dfrac{dY_2}{dt} + R_2 Y_2 |Y_2| = f(X). \end{cases} .. \tag{19}$$

A simulation model of NDOLP:

$$\begin{cases} X = Y_1 + Y_2; \\ \dfrac{d}{dt}\left(Y_1 + \dfrac{K_x}{K_1} X\right) = \dfrac{1}{K_1}(f(x) - R_x X|X| - R_1 Y_1 |Y_1|); \\ \dfrac{d}{dt}\left(Y_2 + \dfrac{K_x}{K_2} X\right) = \dfrac{1}{K_2}(f(x) - R_x X|X| - R_2 Y_2 |Y_2|). \end{cases} .. \tag{20}$$

The principle of solving systems of equations in BO language corresponds to MIMD-parallelism and can be interpreted as a virtual purpose “Functional block - MIMD-process”: each block of the BO language is assigned a MIMD process that precisely performs the operations of the block; we get a set of n processes that are connected with each other by a communication graph, which is synthesized by basing on the connection diagram between the outputs and inputs of the BO-simulator blocks (Fig. 9).

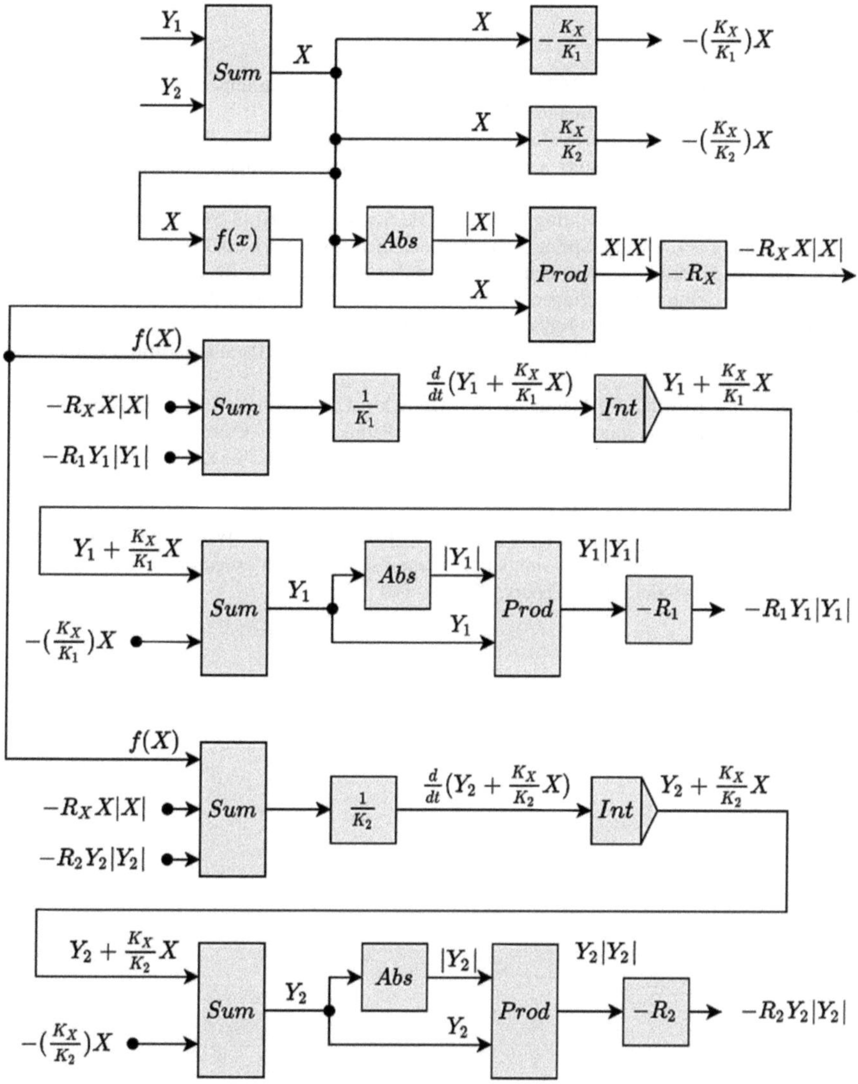

Fig. 9 Stages and means of mathematical modeling of complex dynamical systems

3 Conclusion

This work addresses the challenges of traditional modeling systems by promoting the use of parallel modeling languages and parallel simulators. It holds promise for significantly advancing scientific and engineering endeavors by enabling the efficient simulation of complex dynamic systems.

References

1. Mullen, J., Milechin, L., Milechin, D.: Teaching and learning HPC through serious games. J. Parallel Distrib. Comput. **158**, 115–125 (2021)
2. Bönisch, T., Resch, M., Schwitalla, T., Meinke, M., Wulfmeyer, V., Warrach-Sagi, K.: Hazel Hen – leading HPC technology and its impact on science in Germany and Europe. J. Parallel Comput. **64**, 3–11 (2017)
3. Resch, M., Kaminski, A., Gehring, P. (eds.): The Science and Art of Simulation I: Exploring – Understanding – Knowing. Springer, Berlin/Heidelberg/New York (2017)
4. Knüpfer, A., Hilbrich, T., Niethammer, C., Gracia, J., Nagel, W., Resch, M.: Tools for High Performance Computing. Springer, Berlin/Heidelberg/New York (2019)
5. Gonzalo, P., Rodrigo, P.-O., Östberg, E.E., Antypas, K., Gerber, R., Ramakrishnan, L.: Towards understanding HPC users and systems: a NERSC case study. J. Parallel Distrib. Comput. **111**, 206–221 (2018)
6. Neumann, P., Kowitz, C., Schranner, F., Azarnykh, D.: Interdisciplinary teamwork in HPC education: challenges, concepts, and outcomes. J. Parallel Distrib. Comput. **105**, 83–91 (2017)
7. Chaudhury, B., Varma, A., Keswani, Y., Bhatnagar, Y., Parikh, S.: Let's HPC: a web-based platform to aid parallel, distributed and high-performance computing education. J. Parallel Distrib. Comput. **118**, 213–232 (2018)
8. Czarnul, P., Kuchta, J., Matuszek, M., Proficz, J., Rościszewski, P., Wójcik, M., Szymański, J.: MERPSYS: an environment for simulation of parallel application execution on large scale HPC systems. Simul. Model. Pract. Theory. **77**, 124–140 (2017)

Methods of Biometric Authentication for Person Identification

Daria Polunina, Oksana Zolotukhina, Olena Nehodenko, and Iryna Yarosh

1 Introduction

The methods of biometric authentication are one of the modern technologies of person identification. The term "biometrics" is associated with the use of an individual's unique physiological data, applying biological characteristics or behavioral features to identify a person. Biometric data systems analyze fingerprints, facial geometry, voice recognition, and body contour [1]. The use of biometric authentication in security systems instead of text or graphical passwords is a relevant and modern trend, the development of which is associated with the use of mobile devices and the widespread use of computer information technology.

The purpose of the study is to review and analyze contemporary technologies of biometric authentication for identifying a person by facial features, analyze existing software implementation tools and select methods and algorithms that are most effective in terms of reliability, accuracy, and speed.

The classical identification and authentication procedure consists of the subject undergoing the procedure, its unique characteristics, the selected certification system and its operating principle (biometric authentication), and the mechanism of access control for granting specific access rights to subjects [2].

D. Polunina (✉)
Private Higher Educational Institution "Dnipro Technological University "STEP"", Dnipro, Ukraine
e-mail: polunina_d@itstep.academy

O. Zolotukhina · O. Nehodenko
State University of Telecommunications, Kyiv, Ukraine

I. Yarosh
Donetsk National Technical University, Lutsk, Volyn, Ukraine
e-mail: iryna.yarosh@donntu.edu.ua

M. N. Seyman (ed.), *2nd International Congress of Electrical and Computer Engineering*, EAI/Springer Innovations in Communication and Computing,
https://doi.org/10.1007/978-3-031-52760-9_23

Biometric authentication is the process of confirming the authenticity of data by presenting a biometric image by the user, as well as the process of converting this image in accordance with a predetermined authentication protocol [3].

In the process of biometric authentication, the reference sample and the sample submitted by the user are compared to some predefined tolerances.

There are two main drawbacks that deserve special attention when applying biometric authentication: a spoofing attack on the user interface and leakage from the template database.

A spoofing attack on the user interface involves replacing a real fingerprint or face with a fake biometric image. Spoofing attacks violate the basic principle of operation of recognition systems, and system security is seriously compromised.

A template database leak occurs when information about a user's legitimate template becomes available to an attacker. In this case, it is much easier for the attacker to recover the biometric template by reverse engineering of the template, which increases the risk of forgery. However, the attacker is not able to replace the real template with a fake one, as in the case of a password, and this is an advantage of biometric methods.

The study explores the evolving field of biometric authentication, which harnesses unique physiological or behavioral characteristics for personal identification. Biometric data systems, such as fingerprint recognition, facial analysis, and voice authentication, have gained prominence in the realm of security and access control, driven by the widespread use of technology and mobile devices.

The objective of the research is to assess contemporary technologies in facial feature-based biometric authentication, examining available software tools and selecting methods and algorithms that prioritize reliability, accuracy, and speed. A key aspect of this process is understanding the interplay between the subject, their biometric characteristics, the chosen certification system, and the mechanism for granting access rights.

2 Approaches of Biometric Authentication

In general, methods of biometric identification are divided into two main types: static and dynamic.

Static methods of biometric authentication are based on the physiological (static) features of a person, that is, on unique properties that are given to a person from birth and are inseparable from him or her. Such features include person's fingerprints, the network of blood vessels under the skin, the pattern of the iris, facial contours, and so on [4].

Dynamic methods are based on the analysis of behavioral features that are subconsciously used by a person in the process of demonstrating certain normal behavior (keyboard handwriting, movement patterns, and sequence of performing standard actions). From the point of view of network technologies, authentication methods based on signatures and keyboard handwriting are particularly important. Examples of dynamic methods are given in [5].

Static biometric security systems, which are highly prevalent worldwide, primarily utilize facial recognition technology. This technology has the ability to identify or authenticate individuals using digital images or video frames, which falls under static methods. Various systems incorporate person recognition, but in essence, they operate by comparing specific facial features with an image of a person stored in a database. This approach is often referred to as a biometric application rooted in artificial intelligence, enabling precise identification of individuals by analyzing models based on facial textures and the shape of their head.

Biometric authentication tasks can be classified as tasks of image classification. The main approach to solving the problem of image recognition and classification is considered to be neural networks. Among the main approaches to construction of neural networks are the so-called convolutional or deep neural networks (ConvNets or CNNs). Deep neural networks and deep learning are currently used in computer vision, speech recognition, and image classification systems. A neural network usually includes a large number of neurons working in parallel and located in separate layers, so the expected result is a fairly high performance of methods of this class. Neural networks of this type are adaptive, modify themselves, and learn well even without initial training.

The second method used to solve the classification problem is the Support Vector Machines (SVM) method [5].

The Support Vector Machines (SVM) method belongs to a family of linear classifiers. In this family, the decision on whether an object belongs to the class is made according to the law of a linear decision rule:

$$f(x) = \sum\nolimits_{j=1}^{d} w_j x(j) + b = w^T x + b, f(x) = \begin{cases} +1, \text{if } f(x) \geq 0, \\ -1, \text{if } f(x) < 0. \end{cases} \quad (1)$$

where $b \in R$ – shift parameter and $w_j \in R$ – some weights.

The SVM is based on the construction of an optimal separating hyperplane. The training in the method itself is reduced to solving a quadratic programming problem that has a unique solution. The solution has various properties, including sparsity: the position of the hyperplane depends on a small fraction of the training objects. These very objects are the support vectors that gave the method its name [6].

The third method that has been included in the study is the k-nearest neighbors method. This is also a method for person classification based on their images [6].

When using a k-NN classifier, it is important that the k-nearest neighbors have a relative or absolute majority of images of their own class among other images. Let us consider a simpler case that involves a relative majority. The successful operation of the k-NN classifier is that the condition is met for the k-nearest neighbors:

$$|U_i \tilde{l}_i| > |U_i \tilde{m}_i|, i = 1, 2, 3 \ldots, \quad (2)$$

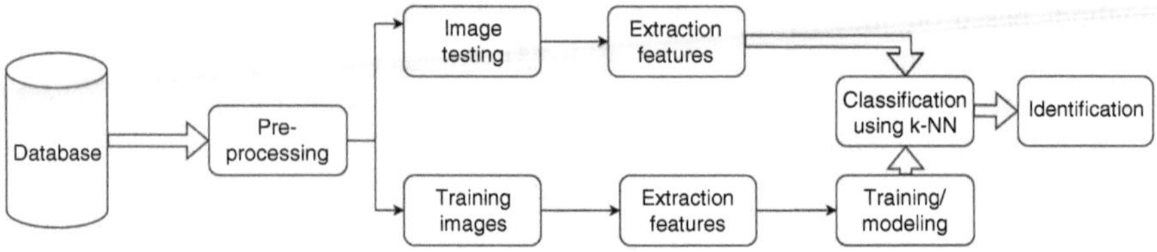

Fig. 1 Scheme of implementation of the method of k-nearest neighbors

where $\tilde{l}_i$, $\tilde{m}_i$ – groups that are formed after the information coverage of classes is reduced. A group means a homogeneous sequence of elements. The scheme of the computational process when using the method of k-nearest neighbors is shown in Fig. 1.

Another common way to solve a classification problem is the application of the Frobenius norm. Informally, the Frobenius norm measures the dispersion or variability (which can be interpreted by size) of a matrix [8]. In our case, we will use it for approximation and calculate it using the formula:

$$||A||_F = \sqrt{\sum\nolimits_{j=1}^{n} \sum\nolimits_{i=1}^{m} A_{ij}^2} = \sqrt{Tr\left(A^T A\right)} \quad (3)$$

where $Tr(A^T A)$ means the trace of the matrix (the trace of a square matrix is defined as the sum of elements of the main diagonal).

Person identification requires the methods used in the process of recognition and classification that meet a number of requirements, namely, speed and high accuracy that is obtained on little data.

The first requirement of a facial recognition system (speed) is that it should provide results in the shortest possible time. For example, in smartphone-based facial recognition, it is usually necessary to use the front-facing camera of the phone to scan the face from different angles for training, and subsequent recognitions should occur in a matter of seconds.

It is worth noting that convolutional neural networks (ConvNets or CNNs), as it turned out, are not quite adequate for face recognition in such systems.

The second requirement is high accuracy on little data, which means that the system must be sufficiently accurate (and therefore safe) when working with a small amount of training data. It turned out that CNN is quite accurate in image classification tasks, but only at the expense of a huge training set available. Therefore, the method does not meet the requirements for this parameter either and, as a result, cannot be used in person identification systems based on the facial features biometric authentication method [7, 8].

In addition, biometric methods can be classified and compared by several factors and indicators, which are described in the following:

FAR (false acceptance rate): It shows how often the system recognizes a person who does not have access as an authorized person, measured as a percentage [9].

Table 1 Levels of parameter values of authentication methods

Authentication method	FAR	FRR	GFRR	FTE	Risk of error	Live pattern recognition
Face recognition	10^{-2}	10^{-2}	1–5	1	Low-middle	Absent

FRR (false rejection rate): It shows how often the system will incorrectly process an authorized person as unauthorized one and deny access, measured as a percentage [9].

GFRR (generalized FRR): It shows, with the help and extensions provided by the SBA study, the FRR of the device in real-world circumstances, including user errors, measured as a percentage [10].

FTE (failure to enroll rate): It shows the number of people who cannot be successfully enrolled in the system and therefore cannot use it, measured as a percentage [10].

Risk of error: It shows how easy it is to bypass the system using a particular method (e.g., instead of an actual fingerprint, showing only a photo of the optical fingerprint scanner).

Live pattern recognition: It shows whether the system can distinguish between a live user and an attempted spoof (Table 1).

Let us explain the selected evaluation factors. False rejection rate (FRR) or generalized false rejection rate (GFRR) is the probability that the system incorrectly determines the match between an input template and a corresponding template in the database. It is measured as a percentage of the deviation of valid input data.

It represents the proportion of incorrect identifications relative to the total number of attempts to identify. For the purposes of this study, seven hundred and fifty face images were registered and confirmed, in which one of the unregistered images was accepted. Therefore, the false rejection rate is calculated using the following formula:

$$\mathrm{FRR} = \frac{\text{Number of false rejected image}}{\text{Number of registered image}} * 100 \tag{4}$$

False acceptance rate (FAR) is the probability that the system incorrectly identifies a successful match between an input template and a non-matching template in the database. This indicator measures the percentage of invalid matches. These parameters are crucial because they are usually used to prohibit certain actions by banned people.

System Recognition Accuracy is the overall percentage of correct system recognitions. Recognition accuracy is defined as follows:

$$\mathrm{RA} = (100 - (\mathrm{FAR} + \mathrm{FRR}))\% \tag{5}$$

Analyzing how the false rejection rate (FRR) and false acceptance rate (FAR) vary across various biometric authentication systems offers valuable insights into the trade-offs between security and user convenience. The optimal scenario would involve achieving both minimal FAR and FRR. However, in practice, biometric authentication systems tend to fall along a continuum, often with a choice between high convenience (resulting in a low FRR) but a lower level of security (resulting in a high FAR), or the reverse.

The failure to enroll rate (FTE or FER) represents the proportion of input data that is deemed invalid and cannot be registered in the system. This occurs when the sensor receives data that is considered of poor quality or invalid.

When a person is unable to register in a biometric system, it is called failure to enroll. FTE occurs when the person using the system does not have enough biometric data. In addition, FTE depends on the design and policy of the biometric systems being implemented. If the FTE indicator is higher, a problematic situation arises.

FTE (FER) is measured by the failure rate as follows:

$$\text{FER} = \frac{\text{Number of unsuccessful enrollment attempts for a person}}{\text{Number of all enrollment attempts for a person}} \tag{6}$$

We will calculate the accuracy of the results using the following formula:

$$\frac{\text{Number of unsuccessful}}{\text{Successful checks}} * 100\% \tag{7}$$

where the number of unsuccessful checks is the number of cases when it should not have authenticated but did; the number of successful checks is the sum of those who should have been authenticated and those who should not have been authenticated, but it was a correct case.

$$\text{Training time (T)} = \text{Final training time–Initial training time (seconds)}.$$

The FRR, FAR, FTE parameters, recognition accuracy, and training time are selected as the main comparative characteristics in the process of analysis of the above recognition methods.

3 Features of Implementation of a System on Facial Recognition

Facial recognition is a method that automatically identifies or verifies a person using a digital image or video frame. One way to do this is to compare the image of a face with examples in a database.

The Georgia Tech face database was chosen for training and testing. No pre-processing of images is performed in the database. It contains a total of 750 images, 15 images per 50 people. The images are in JPEG format, and they all have different sizes. Out of the 15 images per person, four images were taken as the training set and the remaining 11 images were taken as the test images. A total of 200 images are the test images and 550 images are the training set. The images in the database have a resolution of 640 × 480 pixels. The average size of faces in the images is 150 × 150 pixels.

The images show the face from the front, that is, minimally distorting its appearance, as well as the left profile and the right profile [the approaching profile (the tip of the nose extends beyond the cheek, with the nose covering the eyes) and the departing profile] with different facial expressions, as well as with different accessories (glasses, makeup, tattoos, etc.), lighting conditions, and scale.

It should be noted that the human face has numerous distinctive features. These are more than 80 points, some of which can be measured using software. Based on these measurements, a special numerical code called Aface Print is created, and it is this code that represents the face in the database.

The most efficient way of facial biometrics is considered to be 2d identification.

4 Application of the Selected Methods to Solving the Problem of Person Identification

To solve the identification problem, we have developed diagrams of use cases and a class diagram, designed a software application, and selected hardware and software tools for its implementation (Python programming language, Jupyter shell, special libraries). Anaconda development and testing environment with Jupyter Notebook mechanism was installed for machine learning.

The biometric recognition system has two operational modes: identification and verification. Identification involves confirming a person's identity by analyzing a biometric model generated from their biometric characteristics.

In the case of identification, the system learns using the templates of several people. At this stage of training, a biometric template is calculated for each person. The template to be identified is compared with each known template, and the distance describing the similarity between the template and the image is calculated. The system assigns the template to a person with the most similar biometric parameters. To prevent the correct identification of imposter templates (in this case, all templates of persons unknown to the system), the similarity must exceed a certain level. If this level is not reached, the template is rejected.

In the case of verification, the template being checked is compared to an individual's specific template. Similar to identification, it is verified whether the similarity between the template and the image is sufficient to grant access to the secure system.

We utilize parameter estimation to detect the similarity between the template and the biometric template described in the initial section. A higher score indicates a greater degree of similarity between them. Access to the system is only granted if the

Table 2 Characteristics of FAR (%)

Method	Number of images of a person					
	2	4	6	8	10	12
SVM	0.792	0.281	0.373	0.232	0.566	0.098
k-NN	0.192	0.281	0.373	0.232	0.566	0.478
Frobenius norm	0.081	0.001	0.073	0.051	0.093	0.095

Table 3 Characteristics of FRR (%)

Method	Number of images of a person					
	2	4	6	8	10	12
SVM	0.793	0.687	0.475	0.58	0.475	0.273
k-NN	0.981	0.602	0.152	0.013	0.016	0.125
Frobenius norm	0.294	0.182	0.312	0.209	0.046	0.113

Table 4 Recognition accuracy (%)

Method	Number of images of a person					
	2	4	6	8	10	12
SVM	98.415	99.032	99.152	99.188	98.959	99.629
k-NN	98.827	99.117	99.475	99.755	99.418	99.397
Frobenius norm	99.625	99.817	99.615	99.74	99.861	99.792

Table 5 Training time for three methods (sec)

Method	Number of images of a person					
	2	4	6	8	10	12
SVM	50	100	120	140	160	200
k-NN	20	40	60	80	100	120
Frobenius norm	25	50	75	125	150	175

estimation parameters for the trained system (for person identification) surpass a predefined threshold.

The experiment was conducted on a test database containing 50 photos of people with 15 images for each. The training was carried out consistently: from 2 to 12 images, where, for example, two are training images and 13 images are for testing, and so on, increasing the number of images for training process. All results are evaluated by the number of testing images.

In the framework of the study, it was decided to evaluate the FAR, FRR parameters, accuracy, and image recognition time. An additional feature was the use of an algorithm for creating digital makeup during the study.

The results of the study of the main characteristics of image classification algorithms are shown in Tables 2, 3, 4 and 5.

Given the unique requirements and limitations faced by recognition systems, the learning paradigm of CNN is not suitable for identification. This inadequacy of the

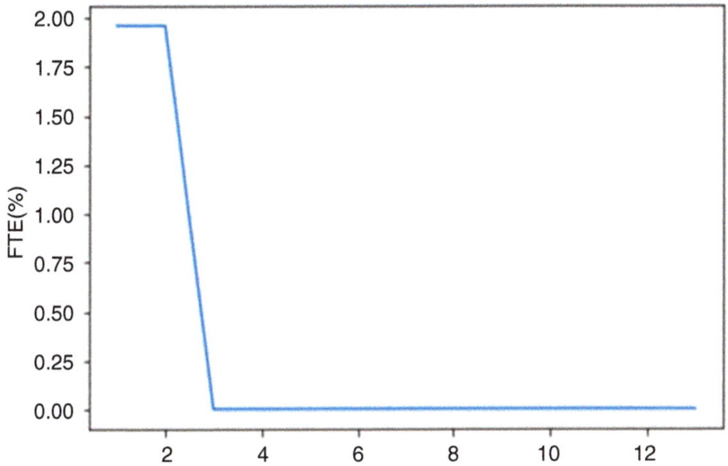

Fig. 2 FTE for three algorithms

method to meet the requirements of speed and accuracy under the condition of small number of operational samples was identified at the initial stages of the study.

The FRR and FAR parameters show a tendency to decrease as the amount of data (images) entering the system increases.

Having analyzed the results, we can conclude that the probability that the system incorrectly identifies a match between the input template and the corresponding template in the database is worst for the k-nearest neighbors method (false positives), while the Frobenius norm and the SVM give close results.

The probability that the system incorrectly determines the match between the input template and the corresponding template in the database has 0.27% of errors for the SVM, which is not a bad indicator; however, compared to the k-nearest neighbors method and the Frobenius norm, which have zero percent error rate, the SVM is not the best solution.

Therefore, in total, according to the three parameters, we can say that the best solution for the classification task is the Frobenius norm.

In terms of recognition accuracy, the best results were found for the SVM and the Frobenius norm. However, if the speed indicator is chosen as the basis, the k-nearest neighbors method will be the fastest.

The three algorithms can also be evaluated by calculating the false rejection rate (FRR), false acceptance rate (FAR), accuracy, completeness, and training time.

The graphs show the obtained results and comparison of each algorithm on three parameters.

A separate parameter that was chosen is FTE (impossibility to register in the system). When testing and checking all algorithms, the same result was obtained, the graph of which is shown in Fig. 2. With a small training sample, there is a small percentage that this error can occur.

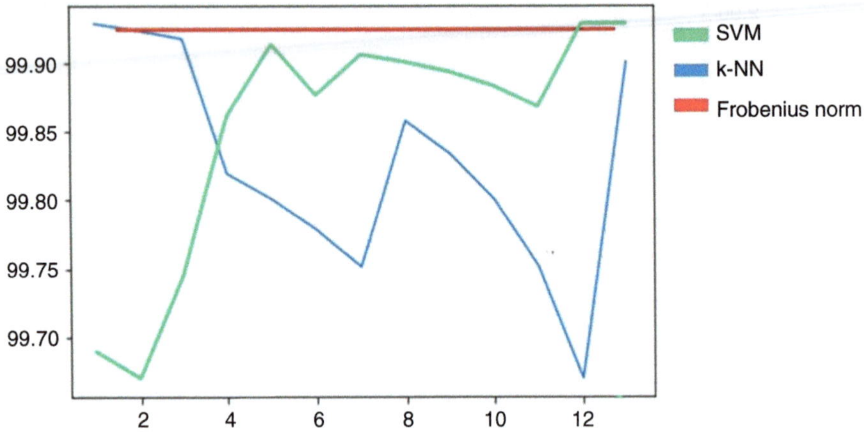

Fig. 3 Comparison of accuracy with make-up for three methods

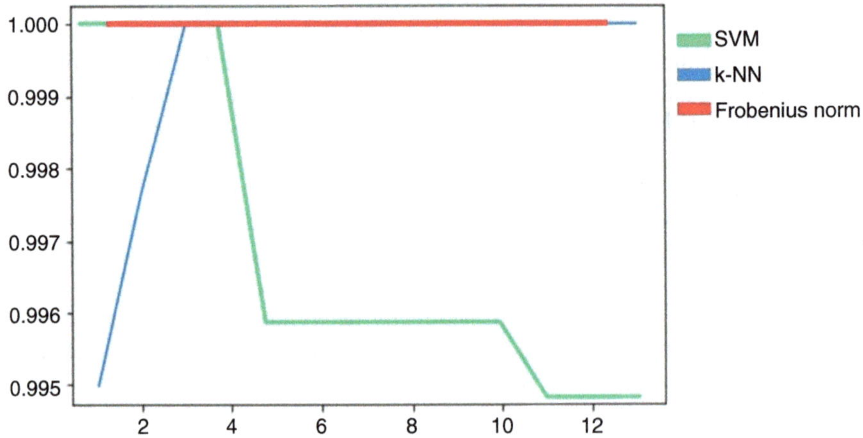

Fig. 4 Comparison of completeness with makeup for three algorithms

A comparison of the accuracy of algorithms on images with makeup is presented in Fig. 3. It can be concluded that the highest accuracy is given by the Frobenius norm.

A comparison of the completeness of three algorithms trained on images with makeup is presented in Fig. 4.

The k-nearest neighbors algorithm and the Frobenius norm have the highest indicators.

FRR calculates the percentage of times the system allows a false rejection, and FAR calculates the percentage of times the system allows a false recognition.

Figure 5 presents a comparison of the results for the three algorithms trained on makeup images.

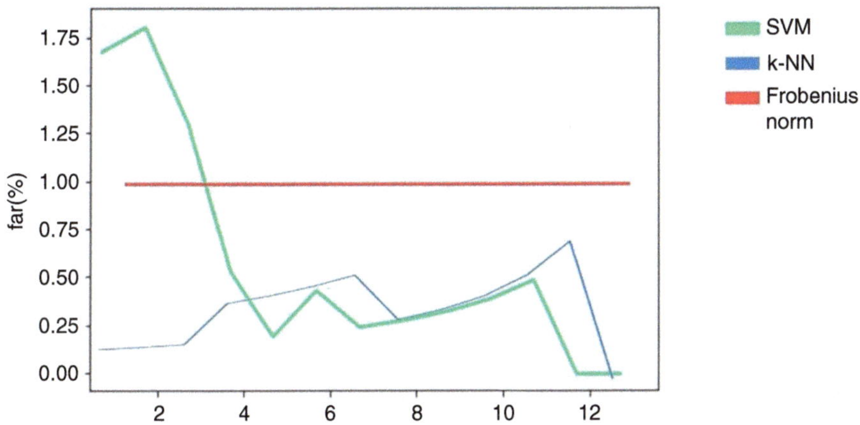

Fig. 5 Comparison of FAR with makeup for three algorithms

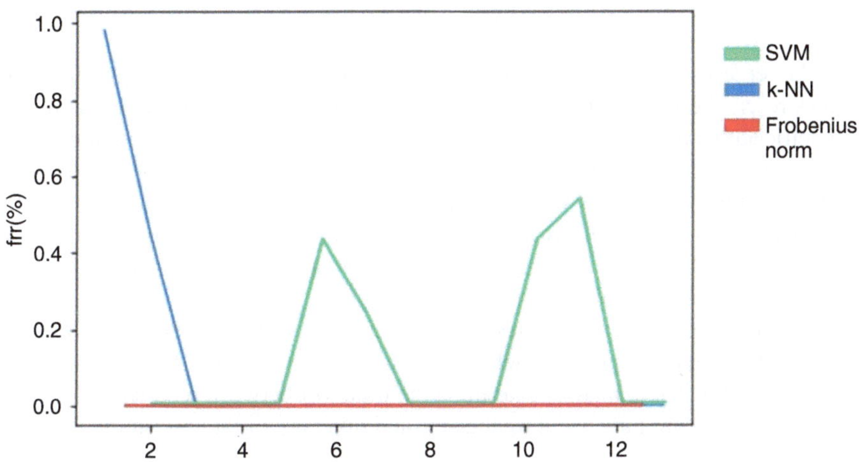

Fig. 6 Comparison of FRR with makeup for three algorithms

Both FRR and FAR show a decreasing trend as more data is fed to the system. A low FRR (Fig. 6) means that the percentage of the system making a false deviation is low and that is a good thing. Also, a low FAR means that the percentage that the system makes a false recognition is low, which is also good.

Figure 7 presents a comparative graph of the training time for the three algorithms on make-up images. Based on these data, it can be concluded that the k-nearest neighbors method is the fastest to learn.

The number of images in the test sample plays a crucial role in the results of biometric authentication studies. It directly influences the accuracy, reliability, and generalizability of the study's findings.

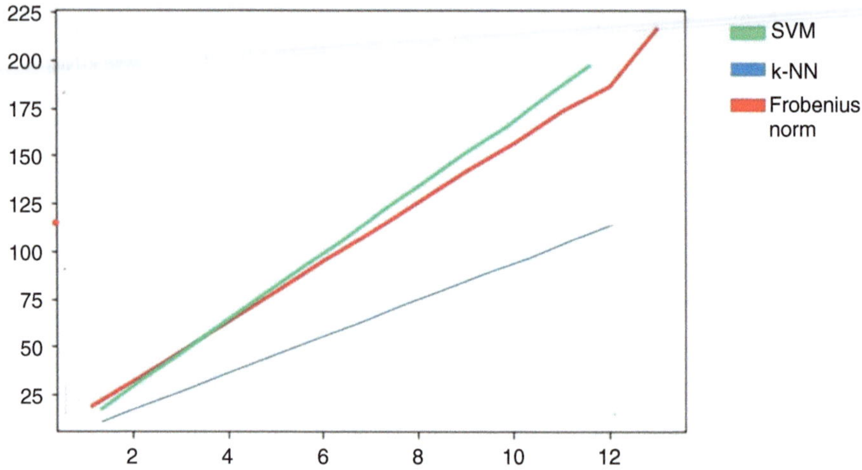

Fig. 7 Comparison of training time with make-up for three algorithms

A larger test sample with a diverse set of images is more statistically significant. It allows researchers to draw more robust conclusions about the effectiveness of the biometric authentication method. Small test samples can lead to skewed or unreliable results due to a lack of diversity.

The number of images directly impacts the accuracy and error rates of the biometric system. A small test sample may lead to overly optimistic results, as it might not reveal the system's limitations. A larger sample provides a more realistic assessment of the system's performance.

In a study, the number of images in the test sample is a key factor in assessing the system's performance metrics, such as false acceptance rate (FAR) and false rejection rate (FRR). These metrics determine the trade-off between security and convenience, and they are influenced by the size and diversity of the test sample.

5 Conclusions

The work analyzes existing biometric authentication methods, including the SVM, k-nearest neighbors, Frobenius norm, and convolutional neural networks.

Testing of the mentioned methods was held. For this purpose, a database with images prepared in advance was selected. It contains various possible variations of a person's face (glasses, head tilt, etc.). The errors that may occur during the system's operation were analyzed. To improve the system's performance and reduce errors, it was decided to use duplicate images with and without the digital makeup algorithm for the purpose of algorithm training and testing. The author's algorithm for images pre-processing by digital makeup was applied.

The results were analyzed.

The convolutional neural network method was excluded as it was unable to meet the basic requirements for person identification using the facial features biometric identification method. CNN did not provide satisfactory results, as the training speed and accuracy were unacceptable.

It was found that the SVM gave satisfactory results for accuracy, false rejection rate, and false acceptance rate; k-nearest neighbors has the best time performance, but lower accuracy; Frobenius norm showed the best results among the stated methods despite the fact that this method is without training.

The general conclusion from the results obtained is that depending on the sphere of system's implementation, it is necessary to choose a method for recognition, since each algorithm has both disadvantages and advantages. It all depends on which indicator should be the main for the system. If the key factor is time, then k-nearest neighbors is the best method, if it is accuracy or a small number of errors, then Frobenius norm is the best.

References

1. Fingerprint vs. Finger-Vein Biometric Authentication [Electronic source]. Access mode: https://www.bayometric.com/fingerprint-vs-finger-vein-biometric-authentication/. Headline from the screen
2. Vorona, V.A.: Systems of Access Control and Management (V.A. Vorona, & V.A. Tikhonov, eds.). Hot Line – Telecom, Moscow, 272 p (2013)
3. Trokielewicz, M.: Cross-spectral iris recognition for mobile applications using high-quality color images. J. Telecommun. Inf. Technol. **3**, 91–97 (2016)
4. Understanding of Convolutional Neural Network (CNN) – Deep Learning [Electronic source]. Access mode: https://medium.com/@RaghavPrabhu/understanding-of-convolutional-neural-network-cnn-deep-learning-99760835f148. Headline from the screen
5. Haykin, S.: Neural Networks: A Complete Course. Dialectics-Williams, 1103 p (2018)
6. Whitehill, J., Serpell, Z., Lin, Y.C., Foster, A., Movellan, J.R.: The faces of engagement: automatic recognition of student engagement from facial expressions. IEEE Trans. Affect. Comput. **5**(1), 86–98 (2014)
7. Gamboa, H., Fred, A.: A behavioral biometric system based on human-computer interaction. Proc. SPIE. **5404**, 381–393 (2004)
8. Scherhag, U., Rathgeb, C., Merkle, J., Breithaupt, R., Busch, C.: Face recognition systems under morphing attacks: a survey. IEEE Access. **7**, 23012 (2019)
9. Keerthi, S.S., Lin, C.-J.: Asymptotic behaviors of support vector machiness for Gaussian kernel. Neural Comput. **15**, 1667–1689 (2003)
10. Draper, B.A., Baek, K., Bartlett, M.S., Beveridge, J.R.: Recognizing faces with PCA and ICA. Comput. Vis. Image Underst. **91**, 115–137 (2003)

Exploring Influencer Dynamics and Information Flow in a Local Restaurant Social Network

Gözde Öztürk, Ahmet Cumhur Öztürk, and Abdullah Tanrısevdi

1 Introduction

Social network analysis (SNA) helps to explore the structural and attribute characteristics of a social network [19]. Identifying the most influential, prestigious, or central social actors in a social network using centrality measurements is one of the common tasks of SNA. Centrality is used to measure the position of a social actor within the overall structure of the social network. In this study, our focus is to identify potential food influencers to promote food businesses. We achieved this by using centrality measurements and examining the effect of food influencers on their followers' restaurant preferences. To obtain the necessary data from Zomato, specifically related to users who write comments on restaurants, as well as their followers, followings, and restaurants located in the Alsancak district of İzmir province, we utilized the BeautifulSoup library in the Python programming language. The scraped data was then converted into a directed graph data using Python, with considering the Zomato's follower network structure. Our social graph consisted of a total of 1755 nodes and 9483 edges, which were imported into Gephi for visualization and centrality calculation of nodes.

After identifying the potential influencers, we conducted ego network analysis for each influencer to investigate whether these influencers had an impact on their followers' restaurant preferences. We examined and compared the difference ratio of the density value of ego networks when influencers remained and when influencers were removed to determine which influencers had the most impact on the information flow within their ego network. Furthermore, we compared the features of potential influencers to observe if these features had an effect on whether

G. Öztürk · A. C. Öztürk (✉) · A. Tanrısevdi
Aydın Adnan Menderes University, Aydın, Turkey
e-mail: gzd_tiga@hotmail.com; cumhur.ozturk@adu.edu.tr; atanrisevdi@adu.edu.tr

M. N. Seyman (ed.), *2nd International Congress of Electrical and Computer Engineering*, EAI/Springer Innovations in Communication and Computing,
https://doi.org/10.1007/978-3-031-52760-9_24

a node was considered an influencer or not. It was observed that the number of photographs uploaded by a user, the number of comments, and the number of followers were determining factors in identifying an effective influencer within a social network.

2 Literature Review

2.1 Influencers, Food Influencers, and Influencer Marketing

Influencers, also known as third-party endorsers, are content creators who cultivate a network of followers by sharing expertise-related content on social media platforms [10]. They possess the ability to influence customer attitudes, behaviors, and opinions, owing to their communication frequency, personal persuasiveness, and their size and centrality within social networks. Influencers consistently generate and share informative and engaging content that captivates their followers including personal touches, with their followers [3].

Influencers engage with their followers by creating content in various domains, such as beauty, health and fitness, travel, and food. In recent years, food-related content generated by influencers has become particularly popular [16]. Influencers who share information and recommendations about food or restaurants with their followers through posts are referred to as food influencers [1]. Influencers inspire businesses to incorporate them as a powerful marketing strategy, known as influencer marketing, as food influencers have the potential to significantly increase audience reach [13]. Influencer marketing intends to enhance the product promotion and increase the brand awareness with the help of the content shared by influential social media users [5]. Food influencer marketing has become a prevalent strategy for attracting and engaging with customers to promote food, food-related products, and restaurants [1]. Identifying suitable food influencers is crucial for achieving influencer marketing goals. According to [1], when restaurants choose the right food influencer, every dollar spent can generate $17.50 in return. To address this research gap, we utilized social network analysis to identify the right food influencers within a digital customer community.

2.2 Social Network Analysis (SNA)

Social network analysis (SNA) helps to examine social structures using network and graph theories. In a social network, nodes depict social actors or entities, such as individuals, groups, or organizations and edges depict the relationships between these social actors or entities. Through the application of social network analysis, one can process large, irregular data obtained from online social networks, reveal the social structure within a specific group of individuals, and create a model for the social relations among these individuals [4].

The edges in a social network can be directional or non-directional, depending on the type of the social network. Directed graphs have edges with a direction. Since a person can be followed by others without necessarily following them, follower networks are an example of directed graphs such as Twitter, Pinterest, and Zomato. Undirected graphs have edges that do not have a direction. Friendship networks, such as Facebook, Yelp, and MySpace, are examples of undirected graphs. In these social networks, the friendship ties established between individuals are typically mutual or reciprocal [19].

In the literature related to SNA in food marketing, there are only a few studies that have been conducted to the best of our knowledge. However, [12] used SNA centrality measures to explore patterns of social relationships in a sampled network of 44 standing out food bloggers that have Twitter accounts. They found a positive association between favorited tweets and Twitter follower relationships. Similarly, [20] identified Social Media Influencers in the Twitter community within the Pizza Hut industry using SNA centrality measurements. The results demonstrated a notable difference among top three influential users. In [20], each node in the graph dataset represented a Twitter user who had written or reposted a tweet with the #pizzahut hashtag. However, the resulting graph was very small and did not provide general information about Twitter influencers, as the dataset only consisted of 23 nodes and 22 edges. Additionally, the authors only applied and analyzed centrality measurements of nodes, without analyzing the effect of interaction through follows, mentions, and replies. In contrast to [20], our social network is a connected graph, meaning that there is always a path exists between each pair of nodes. In our social network, nodes represent social actors and restaurants. Thus, after determining the influential individuals in our network, we examined whether these individuals had an effect on their followers' restaurant preferences using ego network analysis. Additionally, we examined which influencers had an impact on the flow of information in their ego network based on the difference in density values.

2.3 Centrality and Centrality Measurements

One of the fundamental problems in SNA is identifying the most efficient node within the network. Centrality helps to identify nodes in critical positions by measuring their power, effectiveness, and ease of communication. A centrally located social actor has structural advantages, including high popularity, leadership, and prestige. As the centrality of social actors increases, they become closer to the center of the network, and this allows them to quickly and easily access other social actors and control the information flow.

There are several centrality measurements in the literature, such as Degree Centrality (DC), Betweenness Centrality (BC), PageRank (PR), Closeness Centrality (CC), and Eigenvector Centrality (EC). These metrics help to assess the significance of a social actor in the network and provide insight into the concentration of relationships among social actors, giving an idea about their social strength. For instance, [17] used CC, BC, and EC for detecting the influencers on social media in the Malaysian health and beauty sector. They found that EC is the best measurement for identifying influential users in network and this measurement is strongly corelated by both BC and CC. However, [7] compared DC, BC, and EC to detect the network's most central nodes and used a linear threshold model to understand the spreading behavior of social actors. Similarly, [6] employed DC, BC, CC, and EC to predict the probability of authors occupying prominent positions in the co-authorship network on management and accounting in Brazil. In our study, we used the matrix model presented by [18] to classify social actors in our social network. This matrix is two-dimensional where columns represent BC and rows represent EC of individual components. The plane of the matrix is divided into quadrants to determine thresholds in each dimension.

According to [18], potential influencers should have high BC and EC at the same time. In this study, we examined the DC, BC, and EC of social actors to determine which ones are potential influencers. The DC is determined by counting the number of connections a node has with other nodes. DC is related to local centrality and does not consider the global characteristics of the network. There are two variations of DC in directed social networks: in-degree centrality and out-degree centrality. The in-degree centrality value represents the number of incoming connections to a node, while the out-degree centrality value represents the number of outgoing connections from a node. BC is calculated based on the number of shortest paths that pass through a node. A node has a significant and high BC value when it is placed on the only path that other nodes must traverse. Nodes with high BC play vital roles in the network as they connect other nodes and have the capability to control the diffusion of information between them. EC is used to measure the importance of a node by considering both the number and importance of its adjacent nodes. A node with a high EC value has either many neighbors, important neighbors, or both [4].

2.4 Ego Network Analysis

A subgraph S of a graph G is a graph where each node and edge of S are a subset of G. Subgraphs are used to analyze specific portions of a graph. An ego network is a type of subgraph of a given graph G that allows the analysis of nodes and edges surrounding a predefined node. This predefined node is called the ego, and the nodes surrounding the ego are called alters. More formally, an ego network G consists of a node set N and an edge set E, where ego $\subset$ N and ego $\neq N_{\text{Alters}}$. The number of steps it takes for any node to reach the ego determines the size of the ego network [4]. Ego networks can extend any number of steps from the ego. The "1-step neighborhood" ego network is composed of the ego and its directly connected nodes (alters). The "1.5-step neighborhood" ego network expands the 1-step network by including direct connections between1-step alters. The "2-step neighborhood" ego network expands the 1.5-step network by including direct connections between 1.5 step alters.

In social network analysis, ego networks helps in evaluating network design, tie strength, social influence, and conduct comparative analysis. The structure of an ego network provides insights into the information flow within the social network of an individual (ego). Tie strength can be quantified by counting the number of connections between vertices and can provide information about relationships and collaborations among individuals. Social influence can be quantified by analyzing the spread of information and can offer insights into how individuals in a social network are influenced by their immediate social connections. Comparative analysis involves comparing different ego networks in a social network, providing insights into how different communities are formed within the network [4].

There are various measurements for analyzing ego network. In this study, we used size, average degree, density, diameter, average path length, and ego betweenness. In the formulas 1, 2, 3, 4, and 5 explained, n is the sum of nodes and m is the sum of edges in a given graph.

Size of the Graph The total number of edges that will exist if every node in the network has a connection is the size of the graph. The size of an undirected graph is calculated as follows:

$$\text{Undirected Network Size} = \frac{n(n-1)}{2} \tag{1}$$

Graph size is calculated for directed graph as follows:

$$\text{Directed Network Size} = n\,(n-1) \tag{2}$$

Average Degree The average number of edges per node within the graph is called average degree. It is calculated as follows:

$$\text{Average Degree} = \text{Total Edges}/\text{Total Nodes} = m/n \tag{3}$$

Density The density measurement is used for calculating the connectedness of a graph. It is calculated by dividing the total number of connections between vertices to the maximum possible number of connections. The density of an undirected graph is calculated as follows:

$$\text{Undirected Network Density} = \text{Total Edges}/\text{Size} = \frac{m}{\frac{n(n-1)}{2}} \tag{4}$$

The density of a directed graph is calculated as follows:

$$\text{Directed Network Density} = \text{Total Edges}/\text{Size} = m/[n(n-1)] \tag{5}$$

Diameter The length of the shortest path between two most distant connected actors is called diameter. It provides information about the maximum distance between any two actors.

Average Path Length The average distance between any two connected actors in the ego network is called the average path length.

Ego Betweenness The percentage of all shortest paths between pairs of an ego's neighbors that pass through the ego is called the ego betweenness. This measure quantifies how often the ego is positioned on the shortest routes connecting neighbors of ego.

In the literature [9], DC, BC, CC, and EC were used to explore the influencer marketing strategies employed by two popular virtual hotel operators in Indonesia. They utilized centrality metrics to determine the influential actors who disseminate information in the network without providing details of their methodology. However, [21] proposed the following ratio parameter for measuring the collaboration strength among social actors. However, [22] demonstrated that relying solely on the "following ratio" is inadequate for measuring the strength of influence. In contrast to these studies, instead of initially proposing hypotheses related to influencer characteristics, we first identified potential influencers using centrality measurements in our social network. We then applied a one-step neighborhood ego network analysis, which includes all the connections between the ego, the ego's followers, the ego's followings, and the restaurants where the ego visits and writes comments. We also examined the connections among these same partners to assess the influencers' impact on their followers' restaurant preferences. We measured the performance of potential influencers on the flow of information within their ego network and, finally, we found out the characteristics of influential accounts.

3 Methodology

3.1 Data Collection

In this study, we used the BeautifulSoup library of the Python programming language to scrape and filter data from the Zomato website. The Zomato website provides millions of restaurants with descriptive information, making it a valuable source of data on locations, food categories, service facilities, and prices in the market. We chose Python for performing the scraping and filtering of data because it is easy to write and execute code in Python, and the Python libraries are free to use. The Zomato website was used to generate our data repository by following these steps:

1. Zomato provides service in 299 restaurants located in the Alsancak district of İzmir province. We eliminated restaurants that do not contain any user comments related to these restaurants, resulting in a total of 212 restaurants that were examined as of December 2022.
2. We obtained users who wrote comments for these 212 restaurants located in the Alsancak district of İzmir province.
3. We obtained the followers and following bloggers of these users.
4. We eliminated users who did not write any comments for any restaurant in Alsancak within the 212 restaurants. This elimination step was performed because users who do not write comments have no contribution to approving, replying to, or disseminating content created by other users [8].

During the creation of the data repository, in step 3, we obtained a total of 121,189 individual. After step 4, these individuals were reduced to 1543. After the elimination process, we obtained a total of 212 restaurants and 1543 individuals.

3.2 *Data Pre-processing*

The Zomato website has a follower network structure. Therefore, after the data repository was created, the data was converted into a directed graph using the Python programming language (Fig. 1).

In our social network, the nodes represent users and restaurants. If a node represents a user, the outgoing edges of that node represent the act of following

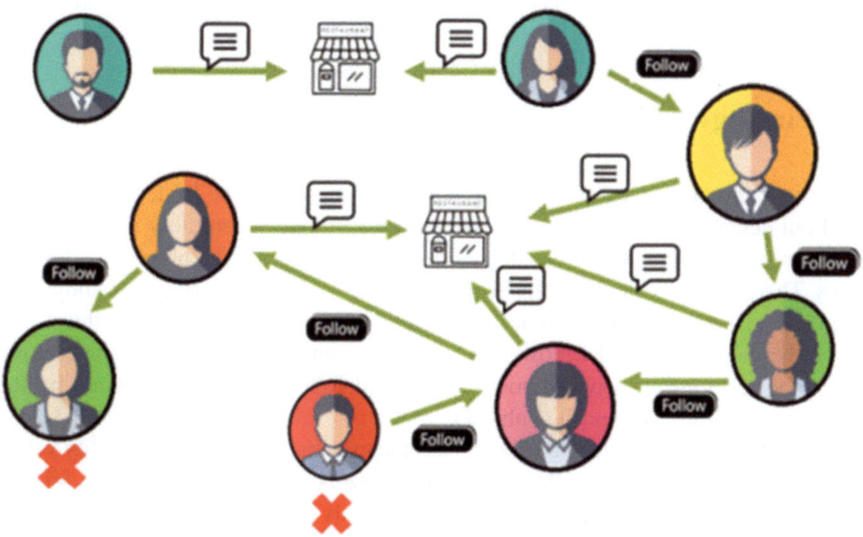

Fig. 1 The graph generation of the social network

other users, while the incoming edges represent followers. If a node represents a restaurant, it only has incoming edges representing users who write comments for that restaurant.

3.3 Data Analyzing

After creating the directed social graph, we imported the graph data into Gephi for visualization and calculation of node centrality values. Gephi is an open-source software application for visualizing and analyzing graphs in various fields [14]. Its visualization module utilizes a specialized 3D rendering engine that allows real-time graph rendering [2]. We chose Gephi for visualizing the directed social graph because it is an open-source software designed for graph and network analysis. It also allows running multiple algorithms simultaneously without blocking the user interface [2]. Upon importing the directed graph into Gephi, we observed that the graph consisted of a total of 1755 nodes and 9483 edges. To create the network visualization, we utilized the ForceAtlas2 layout algorithm. ForceAtlas2 is a force-directed layout algorithm that simulates a physical system to spatialize a network [14]. We adopted for the ForceAtlas2 layout as it effectively illustrates the network's core and periphery. Furthermore, after identifying the influential social actors based on their centrality values, we conducted ego network analysis to examine whether these influential social actors have an impact on their followers' restaurant preferences or not. We applied basic statistics to understand the structure of each ego network, and we compared the difference ratio of the density value of ego networks when influencers remained and when influencers were removed. This analysis aimed to identify which influencer has the greatest impact on the flow of information.

4 Findings

Degree Centrality Degree centrality consists of two variants: in-degree centrality and out-degree centrality. In our social network, in-degree centrality represents the number of followers, while out-degree centrality represents the number of followings. Therefore, we calculated the degree centrality of social actors and examined the top-10 social actors with the highest in-degree centrality values to identify those who have the most number of followers who write comments for any restaurant. However, the relationship between social actors and followers is best captured by the in-degree centrality value [7] (Table 1).

According to the results, "Pisbogaz" has the maximum in-degree centrality value, which means it has the most number of followers who write comments for any restaurant. However, in-degree centrality alone only provides information about the number of an influencer's followers. It does not capture the influencer's ability to effectively spread information. In order to determine the influencers' information

Table 1 Top-10 social actors with high in-degree centrality

Id	In-degree	Out-degree
Pisbogaz	146	19
SeyyahGurme	122	2
Ekndnz	119	13
sucukarmoreno	100	18
eda-t-izmir-16928647	86	18
Mcant	86	62
Gezentianne	86	12
Kanzuk	76	9
Alpartam	75	7
fadanur	72	107

Table 2 Top-10 social actors with high betweenness centrality

Id	In-degree	Out-degree	BC
Pisbogaz	146	19	81103.686644
Gurmeseyyah	66	127	40383.625938
aenderulusoy	59	113	30008.547724
fadanur	72	107	27446.640557
yemekguzel	65	50	26001.146292
selcenocak	26	63	25070.699668
Mcant	86	62	24664.591073
oytun_pinar	38	47	22797.672609
Gezentiobur	53	81	20306.197923
ogrenciKafasi	43	101	20095.149576

spreading capability, we also performed BC, which identifies nodes that can distribute information efficiently.

Betweenness Centrality We calculated the BC of social actors to identify those who can control the information flow between other social actors. We examined the top-10 social actors with high BC (Table 2).

The results indicate that "Pisbogaz" has the highest BC value, meaning that "Pisbogaz" has the ability to spread information further than other users. However, it is important to note that in some cases, users with fewer followers can still distribute knowledge efficiently, as seen with "aenderulusoy," "selcensenocak," and "oytun_pinar." This highlights that the number of followers alone does not determine the effectiveness of information dissemination. Although the nodes that effectively distribute knowledge can be determined by BC, it may not always identify potential influencers.

According to [15], combining both BC and EC measurements would include actors that connect dispersed groups through highly connected actors. Therefore, we conducted EC measurements to determine the most influential node in our social network.

Table 3 Top-10 social actors with high eigenvector centrality

Id	In-degree	Out-degree	BC	EC
Pisbogaz	146	19	81103.686644	1.0
sucukarmoreno	100	18	16571.338627	0.706212
SeyyahGurme	122	2	578.880116	0.699503
ekndnz	119	13	9165.814207	0.694826
eda-t-izmir-16928647	86	18	10396.292603	0.655127
Mcant	86	62	24664.591073	0.63566
kanzuk	76	9	4577.316375	0.620958
gezentianne	86	12	4755.174437	0.600691
ikoo	55	42	5330.812744	0.537024
gurmeseyyah	66	127	40383.625938	0.506083

Eigenvector Centrality We calculated the EC of social actors to determine the most influential social actor. We examined the top 10 social actors with high EC (Table 3). We ranked the social actors according to their EC values (Fig. 2).

The results indicate that "Pisbogaz" has the highest EC value. It is observed that "SeyyahGurme," "kanzuk," "gezentianne," "ekndnz," and "ikoo" have low betweenness centrality values and high EC values. This suggests that although "SeyyahGurme," "kanzuk," "gezentianne," and "ikoo" are located near highly influential nodes, they are not able to quickly disseminate information among their followers. Furthermore, "gurmeseyyah" has a high BC value and a low EC value. As stated in [18], social actors in the network that simultaneously exhibit high values of both BC and EC should be considered potential influencers. Therefore, "Pisbogaz" is the most suitable influencer, as it has the highest BC and EC values. Additionally, "sucukarmoreno," "eda-t-izmir-16928647," and "Mcant" are effective potential influencers in our social network. After identifying the potential influencers in our social network, we examined the ego networks of "Pisbogaz," "Mcant," "eda-t-izmir-16928647," and "sucukarmoreno" to understand whether these influencers have an impact on their followers' restaurant preferences or not.

The results show that "Pisbogaz" has visited and written comments for three restaurants: Reyhan Patisserie, AlsancakDostlar Patisserie, and Kardeşler Fast Food Restaurant (Fig. 3). In this ego network, there are 41 individuals who have visited and written comments for Reyhan Patisserie, 29 individuals for AlsancakDostlar Patisserie, and 7 individuals for Kardeşler Fast Food Restaurant. Additionally, out of the 41 individuals who visited Reyhan Patisserie, two individuals have also visited and written comments for all three restaurants, and five individuals have visited and written comments for both Reyhan and AlsancakDostlar Patisseries. Furthermore, one individual has visited and written comments for Reyhan Patisserie and Kardeşler Fast Food Restaurant. It was discovered that all the individuals who visited and wrote comments for these restaurants are followers of Pisbogaz.

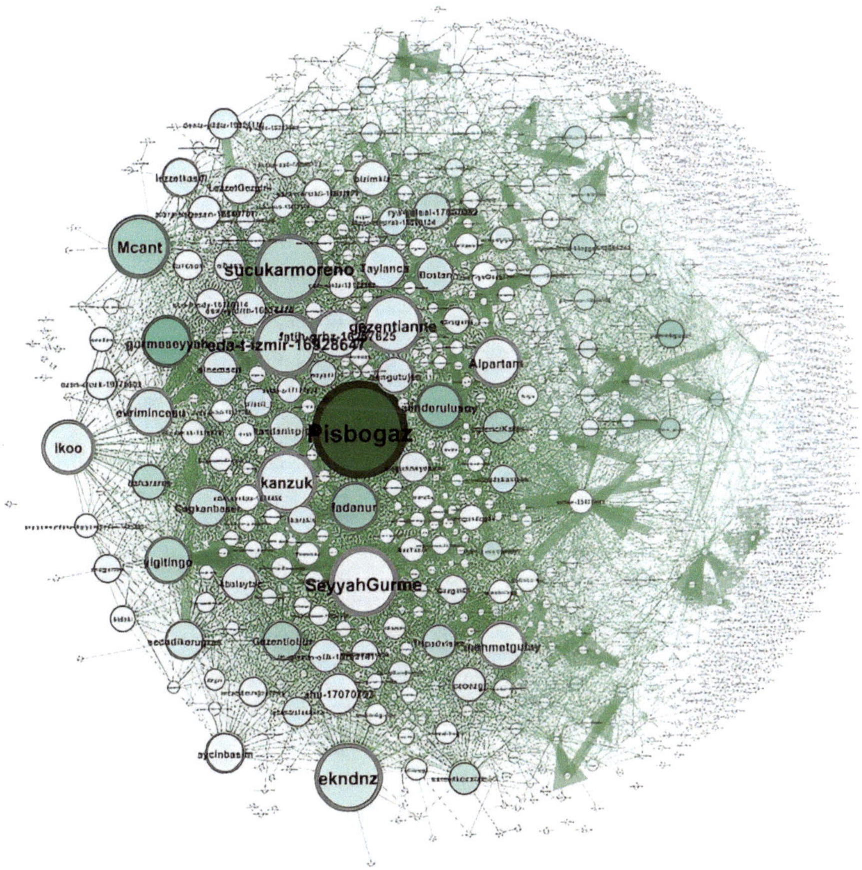

Fig. 2 Social actors graph representation according to their eigenvector centrality

In the ego network of "sucukarmoreno," it is observed that this potential influencer only has visited and written comments for Sevinç Patisserie (Fig. 4). In this ego network, there are 12 individuals who have visited and written comments for Sevinç Patisserie. The results indicated that all the individuals who have visited and written comments for Sevinç Patisserie are followers of sucukarmoreno.

In the ego network of "Mcant," it is observed that this potential influencer has visited and written comments for three restaurants: Reyhan Patisserie, Ristorante Pizzeria Venedik, and Deniz Restaurant (Fig. 5). In this ego network, there are 24 individuals who have visited and written comments for Reyhan Patisserie, eight individuals for Ristorante Pizzeria Venedik, and nine individuals for Deniz Restaurant. Additionally, one individual has visited and written comments for all three restaurants, and two individuals have visited and written comments for Reyhan Patisserie and Ristorante Pizzeria Venedik. It is seen that 78% of the individuals

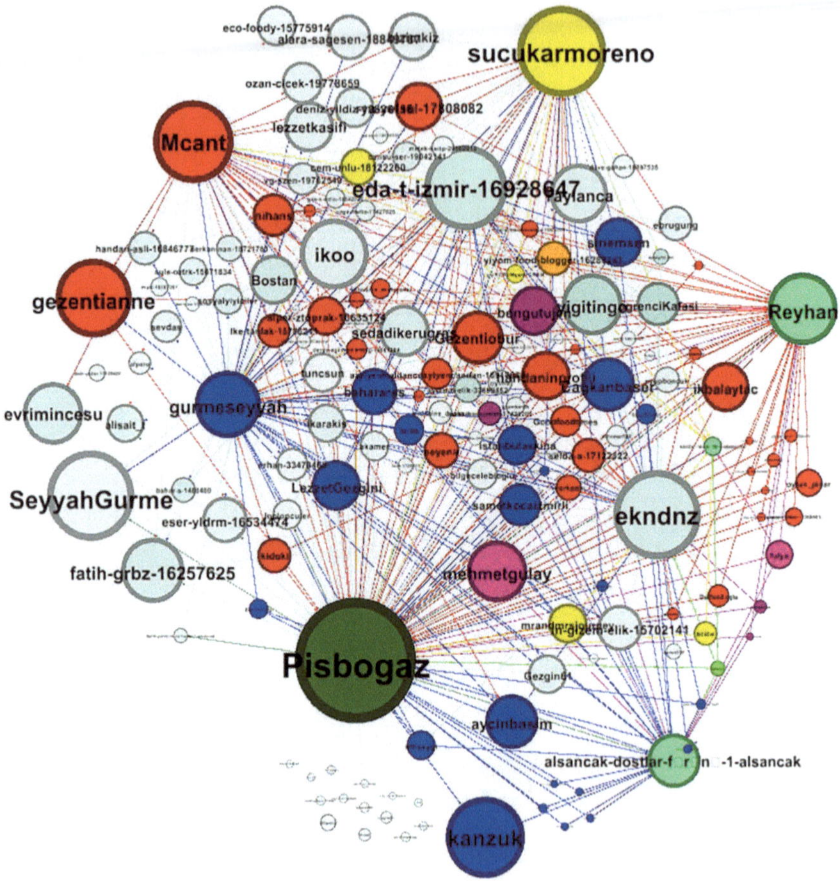

Fig. 3 Ego network of "Pisbogaz"

who have visited and written comments for Reyhan Patisserie, 88% for Deniz Restaurant, and all the individuals who have visited and written comments for Ristorante Pizzeria Venedik are followers of Mcant.

In the ego network of "eda-t-izmir-16928647," it is observed that this potential influencer has visited and written comments only for Sevinç Patisserie (Fig. 6). In this ego network, there are 12 individuals who have visited and written comments for Sevinç Patisserie. The results indicate that 75% of the individuals who have visited and written comments for Sevinç Patisserie are followers of eda-t-izmir-16928647. We applied basic statistics to all potential influencers' ego networks (Table 4). According to these results, "Pisbogaz" has the largest ego network with 23,562 directly connected alters, while "eda-t-izmir-16928647" has the smallest ego network with 8742 directly connected alters. The ego networks of "Pisbogaz," "Mcant," and "sucukarmoreno" have a lower network diameter value compared to "eda-t-izmir-16928647." A low network diameter value indicates that all the individuals in

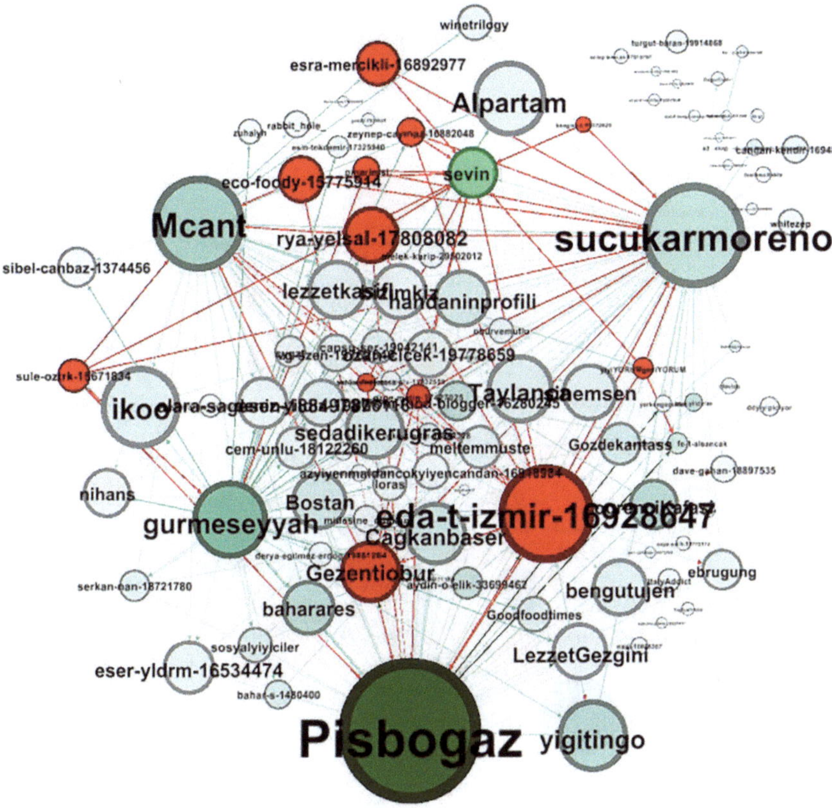

Fig. 4 Ego network of "sucukarmoreno"

their ego network are in close proximity. The mean of the shortest path lengths among all connected pairs in the ego networks of all influencers is close to each other.

In the ego network of "Pisbogaz," each individual has the highest number of neighbors with 15, and "Pisbogaz" has the highest ego betweenness centrality with a value of 4342.553. The ego network of "Pisbogaz" has the lowest density value with 0.103, while "eda-t-izmir-16928647" has the highest density value with 0.148. A higher density value indicates more powerful network effects. The density value enables in discerning distinctions between networks when comparing networks that have equivalent number of nodes and identical type of relationships. However, we cannot assume that "eda-t-izmir-16928647" is more effective in terms of information flow in its ego network solely based on the density values, as the size of "Pisbogaz's" ego network is larger. Therefore, we attempted to identify which influencers have the most effect on information flow in their ego networks.

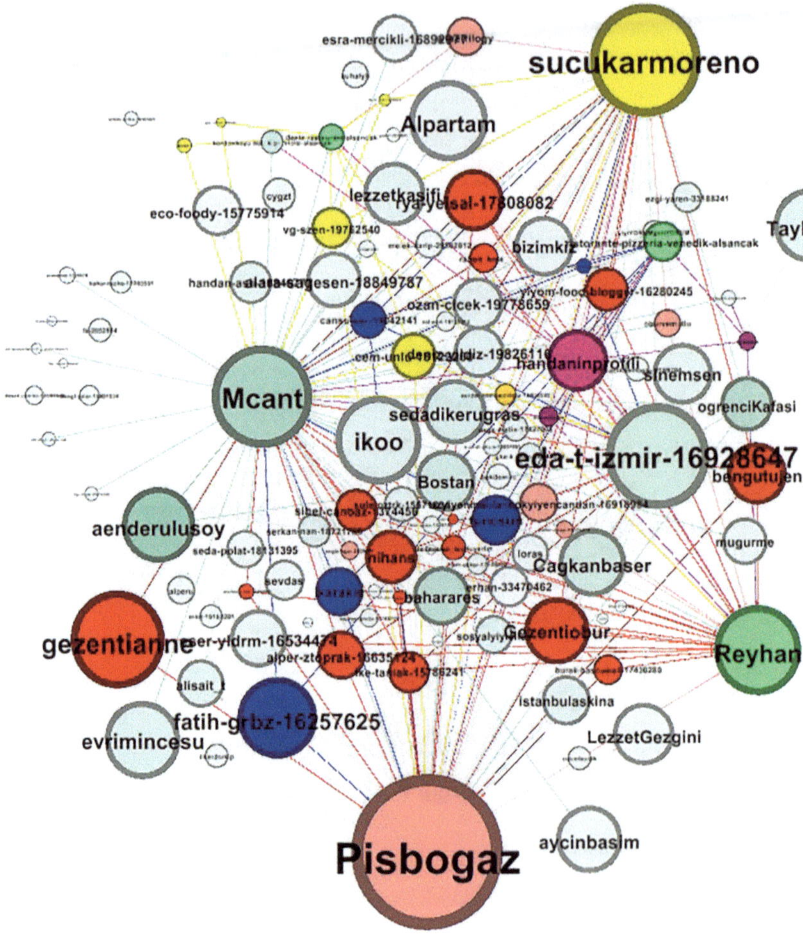

Fig. 5 Ego network of "Mcant"

First, we calculated the density value of each ego network while each influencer remains in their own ego network. Then, we calculated the density value of each ego network after removing the influencers. Finally, we compared the differences between these density values (Table 5). A higher difference between densities indicates that information does not transmit efficiently among individuals, while a lower difference indicates more efficient information flow.

According to these results, when "Pisbogaz" is removed from its ego network, the ratio of density difference (RDD) reaches its maximum value compared to other potential influencers. This indicates that "Pisbogaz" plays a crucial role in the dissemination of information within its ego network, and the removal of "Pisbogaz" has the most significant impact on information flow. Therefore, "Pisbogaz" has a

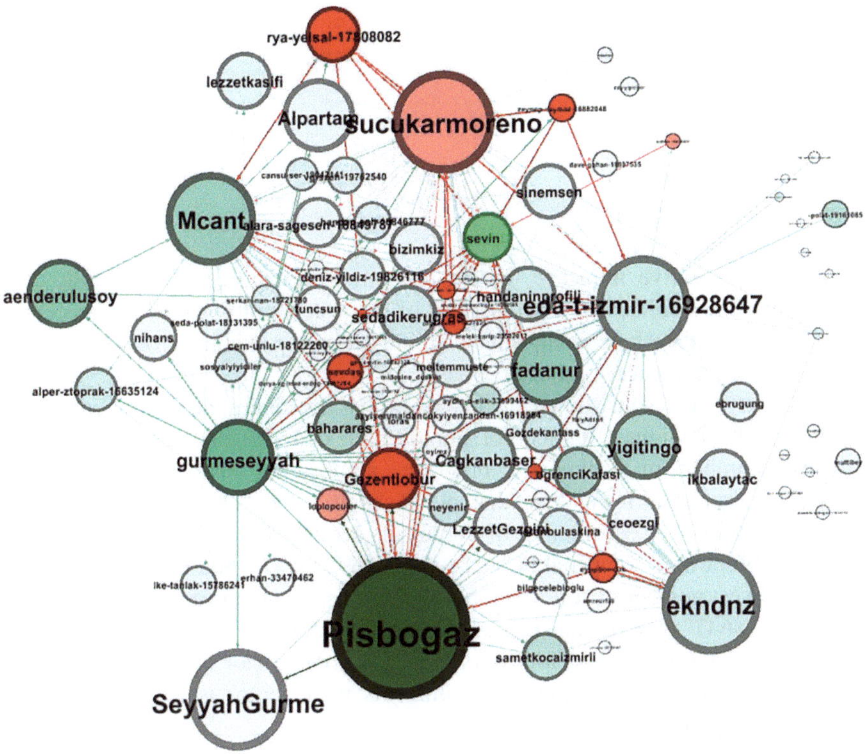

Fig. 6 Ego network of "eda-t-izmir-16928647"

Table 4 Statistics of the ego network of potential influencers

	Pisbogaz	Mcant	sucukarmoreno	eda-t-izmir-16928647
Node	154	112	105	94
Edge	2437	1416	1186	1296
Size	23,562	12,432	10,920	8742
Density	0.103	0.114	0.109	0.148
Network diameter	5	5	5	6
Average path length	2.355	2.173	2.339	2.199
Average degree	15.825	12.679	11.295	13.787
EgoBetweenness	4342.553	2681.28	2331.017	1495.95

Table 5 The ratio of the density difference of potential influencers

	Pisbogaz	Mcant	sucukarmoreno	eda-t-izmir-16928647
Density (ego remained)	0.103	0.114	0.109	0.148
Density (ego removed)	0.092	0.104	0.100	0.139
RDD	10.67%	8.77%	8.25%	6.08%

Table 6 The features of potential influencers

	Pisbogaz	Mcant	Sucukarmoreno	eda-t-izmir-16928647
Comment	1.5 K	488	300	817
Follower	6575	21,033	3238	1700
Photography	13.0 K	1.9 K	1.3 K	5.1 K
Following	51	70	279	44

significantly positive influence on the restaurant preferences of their followers. Furthermore, we conducted a comparison of the features of potential influencers to understand whether these features have an impact on the likelihood of a node being considered an influencer (Table 6).

In Table 6, the results show that "Pisbogaz" has the highest number of comments written and the highest number of photos uploaded. "Sucukarmoreno" has the highest number of followers, but the number of photos uploaded by "Sucukarmoreno" is fewer than that of "Pisbogaz" and "eda-t-izmir-16928647." "Mcant" has the fewest number of comments written, but "Mcant" has a higher number of followers compared to "eda-t-izmir-16928647." "Eda-t-izmir-16928647" has the fewest number of followers, but the number of photos uploaded by "eda-t-izmir-16928647" is fewer than that of "Sucukarmoreno" and "Mcant." According to these results, the number of followers is an important factor in being an effective influencer. Additionally, supporting comments with photos is another important factor in being an effective influencer.

5 Conclusion

One of the main challenges in influencer marketing is to identify the right influencers [11] who can successfully execute influencer marketing campaigns. The identification of suitable influencers in a social network is crucial for businesses to effectively reach their target audiences. In this study, we aimed to determine the potential influencers in a social network within the food industry using social network analysis (SNA). We also examined the impact of these influencers on their followers' restaurant preferences and the flow of information among individuals in their ego network.

In this study, we utilized social network analysis (SNA) centrality measurements to identify the most influential social actors based on the matrix model presented by [18] for classifying social actors. Data for our analysis was obtained from Zomato, an online food ordering and delivery application that contains user-generated content and provides a community-based platform. Since Zomato has a follower network structure, we converted the scraped data into a directed graph data type using Python. The resulting directed graph contained a total of 1755 nodes and 9483 edges. To visualize and identify the most influential nodes in our social network, we imported the directed graph data into Gephi. We calculated and compared the top 10 users with high DC, BC, and EC values. The results revealed that “Pisbogaz” was the most influential user with 1.5 K comments on restaurants, 6574 followers, 51 following, and 13 K photographs uploaded to Zomato. Additionally, we found that “sucukarmoreno,” “eda-t-izmir-16928647,” and “Mcant” were other effective potential influencers in our social network.

Furthermore, we conducted a 1-step neighborhood ego network analysis for each potential influencer to determine whether these influencers had an impact on their followers’ restaurant preferences. The results indicated that “Pisbogaz” had the greatest impact compared to other potential influencers in terms of disseminating information and influencing their followers’ restaurant preferences. Moreover, we compared the features of the potential influencers to understand whether these features influenced how users perceived them as influencers. The results indicated that the number of followers and supporting comments with photos were important factors for being an effective influencer.

Conflict of Interest The authors have no relevant financial or non-financial interests to disclose.

References

1. Anjos, C.J.F., Marques, S., Dias, A.: The impact of Instagram influencer marketing in the restaurant industry. Int. J. Serv. Sci. Manag. Eng. Technol. (IJSSMET). **13**(1), 1–20 (2022)
2. Bastian, M., Heymann, S., Jacomy, M.: Gephi: an open-source software for exploring and manipulating networks. In: Proceedings of the International AAAI Conference on Web and Social Media, ICWSM-09, vol. 3, pp. 361–362. PKP, San Jose, California, USA (2009)
3. Boerman, S.C., Müller, C.M.: Understanding which cues people use to identify influencer marketing on Instagram: an eye-tracking study and experiment. Proc. Int. J. Advert. **41**(1), 6–29 (2022)
4. Borgatti, S.P., Everett, M.G., Johnson, J.C.: Analyzing Social Networks. Sage (2018)
5. Carter, D.: Hustle and brand: the sociotechnical shaping of influence. Soc. Media Soc. **2**(3), 1–12 (2016)
6. Dias, A., Ruthes, S., Lima, L., Campra, E., Silva, M., Bragança de Sousa, M., Porto, G.: Network centrality analysis in management and accounting sciences. RAUSP Manag. J. **55**(1), 207–226 (2020)
7. Dihyat, M.M., Malik, K., Khan, M.A., Imran, B.: Detecting ideal Instagram influencer using social network analysis. Int. J. Eng. Technol. **7**(4.38), 950–954 (2018)
8. Doub, A.E., Small, M., Birch, L.L.: A call for research exploring social media influences on mothers’ child feeding practices and childhood obesity risk. Appetite. **99**(6), 298–305 (2016)
9. Febrianta, M.Y., Yusditira, Y., Widianesty, S.: Application of social network analysis for determining the suitable social media influencers. Int. J. Res. Bus. Soc. Sci. (2147–4478). **10**(6), 349–354 (2021)
10. Gamage, T.C., Ashill, N.J.: #Sponsored-influencer marketing: effects of the commercial orientation of influencer-created content on followers’ willingness to search for information. J. Prod. Brand. Manag. **32**(2), 316–329 (2023)
11. Gretzel, U.: Influencer marketing in travel and tourism. In: Advances in Social Media for Travel, Tourism and Hospitality. Routledge (2017)

12. Hepworth, A.D., Kropczynski, J., Walden, J., Smith, R.A.: Exploring patterns of social relationships among food bloggers on twitter using a social network analysis approach. J. Soc. Struct. **20**(4), 1–21 (2019)
13. Hudders, L., De Jans, S., De Veirman, M.: The commercialization of social media stars: a literature review and conceptual framework on the strategic use of social media influencers. Int. J. Advert. **40**(6), 327–375 (2021)
14. Jacomy, M., Venturini, T., Heymann, S., Bastian, M.: ForceAtlas2, a continuous graph layout algorithm for handy network visualization designed for the Gephi software. PLoS One. **9**(6), e98679 (2014)
15. Litterio, A.M., Nantes, E.A., Larrosa, J.M., Gómez, L.J.: Marketing and social networks: a criterion for detecting opinion leaders. Eur. J. Manag. Bus. Econ. **26**(3), 347–366 (2017)
16. Mainolfi, G., Marino, V., Resciniti, R.: Not just food: exploring the influence of food blog engagement on intention to taste and to visit. Br. Food J. **124**(2), 430–461 (2022)
17. Rum, N.M.S., Yaakob, R., Affendey, L.S.: Detecting influencers in social media using social network analysis (SNA). Int. J. Eng. Technol. **7**(4.38), 950–957 (2018)
18. Scoponi, L., Pacheco Días, M., Pesce, G., Durán, R., Schmidt, M.A., Gzain, M.: Redes de cooperacióncientífico-tecnológica para la Innovaciónenagronegociosen dos universidadeslatinoamericanas. Universidad y agronegocios, Editorial de la Universidad Nacional del Sur, Cap. **3**, 165–211 (2016)
19. Tabassum, S., Pereira, F.S., Fernandes, S., Gama, J.: Social network analysis: an overview. Wiley Interdiscip. Rev.: Data Min. Knowl. Discov. **8**(5), 1–21 (2018)
20. Tan, W.B.: A study on the centrality measures to determine social media influencers of food-beverage products in Twitter. J. Inst. Eng. **82**(3), 19–26 (2021)
21. Teutle, A.R.M.: Twitter: network properties analysis. In: 20th International Conference on Electronics Communications and Computers (CONIELECOMP), pp. 180–186. IEEE, Cholula, Puebla, Mexico (2010)
22. Wibisono, A.I., Ruldeviyani, Y.: Detecting social media influencers of airline services through social network analysis on Twitter: a case study of the Indonesian airline industry. In: 3rd East Indonesia Conference on Computer and Information Technology (EIConCIT), pp. 314–319. IEEE, Surabaya, Indonesia (2021)

Part III
Signal Processing

DSP Runtime Emulator on FPGA: Implementation of FIR Filter Using Neural Networks

Hoda Desouki, Hassan Soubra, and Hisham Othman

1 Introduction

Digital signal processing is still one of the ever-growing fields in the electronics industry as it has various applications such as audio and speech processing, radar, telecommunications, image processing, and speech recognition. Furthermore, a Digital Signal Processor (DSP) is a dedicated microprocessor chip that is highly optimized for digital signal processing. DSPs need to have very high throughput as they operate under real-time constraints which distinguish them from other Reduced Instruction Set Computer (RISC)-based processors. That is why DSPs can execute high-speed multiplications and additions with specially designed hardware called a Multiply Accumulate (MAC) unit integrated into the architecture of the DSP. Furthermore, the architecture of DSPs is Harvard to make use of having two separate busses for data and instruction memories [1].

Digital filters take different forms, notably Finite Impulse Response (FIR) and Infinite Impulse Response (IIR) forms. FIR filters are usually preferable to IIR ones as they are more stable, less sensitive to noise, have a linear phase response, and do not tend to accumulate errors, thanks to their non-recursive nature. However, due to its non-recursive nature, more taps are needed for FIR to get a similar performance to that of IIR which results in longer computation time, less bandwidth, and more memory requirements. This bottleneck can be overcome by choosing a highly parallelized hardware architecture with efficient memory such as a Field-

H. Desouki (✉) · H. Othman
German University In Cairo, Cairo, Egypt
e-mail: hdesouki27@gmail.com; hisham.othman@guc.edu.eg

H. Soubra
Ecole Centrale d'Electronique-ECE, Lyon, France
e-mail: hsoubra@ece.fr

M. N. Seyman (ed.), *2nd International Congress of Electrical and Computer Engineering*, EAI/Springer Innovations in Communication and Computing,
https://doi.org/10.1007/978-3-031-52760-9_25

Programmable Gate Array (FPGA) [2]. In [3], a comparison between FPGA and DSP solutions is presented with a 40-order bandpass FIR filter. It was shown that in terms of execution time, the FPGA offers better performance in terms of execution time and power consumption than DSPs; however, it requires more silicon area.

This paper proposes a real-time emulator for a lowpass FIR filter design by designing a Neural Network (NN) to predict the filter operation on an FPGA. The Nios II soft-core processor is used. The performance of the neural network is benchmarked against a software implementation of the filter operation, showing improvement in the performance [4]. This paper is organized as follows: Sect. 2 presents the literature review related to our work. Section 3 presents our methodology. Section 4 describes the implementation of our case study. Section 5 presents the results. Finally, a conclusion follows in Sect. 6.

2 Literature Review

2.1 Using Machine Learning for DSP Operations

In [5], a 6-tap FIR filter of optimized performance and optimized power consumption is implemented using a Multiplier and Accumulation Control (MAC) unit provided by the Xilinx block set available in the Simulink library, and the resources used are computed using Xilinx ISE software which provides the basic concept of the neural network. The demonstrated design offers better performance than that of a typical FIR filter. The FIR filter is synthesized and implemented in Xilinx FPGA environment utilizing Xilinx ISE platform, then the performance in terms of LookUp Tables (LUTs), Flip Flops (FFs), and power is evaluated. The approach can be divided into three stages: simulating the neural network taking into account the most convenient activation function to get the nonlinear results, the second is the design of the network, and the use of the software tool of Xilinx ISE and System Generator tool for Simulink to implement the hardware, and the third is a convenient utilization of the system. The Multiplier and accumulator (MAC) block where the implementation of MAC can be designed using the Xilinx generator block without the dedicated MAC blocks. Neuron input weights are stored in ROM while temporary storage of data are stored in RAM. LUTs implement the transfer function for the linear operation. The proposed design gives improved performance in terms of LUTs, FFs, and especially power compared to that of the conventional FIR filter.

2.2 *Using Soft-Core Processors to Implement DSP Operations*

In [6], the performance of digital signal processing algorithms is assessed on the Nios II/f soft-core RISC-based processor on the Cyclone IV FPGA on the Altera DE2-115 development board. The operations are then validated and compared against Texas Instruments' TMS320DM64x+ DSP. Furthermore, IF condition, FOR Loop, and 2D convolution are the operations that were chosen as test cases. The timing reports for these operations are used for the comparison between the Nios II processor and the TMS320DM64x+ chip. Moreover, the instructions for the tested operations are demonstrated with their equivalent ones from the TMS320DM64x+ technical reference manual. In addition, they are divided into three categories: arithmetic operations, loading and storing, and comparisons and branching. For the arithmetic instructions, both Nios II and TMS320DM64x+ DSP have the same execution time except for the multiplication with immediate instruction which has a better performance on the TMS320DM64x+ DSP. Whereas for the loading and storing instructions, the performance of Nios II fluctuates which makes TMS320DM64x+ more reliable for performance. Comparison instructions take the same amount of clock cycles for both processors while the branching instructions perform better on Nios II.

3 Approach

The goal is to build a lowpass FIR filter using neural networks on the Nios II soft-core processor in the C programming language. It is divided into four stages:

1. Building the NN model using the machine learning TensorFlow environment for Python. It was chosen as it enables fast experimentation where an NN can be built in just a few lines of code. Unlike the TensorFlow API for C, it is very well documented which facilitates and shortens the time required for development.
2. Importing the model to MATLAB for code generation into C using MATLAB Coder. It can be used for converting neural networks designed in MATLAB to their equivalent C code when used with the Deep Learning Toolbox provided by MATLAB.
3. Building the hardware system architecture using the Quartus Prime Lite Edition which is a programmable logic device design software. The Platform Designer, which is a tool, provided by Quartus is a system integration tool that offers the ability to generate automatic interconnect logic to connect Intellectual property (IP) functions and subsystems.
4. Deploying the code for inference on the FPGA using Eclipse IDE for Nios II.

As illustrated in Fig. 1, the input to our system is an unfiltered digital signal in the form of an audio file in .wav format. Next, the neural network in C does the filtering

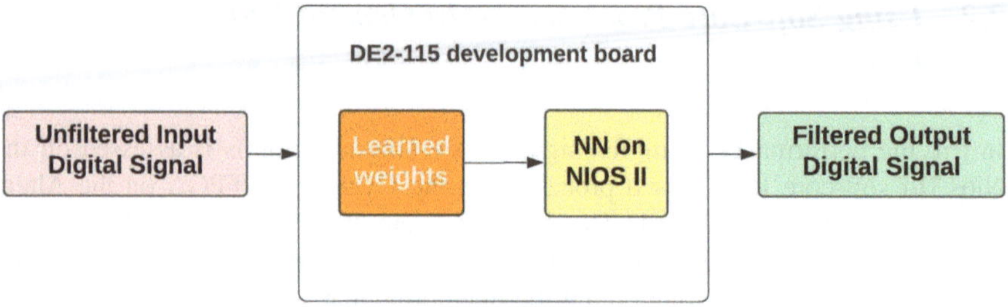

Fig. 1 A model showing an overview of the system

operation with the help of the learned weights and returns the filtered digital signal output samples in an array which can be written back as an audio file.

4 Implementation

4.1 Overview

The FIR filter operation can be described using a simple discrete-time convolution summation:

$$y[n] = \sum_{k=0}^{N-1} h[k] \cdot x[n-k] \tag{1}$$

In Eq. (1), $x[n]$ is the input signal, $y[n]$ is the output signal, $h[k]$ represents the filter coefficients, and N is the filter order. From Eq. (1), a single output sample can be obtained by the multiplication of the N coefficients with their corresponding input samples and then performing a summation of these weighted input samples, so a single output sample requires N input samples: the current input sample in addition to $N-1$ delayed input samples.

Linear regression models are analogous to the linear equation of algebra where the inputs are mapped to the output by multiplying each of these inputs by their respective weights and adding them together to get the output then adding a bias factor. From Eq. (1) and from the equation of the linear regression model, FIR filters can be mapped to the linear regression model equation where the input signal samples represent the input layer, the filter coefficients represent the weights, the output layer represents the output signal, and the bias factor is set to zero.

4.2 Software Implementation

The model was initialized as a sequential model with only one hidden fully-connected dense layer with an input dimension equal to the number of filter coefficients which was chosen to be 13, and the weights were initialized as zeroes. Figure 2 shows the NN architecture.

Furthermore, the model was compiled by setting the optimizer to Adam and choosing the mean squared error as the loss function. A (10000, 1) NumPy array of random floating-point numbers was used as the input for training. However, the input must be in the shape (9988, 13) as explained in the overview section that each output sample is generated from a current input sample in addition to $N-1$ previous samples. This can be achieved by performing a window-sliding technique that chunks an array into each N-consecutive array elements (where N is the fixed window size and is equal to 13 which is the number of learnable weights of the model) by dropping the earliest sample and adding the latest one. This technique reduces the time complexity to $O(n)$ (n is the size of the input array) instead of using the brute force technique with nested-for loops of time complexity $O(n^2)$.

The output label of the input test data is the output of an applied FIR filter to the randomly generated input array. t can be obtained by using the built-in function *firwin* defined in the SciPy Python library which generates the FIR filter coefficients using the window method by specifying some parameters: The number of filter taps, $N = 13$, the cut-off frequency and the sampling frequency which both depend on the target signal, pass_zero = True as the filter is a lowpass one and finally the window method where "blackmanharris" window was chosen as it has a wide main lobe and a narrower side lobe than that of the standard Blackman window due to the extra added cosine terms [7]. After generating the coefficients, the convolution operation is performed to get the output as a NumPy array with dimension (9988, 1). Having the input and the output training data, the model is then trained.

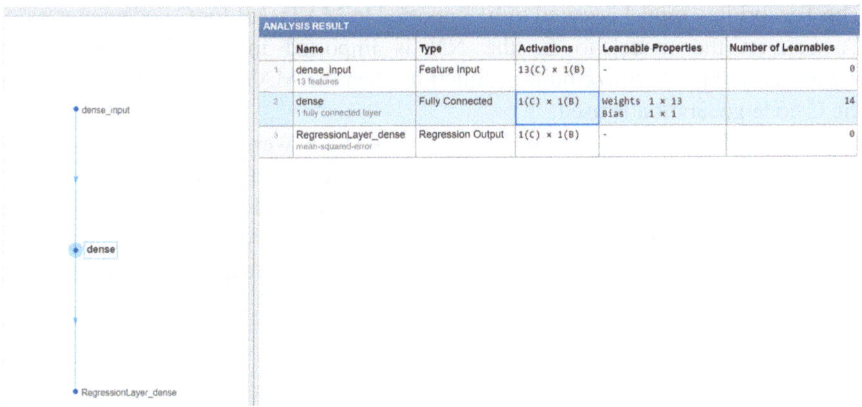

Fig. 2 Neural network architecture

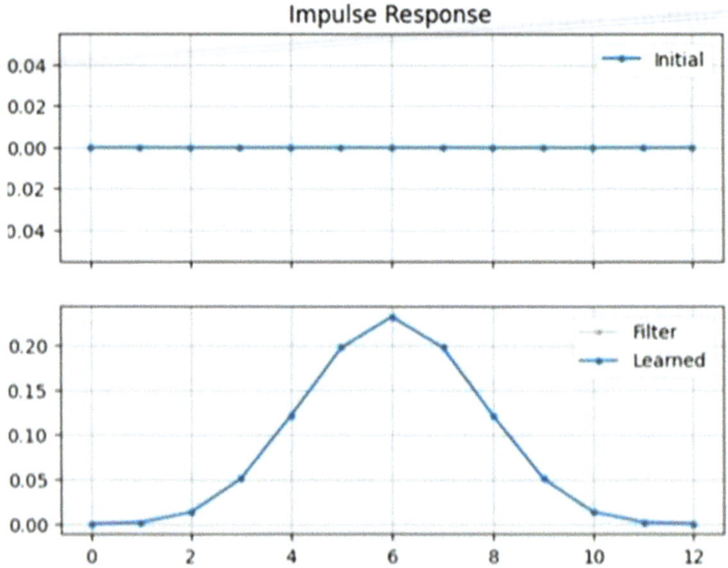

Fig. 3 Impulse response of the filter and the network weights

A plot of the initial network weights and a plot of the learned network weights versus the impulse response of the filter and its frequency domain are shown in Figs. 3 and 4. As it can be shown in both figures, the weights indeed take the shape of a Blackmannharris window.

To evaluate the model performance, the built-in evaluate function was used to get the loss and accuracy of the model. The input test data were in the form of a NumPy array of arbitrary sinusoidal signals and to get the output test data, the convolution is performed. The accuracy score was found to be approximately 99.9%.

The plot of the loss versus the batch is shown in Fig. 5, where the convergence of the plot and the evaluation accuracy indicate that the model is ready for prediction. The TensorFlow model is saved to be imported to MATLAB for C code generation. To convert the model to C, the NN is imported to MATLAB using the *importTensorFlowNetwork* built-in function, then the MATLAB Coder add-on for the C code generation is used.

The test data for the TensorFlow NN evaluation were also evaluated on the MATLAB NN. The MATLAB NN predictions and the actual filtered signal of the FIR filter were plotted in the time domain as shown in Figs. 6 and 7 and in the frequency domain as shown in Figs. 8 and 9. From observing the aforementioned figures, it can be concluded that the model is functioning as expected and is ready for code generation.

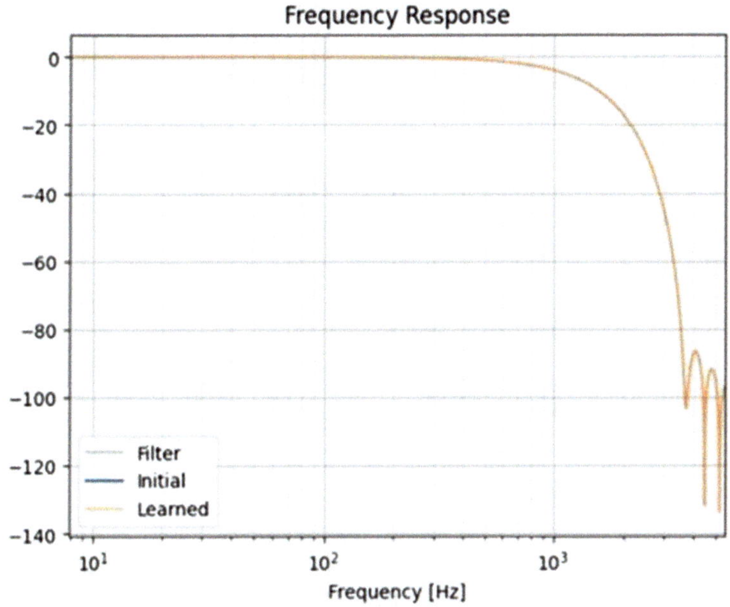

Fig. 4 Frequency response of the filter and the network weights

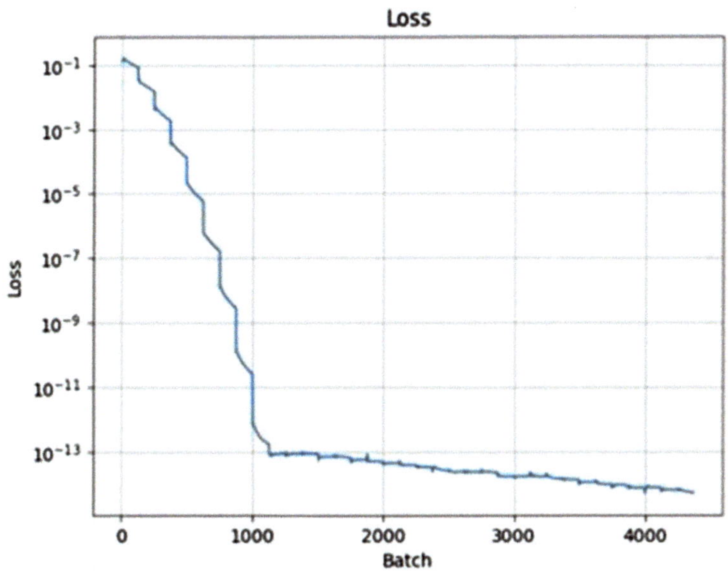

Fig. 5 Loss versus batch after training the neural network

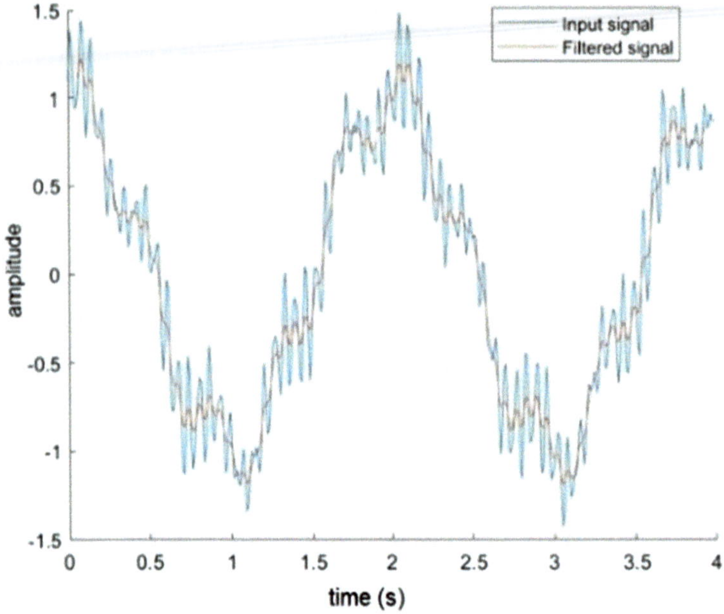

Fig. 6 The input signal versus the filtered signal in the time domain

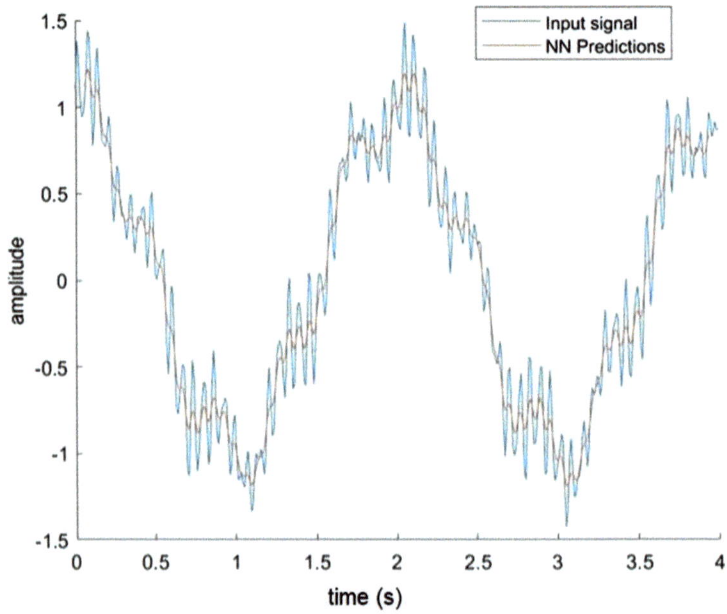

Fig. 7 The input signal versus the neural network prediction in the time domain

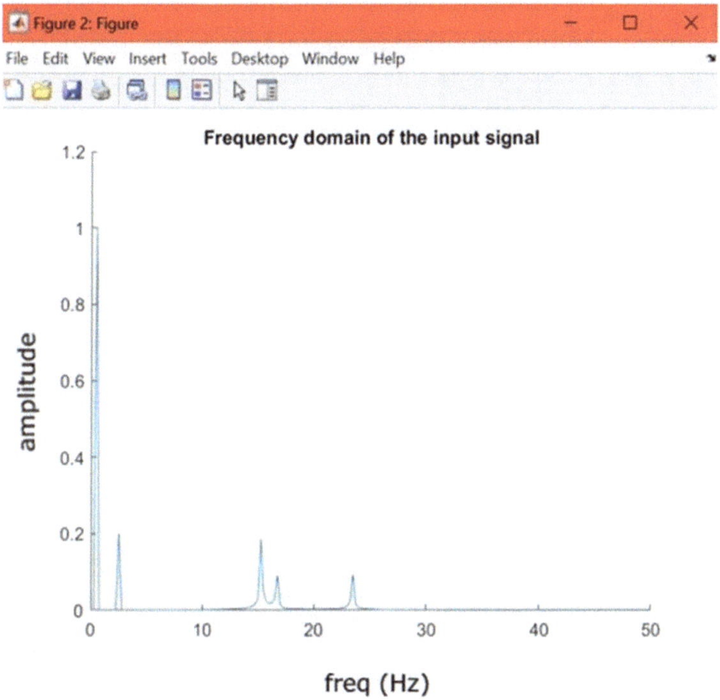

Fig. 8 Input signal in the frequency domain

4.3 Hardware Implementation

The DE2-115 development board is used. The used modules are a 50 MHz clock which is the maximum clock frequency available on the board, the Nios II processor core, SDRAM as the memory module, and a JTAG interface along with a performance counter module.

Nios II Processor It is a soft-core 32-bit RISC-based processor. It is deemed appropriate as the processor of the system for two reasons: (1) It uses Harvard architecture which is crucial for DSP operations as indicated earlier in this paper.

(2) It allows programming the FPGA in the C programming language on the Eclipse IDE where building neural networks in C language is much easier than building the model in HDL from scratch which reduces the development cycle time. The Nios II/f core is used as it is designed to have maximum performance. It has hardware multipliers that can accelerate the NN performance and separate data and instructions caches along with other features.

Memory Module SDRAM is used as the memory module for both data and instructions. It was used instead of using the on-chip memory FPGA IP as it is very limited with a size of 64 KB only whereas the neural network requires a

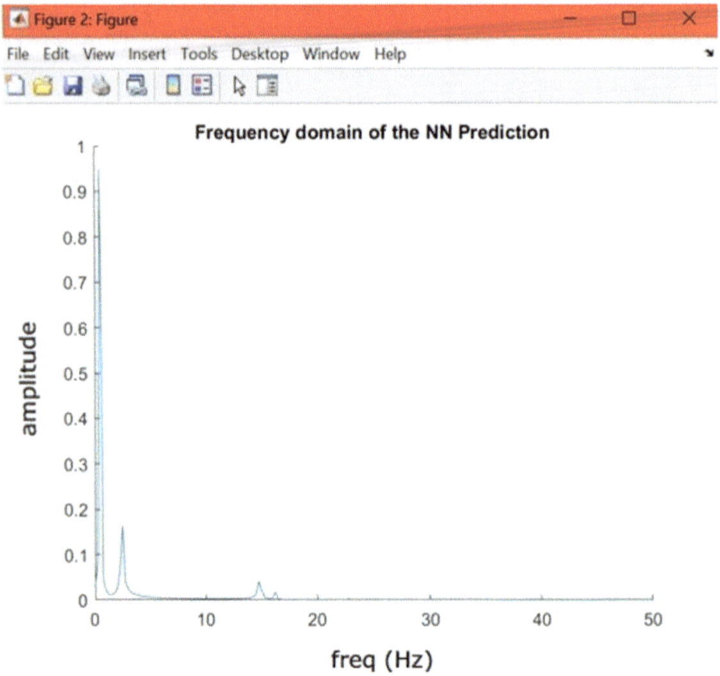

Fig. 9 Neural network prediction in the frequency domain

dedicated memory chip exceeding the on-chip memory capacity. The DE2-115 development board has up to 128 MB SDRAM implemented using two 64 MB SDRAM Devices. It can be managed through the Intel FPGA IP SDRAM controller. Moreover, the clock signal of the SDRAM chip must lead the system clock by 3 nanoseconds. A Phase-Locked-Loop (PLL) circuit needs to be manually created as an IP module to overcome this clock skew between the SDRAM and the system clock [8].

JTAG Interface Module The JTAG interface allows communication between the host computer having the HDL code and the FPGA.

Performance Counter Module The performance counter module is added to measure the execution time of the different code sections both in clock cycles and in seconds.

A full picture of the system after integration is shown in Fig. 10. The Verilog code was then automatically generated.

Later, the proper pin assignments are performed using the Pin Planner tool by referring to the user manual of the board [9]. Then, the system is compiled and the code is flashed to the FPGA. After generating the HDL and synthesizing the system onto the FPGA, a *.sopcinfo* file is created which is used to create the Board Support

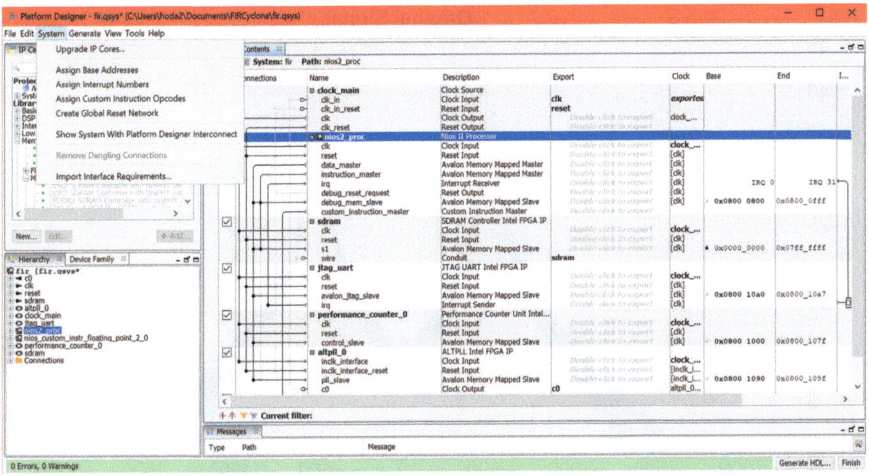

Fig. 10 Integration of the system modules

Package (BSP). After building the BSP, a user application project is created and the generated code is added.

5 Experimental Results

5.1 Time Profiling Approach

Since the goal is to have a runtime emulator for the FIR operation on the FPGA, the execution time of the NN needs to be measured to assess its performance. This was achieved by using the Performance Counter core that was added to the system as a hardware module. It was chosen over other profiling methods as it can be used for real-time profiling without obstructing the code execution. It also requires only a single instruction to start and stop profiling, and no RAM. In addition, it can be accessed in software through the Performance counter API provided by Altera which provides routines to access the hardware core. The API consists of functions, macros, and constants that define the low-level access to the hardware and provide control and reporting functionalities [10].

5.2 Testing the Neural Network on the FPGA

The NN performance on the FPGA is tested using the sample audio file, created by [11]. The frequency analysis of the input file sample and the NN output were plotted

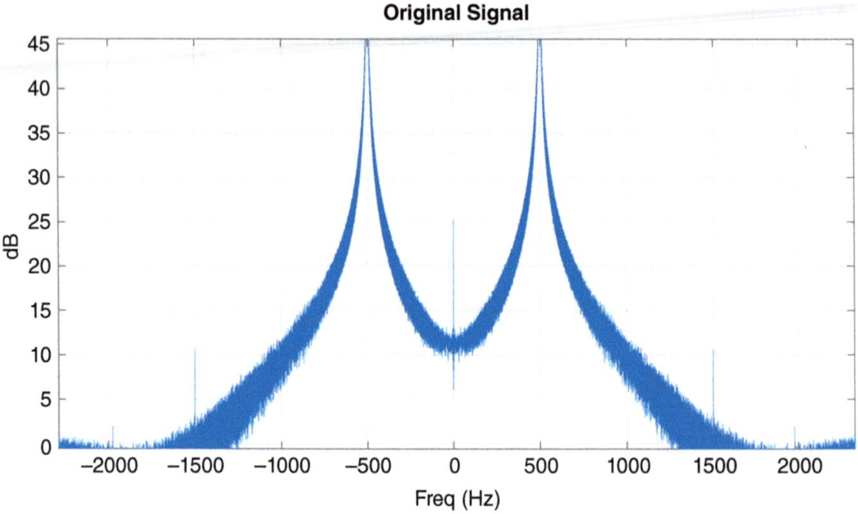

Fig. 11 Frequency analysis of the original signal

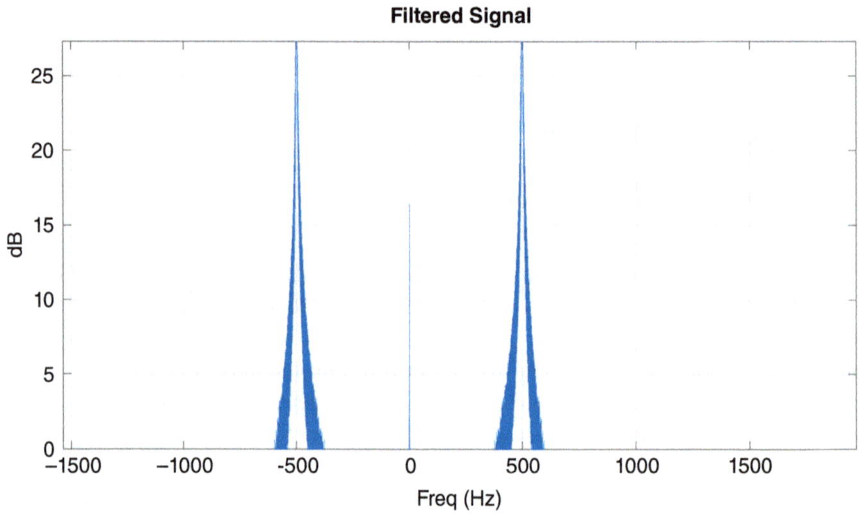

Fig. 12 Frequency analysis of the filtered signal

in MATLAB to visualize the accuracy of the NN in C. Figure 11 shows the signal before filtering, whereas Fig. 12 shows the output of the NN in C. It can be shown that the frequencies higher than 500 Hz were attenuated. Figure 13 shows the NN profiling results where it took the NN approximately 6.759 s and 337,968,675 clock cycles to predict the output.

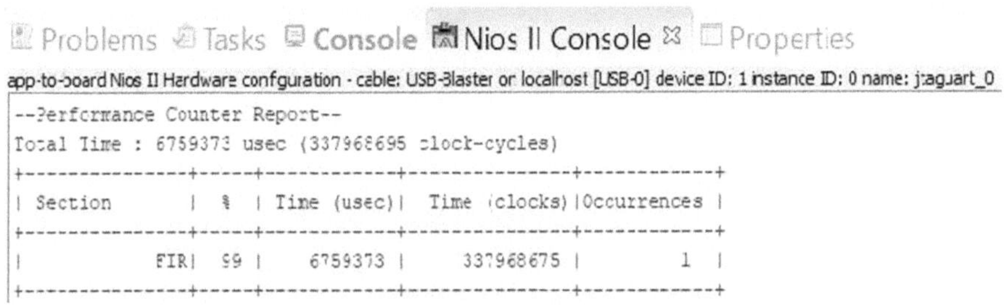

Fig. 13 The execution time report of the neural network

5.3 *Performance Optimization*

As a real-time emulator, the performance of the neural network must be optimized as much as it could be. In general, there are two approaches for optimizing performance: software optimization and hardware optimization. The main objective of software optimization regarding performance is to reduce the time complexity of the algorithms, either by replacing them with more optimal ones or by analyzing the code thoroughly to check for unnecessary time-consuming code sections, e.g., removing loops that could be eliminated without altering the code functionality. On the other hand, hardware optimization can range from using a higher frequency oscillator, modifying the hardware architecture, or moving the software to another hardware platform with higher performance.

Software Optimization Taking a closer look at how the code functions, it was found that two calls to the time-consuming *memcpy* C function could be eliminated by passing the pointer to the array having the final output directly to the *FullyConnectedLayer_predict* function instead of having to do it on two intermediate steps. This improved the performance by 19% as the NN profiling results show an execution time of 5.481 s, and 274,096,895 clock cycles to predict the output.

Hardware Optimization The Nios II Floating-Point Custom Instruction was integrated into the system using the Platform Designer to accelerate the NN execution time. The NN performs floating-point arithmetic frequently, especially the multiplication and addition operations; hence, it is very convenient to have a dedicated hardware unit that optimizes these operations. The basic operations of floating-point custom instructions include single-precision floating-point addition, subtraction, and multiplication which when integrated into the hardware, the Nios II Software Build Tools (SBT) for Eclipse compiles the code to use the custom instructions for floating-point operations. In this project, the Floating-point hardware 2 unit (FPH2) was used as recommended by Intel for getting the most optimal performance and for minimal device footprint [12]. Figure 14 shows its integration into the system. Next, the HDL is regenerated and flashed to the FPGA, the BSP is rebuilt, and the code is run. The time-profiling report indicated that the NN execution time is

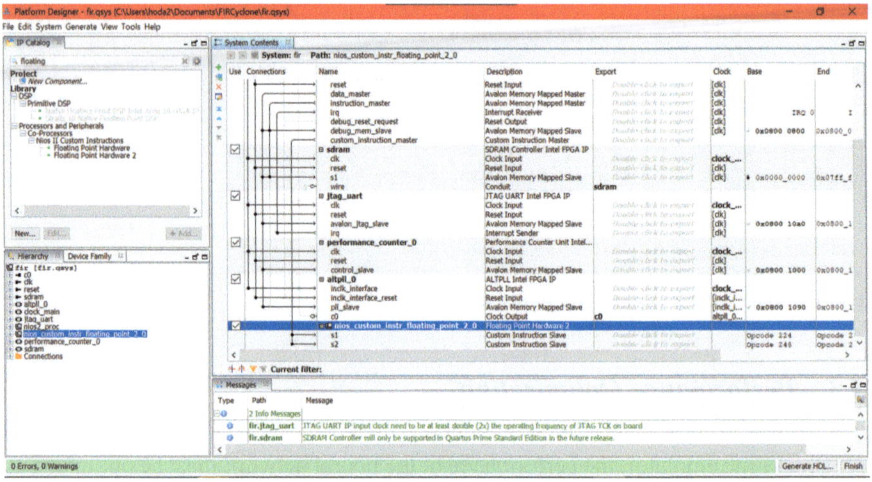

Fig. 14 Integrating the FPH2 to the system

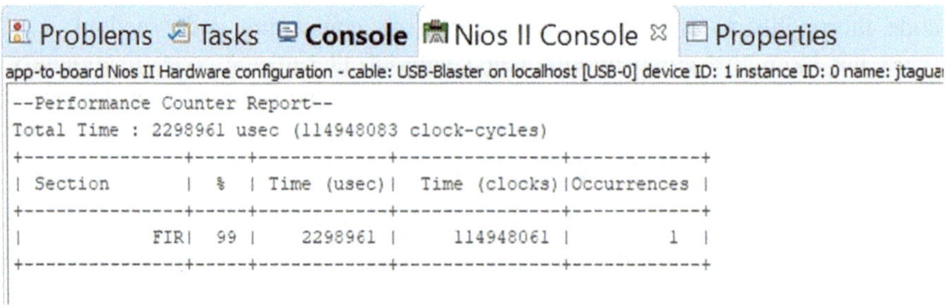

Fig. 15 The execution time report of the NN after adding the FPH2

2.298 s, and 114,948,083 clock cycles as shown in Fig. 15 which is approximately 66% improvement to the performance of the original code before any optimizations.

5.4 *Results Analysis*

The software implementation of the FIR filter by [4] was chosen to be used as a benchmark for our approach. The FIR filter in [4] has a 32-bit datapath and constant coefficients. The software was implemented based on the soft-core Nios II on a Cyclone III FPGA Development Kit equipped with an EP3C120 FPGA device. All of these factors made their software C++ implementation a suitable candidate to compare its performance against our approach. The same implementation steps were

Table 1 FIR evaluation

Implementation	Execution time (s)[a]	Logical cells
SW [4]	3.827	4309
NN (DE2-115 board)	**2.618**	5665
NN (DE10-Lite board)	3.434	5908

[a]Execution time for 61,234 samples

performed again to design a neural network for the 15-tap FIR filter and the generated C code of our earlier chosen sample created by [11], was tested on the DE2–115 board (Cyclone IV FPGA family) and DE10-Lite board (Max 10 FPGA family). From Table 1, it can be shown that the performance of the NN on both the DE2–115 board and the DE10-Lite board is better than that of the direct software implementation.

6 Conclusion

The goal of this paper was to propose a real-time emulator that offers more agility than that of conventional DSPs for the FIR filter operation using a neural network on the FPGA-based Nios II/f soft-core processor while optimizing the performance. This is achieved in two stages: software implementation and hardware implementation. Software implementation involved designing the neural network and then importing it to MATLAB for C code generation while the hardware implementation involved identifying and integrating the needed modules for the system. A sample audio file was used to test the emulator which proved its accuracy. Time profiling was performed, and its results were recorded before and after applying optimization techniques.

The results presented in this paper confirm that the use of the emulator is preferable to implementing the FIR operation in software in terms of speed. In addition, the use of the DE2-115 board was found to be more optimal than using the DE10-Lite board as the data width of the SDRAM on the DE10-Lite board is 16-bit while the Altera Avalon bus connecting the architecture is a 32-bit bus.

Using Nios II is more flexible yet not the best choice for acceleration compared to tailored hardware circuits as in [13]. However, our approach is more resource efficient than using hardware circuits because it only uses 5% of the Cyclone IV FPGA, and this percentage remains constant regardless of the number of taps in the filter. In contrast, in [13], the number of DSPs, LUTs, and FFs increases as the number of filter taps increases.

References

1. Frantz, G.: Digital signal processor trends. IEEE Micro. **20**(6), 52–59 (2000)
2. Austerlitz, H.: Chapter 10 – data processing and analysis. In: Austerlitz, H. (ed.) Data Acquisition Techniques Using PCs, 2nd edn, pp. 222–250. Academic (2003)
3. Diouri, O., Gaga, A., Ouanan, H., Senhaji, S., Faquir, S., Jamil, M.O.: Comparison study of hardware architectures performance between FPGA and DSP processors for implementing digital signal processing algorithms: application of FIR digital filter. Res. Eng. Des. **16**, 100639 (2022)
4. Possa, P., Schaillie, D., Valderrama, C.: FPGA-based hardware acceleration: a CPU/accelerator interface exploration. In: 2011 18th IEEE International Conference on Electronics, Circuits, and Systems, pp. 374–377. IEEE, Beirut, Lebanon (2011)
5. Chauhan, A., Kumar, P.S.: Implementation of FIR filter & Mac unit by using neural networks in FPGA. In: 2018 International Conference on Advances in Computing, Communications and Informatics (ICACCI), pp. 2496–2501. IEEE, Bangalore, India (2018)
6. Shamseldin, A., Soubra, H., ElNabawy, R.: Performance of DSP operations implemented using a soft microprocessor: a case study based on Nios II. In: 2021 International Conference on Microelectronics (ICM), pp. 66–69. IEEE, New Cairo City, Egypt (2021)
7. Kaur, M., Kaur, S.: FIR low pass filter designing using different window functions and their comparison using MATLAB. Int. J. Adv. Res. electr. Electron. Instrum. Eng. **5**(2), 753–758 (2016)
8. Intel: using the SDRAM on Intel's DE2-115 board with verilog designs. https://ftp.intel.com/Public/Pub/fpgaup/pub/Teaching_Materials/current/Tutorials/Verilog/DE2-115/Using_the_SDRAM.pdf (2019, March)
9. Finlayson, M.A.: De2-115 user manual world leading FPGA based products and design services (2017)
10. Performance Counter Core. Available at https://pages.mtu.edu/~saeid/multimedia/labs/Documentation/qts_qii55001_Performance_Counter_Core.pdf (2009)
11. Franco, T.H.: Lowpass FIR filter on.wav file with windowing. https://www.mathworks.com/matlabcentral/fileexchange/20986-lowpass-fir-filter-on-wav-file-with-windowing (2008, August)
12. Intel: introduction to Nios II floating point custom instructions. https://www.intel.com/content/www/us/en/docs/programmable/683242/current/introduction-to-floating-point-custom.html (2020, April)
13. Mirzaei, S., Hosangadi, A., Kastner, R.: FPGA implementation of high-speed FIR filters using add and shift method. In: 2006 International Conference on Computer Design, pp. 308–313. IEEE, San Jose, CA, USA (2006)

A Real-Time Adaptive Reconfiguration System for Swarm Robots

Nora Kalifa, Hassan Soubra, and Nora Gamal

1 Introduction

Swarming, the idea of forming large groups to complete a task, originally comes from nature. A single ant, bee, fish, bird, or even bacteria cannot complete a task on its own. However, with the power of a swarm and collaboration, the highly complex task becomes simple. The idea of swarm robots was first proposed in 1993 [1]. Swarm robots consist of simple robots that work together to accomplish a complex task by communicating and coordinating with each other. With the technological advances today, swarm robots proved to be highly beneficial, especially in hazardous zones, in numerous fields, e.g., agriculture, military, and industrial. Another advantage of swarm robots is their ability to adapt to the environment using different sensors. The communication between the robots also helps in the adaptation as they can share information about the surroundings.

Although the field of adaptive swarm robots is relatively new, a lot of advances are yet to be made, numerous algorithms have been developed for swarm robots. Algorithms include ways in which the task can be divided among the swarm, ways they can optimally reach a goal, and different communication protocols. Nevertheless, reconfiguration systems in the context of Swarm robotics are still scarce. One or more robots of the swarm can be lost or damaged as a consequence of different failures and errors, e.g., software crashes, hardware errors, and low energy levels. This can cause a task to fail because of communication problems or because of an

N. Kalifa (✉) · N. Gamal
German University in Cairo, Cairo, Egypt
e-mail: Nora.khalifa@student.guc.edu.eg; Nora.mohammed@guc.edu.eg

H. Soubra
ECE-Ecole Centrale d'Electronique, Lyon, France
e-mail: hsoubra@ece.fr

M. N. Seyman (ed.), *2nd International Congress of Electrical and Computer Engineering*, EAI/Springer Innovations in Communication and Computing,
https://doi.org/10.1007/978-3-031-52760-9_26

insufficient number of remaining robots working on the task. Furthermore, a damaged robot could halt the execution of a crucial sub-task necessary for completing the main task. A reconfiguration system needs to be put in place to ensure the completion of the main task even when a robot is damaged.

This paper presents a reconfiguration system for adaptive swarm robots based on the energy level of the robots in the swarm. The reconfiguration focuses on homogenous swarms that execute a task by dividing it into two sub-tasks, in which, each sub-task would be executed by a sub-swarm of the whole swarm.

The rest of this paper is structured as follows: The literature review section discusses previous work and related research. Our approach section presents the methodology used. Moreover, our system proposed section provides details about how the proposed system was implemented in both the simulation and real-life implementation. The testing section discusses the test cases conducted as well as their results. The results and discussion section discusses the system test results with a brief analysis of those results. The future work section identifies the limitations faced and suggestions for future work. Finally, the conclusion section gives a summary of the paper as well as the results obtained.

2 Literature Review

Pini et al. [2] discussed the idea of dividing a task into two subtasks to be able to reduce the task's complexity. The robot would decide whether it is more beneficial to divide the task or not. If the task is divided into two sub-tasks, then the partitioning of the task is done automatically by the self-organizing feature of swarm robots. Wei et al. [3] developed a further method that dynamically divides the tasks into sub-tasks and then divides them among the robots in the swarm.

Hanlin et al. [4] developed a task-swapping approach for homogenous swarm robots. The swarm robots were given the task of creating a shape by the human operator while having a deadlock-free guarantee as well as an absolute collision avoidance guarantee. The robots would pick a goal which would be their location in the shape according to the information sensed and gathered from local robots. If two robots select the same location, then one of the robots would swap its task and move to another free location.

Liu et al. [5] created a bilateral matching approach for task-based network reconfiguration in heterogeneous UAVs. In the system of UAVs, each slave communicates with a master, and the master sends the message to ground control. This decreases network traffic. The main problem faced is when one of the masters suffers damage. The slave UAV would be unable to send data obtained to ground control. A reconfiguration in the slave-master connections is done using a bilateral matching approach.

While both approaches of using heterogeneous and homogeneous swarm robots exist on their own, Pinciroli et al. [6] combined both approaches. The goal was to show that minimal communication among heterogeneous sub-swarms is sufficient to create a coherent global behavior. The first homogeneous sub-swarm consisted of wheeled robots, while the second consisted of flying robots. The approach proved to be effective in a way that makes teams able to operate despite communication errors.

Lee et al. [7] proposed an approach to divide the robots among the swarm into different foraging subtasks according to the task demands. Those robots depend on their knowledge of the task demands and sensing of their neighboring robots. Using their local information, the robot uses a task selection function to decide whether they would switch the task or not. Since each robot responds differently to this function, a self-organizing aspect is shown. Therefore, the system converges to the optimal task distribution. Simulation results verify that the robots re-distribute themselves among the tasks using the proposed approach.

UAV battery levels run out quickly; hence, Timothy et al. [8] proposed an approach to efficiently send out search UAVs into an unknown environment. While his approach successfully minimized the energy consumed by the UAVs to finish the task, however, it took a longer time. Chen et al. [9] proposed another approach that would minimize the energy consumed by the swarm robots and therefore reduce the costs. The swarm robots would be sent out into the environment to search, and when it detects that its energy level is below the energy capacity threshold, it begins retreating to the nest to charge. The results of the proposed system show an overall improvement in minimizing the energy levels consumed. Sheth [10] proposed an algorithm for swarm robots to be able to manage tasks distributed along huge distances such as search and rescue missions. The algorithm optimized the distance traveled by each robot in the swarm so it would consume less energy. The need for such an algorithm was put in place because robots run out of battery fast.

Diehl et al. [11] proposed a swap algorithm for UAVs on the field with low battery levels. Agents with low batteries pause the task they are doing and return to the launch site. A human operator then picks a UAV with fully charged batteries from the launch site and sends it to resume the task. However, swaps may become slow if the human operator has to handle several parallel swaps at the same time. The results of the field experiments confirm an increase in the mission's success and outcome consistency.

Although many studies proposed different types of swarm robots by implementing different algorithms for doing tasks or dividing tasks, works on reconfiguration when one of the robots goes off the grid due to some error, failure, or damage for homogeneous swarm robots are still scarce. Moreover, real-life implementations for such reconfiguration systems have not been proposed to the best of our knowledge.

3 Our Approach

3.1 Reconfiguration System

As depicted in Fig. 1, the reconfiguration system is as follows: each sub-swarm starts executing with the maximum number of robots for its specific task to obtain optimal

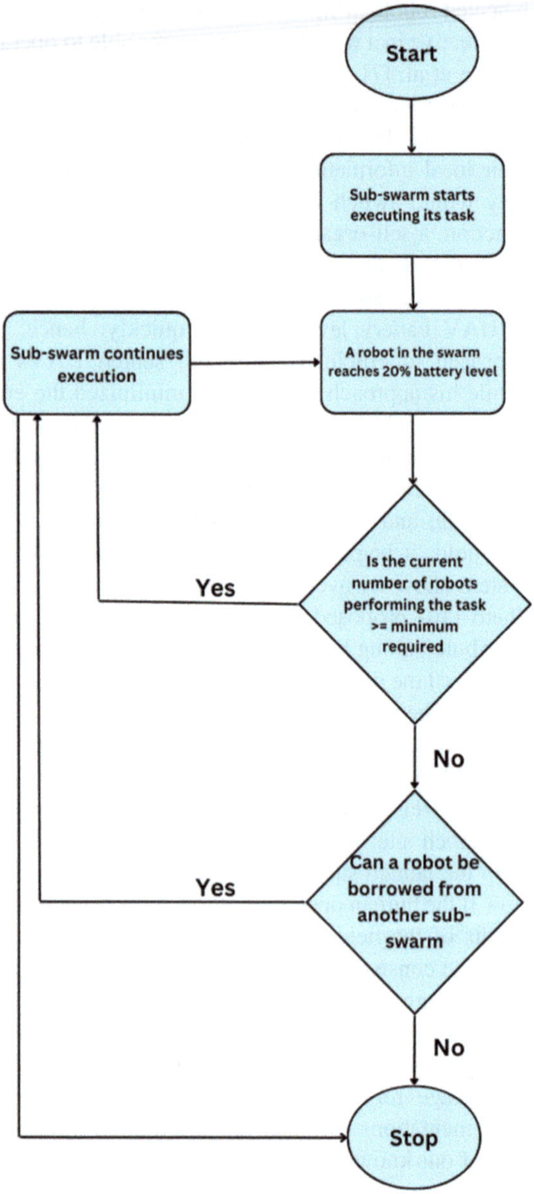

Fig. 1 Reconfiguration system overview

results. Moreover, each sub-swarm has an optimal minimum number of robots determined before the start of execution to guarantee the proper execution of a task. Once a robot reaches 20% of its battery level, it is excluded from the sub-swarm, and if the number of robots falls below the minimum number required, then if possible, it will borrow a robot from the other sub-swarm. The robot borrowed would straightforwardly switch the task it is doing to replace the excluded one. This is possible since all the robots are homogeneous. They might be doing different tasks in different sub-swarms, but in the end, they all contain the same set of software and hardware features. A controller is used to keep track of which robot is low on battery and to reconfigure the robot's task. After the task has been reconfigured, the robots will go back to executing their tasks. However, the controller will not reconfigure the robot's task if the number of robots in the other sub-swarm is exactly equal to the minimum required, thereby leading to the termination of the algorithm. The other case in which the algorithm stops is when the sub-swarm finishes execution. The main difference between the two cases that stop the algorithm is one would stop it due to the failure of completing the task and the other one stops it due to the completion and success of the task.

3.2 Swarm Communication

Swarm communication is usually direct between the robots in the swarms or explicit with a portion of the swarm. However, the communication for reconfiguration is set up differently than the way swarm robots would normally communicate with each other. As shown in Fig. 2, there is no direct communication between the robots in the swarm. The robots would communicate with the controller and the controller would make a decision based on the message sent. Finally, the controller would send the message to a specific robot or broadcast the message to all robots according to the decision it made.

A radio communication module is used in the real-life implementation, and a radio transceiver and receiver are used in the simulation. Those modules broadcast messages to all recipients. Although for each implementation the communication is set up using different modules, they both follow the same procedures as they both communicate through radio waves. Therefore, abbreviated messages are sent for clarity and efficiency.

To set up the communication, first, the controller sends a message to all robots and waits to receive an acknowledgment from them to make sure that the

Fig. 2 Communication flow

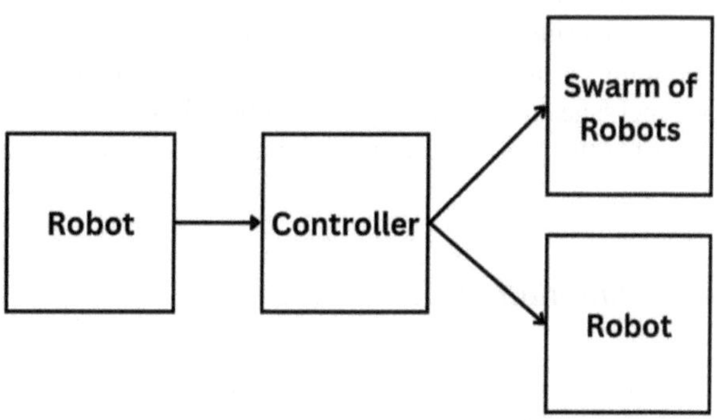

communication between them is established. The controller then keeps listening for any message coming from the robots, indicating that their battery level is at 20%. Then, the controller decides to either reconfigure the robots or to announce the failure of the task. If it decides to reconfigure, then it picks a robot to send a message to the new task number they will be performing. If it decides to announce task failure, then a message is broadcast to all robots to end performing their tasks.

The system starts working by having each sub-swarm execute its tasks as they usually do. Now when a robot's battery level is at 20%, it would send to the controller a message providing the robot's identification. Then, the controller would check the task number that this specific robot is executing. It would then check if the current number of robots, excluding the one that sent the message, greater than or equal to the minimum required for the task. If it is, the controller does not send anything and the swarms continue executing their tasks as they were doing before. If the number of robots is less than the minimum required, then the controller would have to check the number of robots in the other sub-swarm. If the other sub-swarm has more robots than the minimum required then the controller would randomly choose and pick one of the other robots in the other sub-swarm to switch its task and the execution would continue. However, if the other sub-swarm is exactly at the minimum required, then the controller would broadcast to all the robots to stop execution and declare mission failure. The other case would be both tasks finish executing their tasks and the controller would declare mission success.

4 Our Proposed System

4.1 Task Descriptions

For both the simulation and real-life application, the robots are given the same task. The robots in the first sub-swarm would be doing task A which is moving from right to left in a straight line. The robots in the second sub-swarm would be doing task B which is moving from left to right. Both software and hardware implementations of the simulation and hardware are similar to a certain extent. However, there is a difference in the modules.

Considering that the most important aspect of reconfiguration of the robots is communication, an efficient and clear communication system must be put in place. All messages are broadcast to all robots in the simulation and real-life application. To solve the problem, abbreviated and specific messages are used (Table 1).

4.2 Simulation

The simulation environment was set up with a total of 10 robots and one controller, as shown in Fig. 3. The orange robot represents the controller, the five green robots

Table 1 Communication abbreviation

Abbreviation	Meaning
RoboN	The initial message sent from the controller to robot N where $0 \leq N \leq 9$
RoboEN	The acknowledgment message sent from robot N where $0 \leq N \leq 9$
RoboNX	The acknowledgment message sent from robot N where $0 \leq N \leq 9$
RoboLN	Message sent from robot N to the controller where $0 \leq N \leq 9$ when its battery reaches 20%

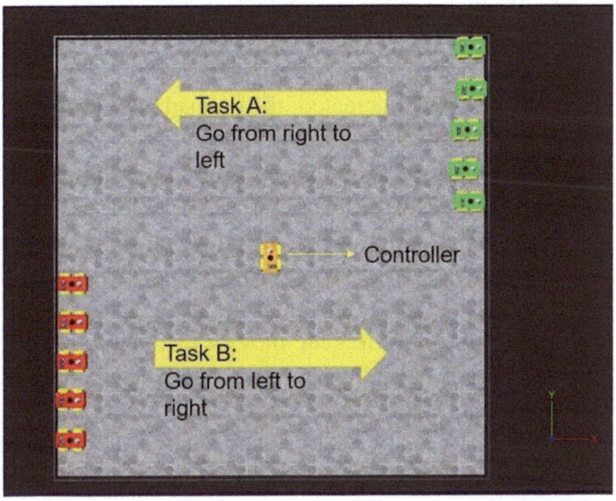

Fig. 3 Simulation environment

represent the robots doing task A, and finally, the five red robots represent the robots doing task B. While the tasks are simple, they can also be developed into more complex tasks. The idea to use simple tasks was done to focus on the reconfiguration more than the task itself. Although the red and green robots appear to be heterogeneous by doing different tasks, the reality is that they both contain the same set of features which makes them homogeneous. Their homogeneity allows the reconfiguration to happen.

Each robot is equipped with a receiver and a transmitter to allow communication between the robots, unlike the hardware implementation which had a transceiver. They are also equipped with an Arduino Nano. While it is not functional, it brings the design closer to reality. The robot itself is made out of solid shapes. A solid cube with dimensions 0.1 cm × 0.2 cm × 0.5 cm was used to create the body of the car. Four cylinders with a radius of 0.04 cm and a height of 0.02 cm were used to create the wheels of the robot. Each cylindrical wheel was connected to a rotational motor.

To set up the communication, first, the controller sends a message to all 10 robots and waits to receive an acknowledgment from them to make sure that the communication between them is established. Then, the algorithm works as mentioned before. The controller then keeps listening for any message coming from the robots, indicating that their battery level is at 20%. Then, the controller decides either to reconfigure the robots or to announce the failure of the task. If it decides to reconfigure, then it picks a robot to send a message to the new task number they will be performing. If it decides to announce task failure, then a message is broadcast to all robots to end performing their tasks.

4.3 Real-Life Implementation

A real-life POC (proof of concept) of our system has been implemented. A total of three robots, as shown in Fig. 4, and one controller are used to be able to have a POC that would allow the testing and assessment of our reconfiguration system proposed. Two sub-swarms are created: the first sub-swarm has one single operating robot and the second has two operating robots.

The robots are implemented using an Arduino Nano to integrate all the robot's modules, Fig. 5 and Tables 2 and 3. A 12 V battery is connected to the Arduino VIN pin to power it up. The NRF24L01, which is a radio transceiver module, is used for the communication between the different robots in the swarms. The L289N module, which is also referred to as an H-Bridge, is used to be able to control the direction and speed of the four gear motors in the robots. Moreover, the 5-channel lane tracking sensor is used to be able to detect a lane accurately which marks the end of the task.

The NRF24L01 module is connected to the Arduino by 7 pins. The VCC pin is connected to the 3.3 V pin of the Arduino, and the GND pin is connected to the GND of the Arduino. A 100 microfarad capacitor is also used to reduce the noise in

Fig. 4 The three robots used in our real-life POC

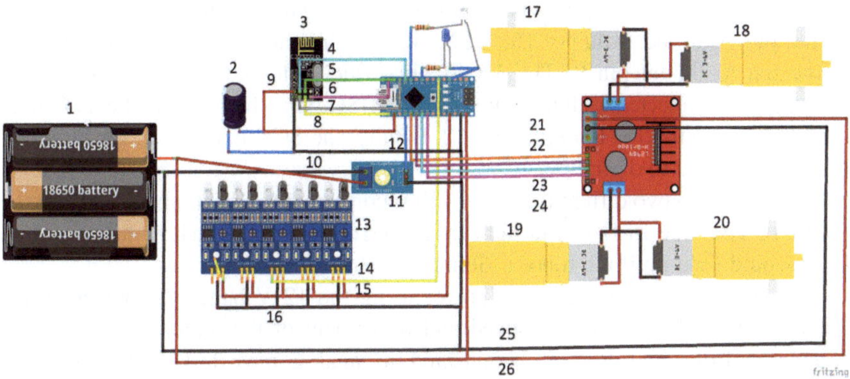

Fig. 5 Hardware connections diagram

Table 2 List of hardware components used

Number	Component name
1	12 V Battery
3	NRF24L01
11	Voltage Sensor
13	5-Channel Line Tracking Sensor
17	Gear Motor 1
18	Gear Motor 2
19	Gear Motor 3
20	Gear Motor 4
27	Arduino Nano
28	H-Bridge with L298N Driver

Table 3 List of connections

Connection number	Connection
4	Chip Select Not (CSN)
5	Master Out Slave In (MOSI)
6	Master In Slave Out (MISO)
7	Chip Enable (CE)
8	Serial Clock (SCK)
12	Input 1
14	Input 2
21	Output 1
22	Output 2
23	Output 3
24	Output 4
9, 15, 26	Power Supply (VCC)
10, 16, 25	Ground (GND)

communication. The CE pin, which allows SPI communication, is connected to the digital pin 10 in the Arduino. The CSN pin, which is always set on active high, is connected to digital pin 9. The MOSI pin, which allows the module to receive messages, is connected to digital pin 11. The MISO pin, which allows the module to send data, is connected to digital pin 12. Finally, the SCK that allows synchronization is connected to digital pin 13.

There are also two different colored LEDs connected to digital pins 8 and 7 in the Arduino. The blue LED would light up if the robot is doing task A and the white LED would light up if the robot is doing task B.

The H-Bridge is supplied with power using the 12 V battery to power up the 4-gear motors. To control the speed and movement of the motors, its input pins connected from pins 1–4 are connected to analog pins 1–4 in the Arduino, respectively.

The 5-channel line tracking sensor is used to detect the finish line. If the robot reaches the finish line, then it has completed its task. The 5-channel sensor is made up of 5 infrared sensors. It cannot detect the color black; therefore, if it sends a low signal to the Arduino through the digital pin 2, then it is detecting the finish line. It is also connected to the 5 V pin in the Arduino as well as the GND pin.

The voltage sensor is connected to the GND of the Arduino. It also has 1 input pin connected to an analog pin in the Arduino to be able to send the voltage measured. The 12 V battery is also connected to the sensor, so its current voltage can be read. Since the Arduino needs at least 7 V to power it up, then when the energy level measured is 7 V, it is considered to be at 0% battery. The 12 V is the full 100% battery.

The communication setup of the real-life robots is similar to the communication in the simulation. First, the controller would broadcast a message to all robots in the system. Then, when the robots receive the message, they reply with an acknowledgment. When the controller receives all acknowledgments, it broadcasts a message for all robots to start working on their defined tasks. The controller then goes into active listening mode. It would be waiting for the robots to send a message indicating that it has reached a 20% energy level. When this instance occurs, the controller would check whether the sub-swarm that has lost a robot would be a robot to be reconfigured from the other sub-swarm or will be able to continue the task as normal.

5 Testing

5.1 Simulation Testing

A simulation was set up to implement the proposed method as well as test it. The testing would then take the algorithm to be implemented in real life. Five test scenarios were tested with a total of nine test cases (Table 4). The first test scenario is the ideal scenario which is each sub-swarm finishing its task without losing any robot or in other words with no need for reconfiguration. It always ends up with the

Table 4 Simulation test cases

Test case no.	Description	Outcome/ mission status
1	Both swarms can finish their tasks without losing any robot due to the low battery	No reconfiguration Success
2	Sub-swarm A did not lose any of its robots due to low battery, however, B lost 3 robots	1 robot reconfigured Success
3	Sub-swarm B did not lose any of its robots due to low battery; however, Sub-swarm A lost 3 robots	1 robot reconfigured Success
4	Sub-swarm A did not lose any of its robots due to low battery; however, Sub-swarm B lost 4 robots not particularly at the same time	2 robots reconfigured Success
5	Sub-swarm B did not lose any of its robots due to low battery, Sub-swarm A lost 4 robots not particularly at the same time	2 robots reconfigured Success
6	Sub-swarm A loses 1 robot due to low battery and Sub-swarm B loses 4 robots, not particularly at the same time	Failure
7	Sub-swarm B loses 1 robot due to low battery and sub-swarm A loses 4 robots, not particularly at the same time	Failure
8	Stage 1: Sub-swarm A does not lose any robot due to low battery; however, sub-swarm B loses 4 robots not particularly at the same time. Stage 2: Before the task is completed by sub-swarm A and sub-swarm B, 1 of A's robots is lost due to low battery	Failure
9	Stage 1: Sub-swarm B does not lose any robots due to low battery; however, sub-swarm A loses 4 robots not particularly at the same time. Stage 2: Before the task is completed by sub-swarm B and sub-swarm A, 1 of B's robots is lost due to low battery	Failure

ideal success of the task. The second test scenario is having the first sub-swarm reach a minimum number of robots first and requesting robots from the second sub-swarm. The third test scenario is quite similar to the second in which the second sub-swarm reaches the minimum number of robots required first and requests robots from the first sub-swarm. Both the second and third test scenarios can end up with either the success of the task or its failure. The fourth test scenario is an extension of the first where after the first sub-swarm requests a robot from the second sub-swarm, the second sub-swarm would request a robot from the first sub-swarm. The fifth and final test scenario is very similar to the fourth and also an extension of the third test scenario. It happens when after the second sub-swarm requests a robot from the first sub-swarm, the first sub-swarm requests a robot from the second sub-swarm. Both the fourth and fifth test scenarios would end up with task failure.

The nine test cases were tested with constant variables that are constant through all of them. All sub-swarms start with five robots each and the required minimum for each sub-swarm is three robots. It can also be noticed that test cases starting from test case 6 are considered to be extreme cases.

Table 5 Real-life application test cases

Test case no.	Description	Outcome/ mission status
1	Both sub-swarm A and sub-swarm B can complete their task without the loss of any robot due to low battery	No reconfiguration Success
2	Sub-swarm B did not lose any of its robots due to low battery; however, sub-swarm A lost the single robot it had	1 robot reconfigured Success
3	Sub-swarm B lost 2 robots due to low battery while sub-swarm A did not lose any of its robots	Failure
4	Sub-swarm B lost one of its robots followed by sub-swarm A losing one of its robots	Failure

5.2 Real-Life Application Testing

Similar to the simulation, five test scenarios could be tested. Due to the fact that a limited number of robots are used, not all test scenarios were tested, resulting in fewer test cases as well. However, four test cases were conducted to show the effectiveness of the reconfiguration system (Table 5). The four test cases were tested with constant variables that are constant through all of them. Sub-swarm A started with only one robot and Sub-swarm B started with a total of two robots. The minimum number of robots required for each sub-swarm is one.

6 Results and Discussions

The first test case in both the simulation and real-life application is the optimal case and would always end up in task completion with or without reconfiguration. As for test cases 2–9 in the simulation, they would have always ended up in task failure if it were not for the reconfiguration system. However, four out of eight test cases ended up to be successful due to the reconfiguration system. This shows that the percentage of failure in the simulation test cases 2–9 decreased from 100% to 50%. Similar to the simulation, test cases 2–4 in the real-life implementation would also always end up in task failure. However, due to the reconfiguration system, one out of the three test cases ended up to be successful. This shows a decrease in the percentage of failure in the real-life implementation test cases 2–4 from 100% to approximately 66.67%.

7 Future Work

The reconfiguration system is still in its early stages of development; thus, numerous enhancements and developments could be done for future work. First, a more reliable communication module that would be based on Wi-Fi or Bluetooth could be used. It would establish a more stable connection between the controller and the robots, and it would also allow an increase of the swarm used. Furthermore, instead of using simple tasks, real-life tasks could be integrated with the reconfiguration system and tested.

Enhancements that could be done to the system itself would be to include the human operator in choices such as entering the minimum number of robots for each sub-swarm, and which robots would be performing which tasks. The human operator could later on pick which robot to reconfigure according to some factor that is measured. The reconfiguration system could also extend to several other factors other than loss due to a low battery like CPU usage, hardware damage, software damage, and overheating.

A scalability issue was faced while trying to increase the number of robots as one human operator is not able to handle more than 10 robots at a time. The scaling issue was even more apparent when testing the robots in real life. Being able to follow the motion and make sure that the reconfiguration system would be followed by all robots stresses the cognitive load of the single human operator. Therefore, even fewer robots were used in the real-life implementation. Finally, the dynamic adaptation of access control policies could be proposed when new security requirements are introduced at runtime, similar to the one proposed in [12] for connected autonomous vehicles.

8 Conclusion

In conclusion, the reconfiguration system would decrease the possibility of task failure. There are many possibilities of something going wrong while working on a huge number of robots, but the factor that was focused on is the energy level of the robots. A reconfiguration system was first built on a simulation and tested until it proved to be successful, and then, it was implemented and tested on real-life robots. In the simulation, the percentage of failure decreased from 100% to 50% and in the real-life implementation, it decreased from 100% to approximately 66.67%. Since the percentage of failure decreased by 50% in the simulation which was conducted using 10 robots and by approximately 33.33% in the real-life implementation which was conducted using three robots, it can be concluded that the higher the number of robots, the higher chance of the reconfiguration system to be more successful.

References

1. Cheraghi, A.R., Shahzad, S., Graffi, K.: Past, present, and future of swarm robotics. In: Arai, K. (ed.) Intelligent Systems and Applications. IntelliSys 2021 Lecture Notes in Networks and Systems, vol. 296. Springer, Cham (2021)
2. Pini, G., Brutschy, A., Pinciroli, C., Dorigo, M., Birattari, M.: Autonomous task partitioning in robot foraging: an approach based on cost estimation. Adapt. Behav. **21**(2), 118–136 (2013)
3. Wei, Y., Hiraga, M., Ohkura, K., Car, Z.: Autonomous task allocation by artificial evolution for robotic swarms in complex tasks. Artif. Life Robot. **24**, 127–134 (2019)
4. Wang, H., Rubenstein, M.: Shape formation in homogeneous swarms using local task swapping. IEEE Trans. Robot. **36**(3), 597–612 (2020)
5. Liu, D., Du, Z., Liu, X., Luan, H., Xu, Y., Xu, Y.: Task-based network reconfiguration in distributed UAV swarms: a bilateral matching approach. IEEE/ACM Trans. Networking. **30**(6), 2688–2700 (2022)
6. Reina, A., Pinciroli, C., Ferrante, E., Turgut, A.E., O'Grady, R., Birattari, M., Dorigo, M.: Closed-Loop Aerial Robot-Assisted Navigation of a Cohesive Ground-Based Robot Swarm Technical Report IridiaTr2011-020, IRIDIA. Faculté des Sciences Appliquées, Université Libre de Bruxelles (2011)
7. Lee, W., Vaughan, N., Kim, D.: Task allocation into a foraging task with a series of subtasks in swarm robotic system. IEEE Access. **8**, 107549–107561 (2020)
8. Diehl, G., Adams, J.A.: Battery variability management for swarms. In: Distributed Autonomous Robotic Systems: 15th International Symposium, pp. 214–226. Springer (2022)
9. Stirling, T., Floreano, D.: Energy-time efficiency in aerial swarm deployment. In: Distributed Autonomous Robotic Systems: The 10th International Symposium, pp. 5–18. Springer Berlin Heidelberg, Berlin/Heidelberg (2013)
10. Sheth, R.S.: A Decentralized Strategy for Swarm Robots to Manage Spatially Distributed Tasks. Doctoral dissertation, PhD thesis, Worcester Polytechnic Institute (2017)
11. Chen, A., Harwell, J., Gini, M.: Maximizing energy battery efficiency in swarm robotics. arXiv preprint arXiv:1906.01957. (2019)
12. Loulou, H., Saudrais, S., Soubra, H., Larouci, C.: Adapting security policy at runtime for connected autonomous vehicles. In: 2016 IEEE 25th International Conference on Enabling Technologies: Infrastructure for Collaborative Enterprises (WETICE), pp. 26–31. IEEE (2016)

Optimal Feature Selection Using Harris Hawk Optimization for Music Emotion Recognition

Osman Kerem Ates

1 Introduction

Modern datasets are frequently characterized by an abundance of information, encompassing numerous attributes and variables. While this wealth of data offers potential advances, it also poses challenges. The inclusion of irrelevant, redundant, or unnecessary features can not only prevent computational efficiency but can also lead to low classification performance. Excessive features can also lead to overfitting challenges in learning models. One of the most effective solutions to this problem is to select the subset of attributes that best represents the data. Feature selection, as a powerful technique, has the potential to enhance learning performance, construct more generalizable models, reduce computational complexity, and minimize storage requirements.

There are a few feature selection strategies in the literature to reduce feature dimension. Many of the latest techniques for selecting features also involve optimization algorithms to decide the best subset of attributes from the given datasets. Genetic algorithm [1], particle swarm optimization [2], symbiotic organism search algorithm [3], and ant colony optimization [4] are some of the feature selection algorithms used in different fields and for different purposes. For feature selection, a new method named as binary teaching learning-based optimization algorithm was presented that obtains a subset of optimal features in the dataset. Allam and Nandhini concluded that a minimal number of features from data give higher accuracy. In their paper, quadratic binary HHO was proposed to select features in the classification process. They evaluated 22 datasets from the UCI machine learning archive and

O. K. Ates (✉)
Department of Electrical and Electronics Engineering, Karadeniz Technical University, Trabzon, Turkey
e-mail: osmankeremates@ktu.edu.tr

M. N. Seyman (ed.), *2nd International Congress of Electrical and Computer Engineering*, EAI/Springer Innovations in Communication and Computing,
https://doi.org/10.1007/978-3-031-52760-9_27

compared with other algorithms to test the performance of this algorithm. The results showed that the proposed one had better classification performance [5]. Al-Tashi et al. also proposed a method using hybrid gray wolf optimization to find the best feature subset. For this purpose, they used 18 datasets and k-nearest neighbors algorithm as classifier. The results showed the superiority of this method in terms of accuracy and number of features selected [6]. There are also numerous research on music emotion recognition in the literature. Widiyanti and Endah compared three feature selection algorithms to decide very influential features. They obtained highest accuracy as 43%, selecting 6 out of 13 features through the sequential backward selection algorithm. This method increased accuracy up to 8% [7]. Byun et al. said that feature selection is important for EEG classification performance. In their paper, they analyzed a relationship between emotional state and music. To obtain the feature vectors, Relief algorithm and Bhattacharya distance were used, and they decided that power of signal had the best performance for emotion recognition [8]. Generally, different and particular parameters are used by optimization algorithms, so these parameters have powerful effect on the performance of machine learning models.

In this paper, we used the HHO-based feature selection method for deciding an optimal subset of all features to obtain a better classification result with the music emotion recognition dataset. In total, 50 features were extracted with the participation of 13 people. To identify the most effective classification features, we applied this optimization algorithm. Ultimately, we achieved an 85% classification accuracy with selected features using the k-NN classifier, which is 5% higher than the accuracy obtained when all features are considered. The main contribution of this paper is to test the success of an HHO-based method on music emotion recognition data.

The rest part of this paper is organized as follows. In the second section, we presented some information about the dataset used. Section 3 describes the HHO algorithm and explains its phases and parameters. Section 4 also explains how HHO algorithm is used for feature selection. Finally, Sect. 5 provides experimental results and makes an assessment about this work.

2 Dataset Description

Various forms of Turkish music were chosen to prepare the data [9]. There are four classes in all, designed as a discrete model: happy, sad, angry, and relax. To determine the feelings of these pieces of music, the 13 participants were asked to label the selected music according to the above classes.

In this experiment, participants listened to a 30-second randomized segment of the track before selecting the class based on their emotions. The emotion selected by the participants in a work was allocated as the item's label. For example, if nine participants described a piece of music as "happy" and four others labeled it as "relax," the song was assigned to the "happy" class based on the majority vote.

The trials took place over three sessions, with each participant listening to a total of 500 pieces of music. Each class received 100 samples to have an equal number of samples. There are 400 samples, each lasting 30 s. MIR Toolbox was used to choose features. To examine the emotional content of music signals, Mel Frequency Cepstral Coefficients (MFCCs), Tempo, Chromagram, Spectral, and Harmonic aspects were used. A feature vector was constructed after extracting 50 features.

3 Harris Hawk Optimization

Harris hawk optimization (HHO) is a population-based method created in 2019 by Heidari. The hunting methods of Harris hawks, an intelligent bird in nature, inspired this optimization algorithm. Harris hawks usually move in flocks when hunting rabbits. First, the flock leader and other members organize reconnaissance flights. After the detection of the prey, they carry out their hunts in accordance with the hunting model called surprise attack in nature. In the algorithm, Harris hawks represent potential solutions. The prey (rabbit) represents the global best solution. In general, it consists of two phases: exploration and exploration. The energy of the escaped prey plays an important role in advancing these two phases. Mathematically, the energy of the escaped prey, E, is as in Eq. 1:

$$E = 2E_0\left(1 - \frac{t}{t_{\max}}\right) \tag{1}$$

$$E_0 = 2\text{rand} - 1 \tag{2}$$

In the above equations, t is the number of iterations, $t_{\max}$ is the maximum number of iterations, E_0 is the initial energy of the prey [−1,1], and rand is a randomly generated number between [0,1].

3.1 Exploration Phase

In this stage, the exploration strategies of Harris hawks are modeled. The hawks in the algorithm are candidate solutions. In each cycle, the hawk with the closest position to the prey indicates the optimal solution. This modeling is shown in Eq. 3.

$$X(t+1) = \begin{cases} X_k(t) - \text{rand}_1 \left|X_k(t) - 2\text{rand}_2 X(t)\right|, q \geq 0.5 \\ X_{rb}(t) - X_m(t)) - \text{rand}_3(lb + \text{rand}_4(ub - lb)), q < 0.5 \end{cases} \tag{3}$$

$$X_m(t) = \frac{1}{N}\sum\nolimits_{i=1}^{n} X_i(t) \tag{4}$$

In Eqs. 3 and 4, the parameter X is the position of the hawk, X_k is the random hawk position, X_{rb} is the global best solution, ub and lb are the upper and lower limits, respectively, and X_m is the average position of the hawks. rand_1, rand_2, rand_3, rand_4 and q are randomly chosen values in the range [0,1].

3.2 Exploitation Phase

The Harris hawk strikes its target in an unexpected attack during this phase. Their positions are updated based on four distinct behaviors. During the hunting process, the energy of the escaping prey (***E***) and the chances of the prey escaping (***R***) are used to adjust these behaviors.

Soft Besiege In soft besiege phase, the Harris hawk makes moves aimed at reducing the energy of its prey. In this process, called soft encirclement ($R \geq 0.5$ and $|E| \geq 0.5$), the hawk's position is updated as in Eq. 5:

$$X(t+1) = \Delta X(t) - E|JX_{rb}(t) - X(t)| \tag{5}$$

In this equation, ΔX is the difference between the global best solution and the corresponding hawk, and J is the jump power. ΔX and J are as given in Eqs. 6 and 7.

$$\Delta X(t) = X_{rb}(t) - X(t) \tag{6}$$

$$J = 2(1 - \text{rand}_5) \tag{7}$$

rand_5 is a random vector distributed in the interval [0,1].

Hard Besiege When ($R \geq 0.5$ and $|E| < 0.5$), the energy of the prey has decreased considerably, and the position of the Harris hawk is updated as in Eq. 8:

$$X(t+1) = X_{rb}(t) - E|\Delta X(t)| \tag{8}$$

Soft Besiege with Progressive Rapid Dives When $R < 0.5$ and $|E| \geq 0.5$, by making progressively faster dives, the hawk uses the soft besiege approach. In this case, the (solution) position of the new hawk is calculated using Eqs. 9 and 10:

$$Y = X_{rb}(t) - E|JX_{rb}(t) - X(t)| \tag{9}$$

$$Z = Y + \alpha(D) \times LF(D) \tag{10}$$

Y and Z are the two newly generated solutions, D is the number of dimensions, and α is a random vector distributed in the interval [0,1]. LF is the levy function and is given by Eq. 11:

$$\mathrm{Levy}(x)=\frac{\mu\times\sigma}{|v|^{1/\beta}} \tag{11}$$

μ and v are random numbers between (0,1). σ is defined by (12):

$$\sigma=\left(\frac{\Gamma(1+\beta)\times\sin\left(\frac{\pi\beta}{2}\right)}{\Gamma\left(\frac{1+\beta}{2}\right)\times\beta\times 2^{\left(\frac{\beta-1}{2}\right)}}\right)^{\frac{1}{\beta}} \tag{12}$$

β is taken as constant 1.5. Finally, the hawk's position is updated as in Eq. 13:

$$X(t+1)=\begin{cases} Y \text{ if } F(Y)<F(X(t)) \\ X \text{ if } F(Z)<F(X(t)) \end{cases} \tag{13}$$

$F(.)$ in Eq. 13 is the objective function.

Hard Besiege with Progressive Rapid Dives When $R < 0.5$ and $|E| < 0.5$, the hawk performs the hard besiege strategy, this time making progressively faster dives. In this case, two new solutions are found as in Eqs. 14 and 15:

$$Y=X_{rb}(t)-E|JX_{rb}(t)-X_m(t)| \tag{14}$$

$$Z=Y+\alpha(D)\times LF(D) \tag{15}$$

Finally, the hawk changes its position as shown in Eq. 16:

$$X(t+1)=\begin{cases} Y \text{ if } F(Y)<F(X(t)) \\ X \text{ if } F(Z)<F(X(t)) \end{cases} \tag{16}$$

4 Using HHO for Feature Selection

This section describes how HHO is handled and used in feature selection. For the feature selection problem, a k-nearest-neighbor (k-NN) classifier is used as an evaluator. In the k-NN algorithm, the classification of the test data is done by examining the training data and employing specific distance measurement methods [10]. This study is based on the Euclidean distance as a measure of distance.

For the feature selection process, a threshold value must be selected to decide whether feature is selected or not. In this study, this value is chosen as 0.5. In the HHO algorithm, every hawk represents a potential solution. If a solution is greater than the threshold value (>0.5), then this related feature is to be selected. Also, there are two main objectives to be evaluated that are minimum classification error and number of selected features. Thus, it is necessary to minimize the objective function during the feature optimization. The objective function used in this algorithm is presented in Eq. 17:

$$\downarrow f = \omega\Delta + (1-\omega)\frac{|S|}{|T|} \tag{17}$$

where f is the objective (fitness) function, Δ is the classifier error, $|S|$ is the number of features to be selected, $|T|$ is the total number of extracted features in the dataset, and ω; [0,1] is a weighting factor related to classifier performance.

There is no uniform rule for selecting the appropriate parameters for HHO because they are dependent on problem factors such as dimensionality and complexity. However, using large number of solutions or iterations can increase the accuracy. Also, choosing higher weighting factor may reduce the number of features to be selected. All parameters in the algorithm were set to minimize the fitness function. The study's initial parameters are shown in Table 1.

5 Experimental Results and Discussion

The classification accuracies of the features optimized (minimized) by HHO are given in the tables below. In this study, where the effects of the number of iterations and solutions on the performance were observed as the main objective, the processing times of each run were also calculated for comparison purposes.

Initial experiments were carried out with the initial parameters given in Table 1 and the results were obtained. These results were not sufficient, and the values of the parameters were changed. As a result of the evaluations, choosing the threshold value of 0.9 instead of 0.5 gave better results both in terms of speed and performance. In addition, although the value of β in the levy function, which was chosen in the original study, is usually chosen as a constant 1.5, it was taken as 1.0 in this study, which increased the accuracy percentage.

Tables 2 and 3 show, respectively, the results for 100 and 500 iterations. Gradually increasing the number of solutions increased the success even more, but after 50 solutions, it did not have much effect on the success. Although it seems to be a

Table 1 Initial values of the algorithm

Parameter name	Parameter value
Threshold value	0.50
Lower-upper bound (lb-ub)	0–1
Levy value (β)	1.50
Solution (hawk) number	10
Number of iterations	100
k value (for k-NN)	5
ho (holdout ratio) – validation data rate	0.20
ω (weighting factor)	0.99

Table 2 Accuracy values for iteration number 100

Solution number	Accuracy (%)	Feature number	Selected features	Processing time (sec)
10	60.00	5	14,18,20,23,48	7.90
10	70.00	6	24,29,33,35,41,42	7.01
10	65.00	2	32,49	7.27
10	65.00	6	11,25,39,42,44,45	7.17
10	65.00	4	6,24,29,38	6.91
50	82.50	11	6,7,11,13,15,22,30,31,34,39,42	34.37
50	77.50	10	1,10,11,12,13,24,40,45,47,50	33.94
50	77.50	6	2,11,26,32,45,46	34.91
50	80.00	7	3,8,11,16,35,40,44	37.01
50	82.50	10	3,5,6,7,8,14,22, 39,47,49	34.84

Table 3 Accuracy values for iteration number 500

Solution number	Accuracy (%)	Feature number	Selected features	Processing time (sec)
10	70.00	6	1,2,16,24,45,46	34.91
10	60.00	3	4,23,27	34.55
10	75.00	3	32,44,46	33.25
10	67.50	9	6,7,8,9,19,29,41,42,43	35.38
10	67.50	3	2,25,44	36.14
50	80.00	4	8,32,46,50	163.08
50	85.00	11	3,10,11,22,24,26,33, 34,38,39,42	179.99
50	82.50	14	3,6,10,16,19,26,33,34,38,40,41,42,43,49	170.84
50	85.00	9	6,11,13,16,17,25,26,31,49	177.81
50	80.00	6	3,9,16,24,25,38	160.87

problem that searching for the result with more solutions increases the processing time five times, it is also seen from the results that it provides a 15% increase in the accuracy results. In other words, optimum results were obtained when the number of solutions was high. In addition to the number of solutions, another factor affecting the success was the number of iterations. Increasing the number of iterations also increased the success.

Each example was run five times to ensure that the outcomes were consistent, and the maximum classification accuracies were discovered when 50 solutions were searched and 500 iterations were conducted. The greatest results of this investigation are 85% twice, as shown in Table 3. However, when the table is examined again, 11 of the 50 features in the feature vector were selected in the first case, while this result was accomplished in the second case by reducing the number of features to 9.

Figure 1 also presents the convergence analysis on dataset. The HHO algorithm converged fast in order to find valuable solutions as seen in Fig. 1.

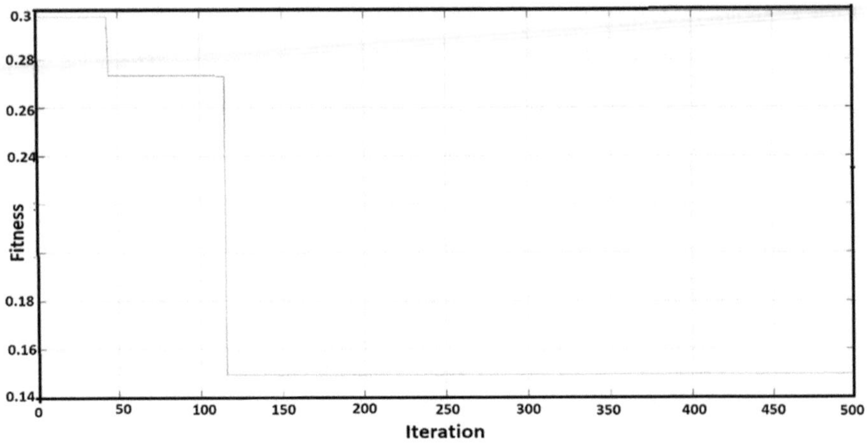

Fig. 1 Convergence curve of dataset

6 Conclusion

In this research, we applied an HHO-based feature selection strategy to determine the best subset of all features to improve classification results with the music emotion recognition dataset. Optimization algorithms usually require specific parameters, and these parameters have a major effect on the performance of machine learning models. In HHO, selection of these parameters like threshold value and solution number was particularly important. It was discovered that selecting smaller solutions had less success rate (i.e., 10 solutions instead of 50 solutions). By increasing the number of solutions, the results were boosted. While operating speed and performance are important for this study, it is more important for this study to have a higher value of accuracy, which is our main focus. Comparisons have also been made in terms of iteration and parameter choices, as already shown in the tables above.

To demonstrate the consistency of the study, the study was run several times. As an overall conclusion, the total of 50 features was reduced to nine by utilizing the k-NN classifier as an evaluator. With these nine features, an accuracy of 85% was reached. As a result, the HHO presented in this study is projected to play a very stable and effective function in feature selection difficulties.

Conflict of Interest The authors have no relevant financial or non-financial interests to disclose.

Data Availability Training and testing processes have been carried out using the music emotion recognition dataset.

References

1. Maleki, N., Zeinali, Y., Niaki, S.T.A.: A k-NN method for lung cancer prognosis with the use of a genetic algorithm for feature selection. Expert Syst. Appl. **164**, 113981 (2021)
2. Abualigah, L.M., Khader, A.T., Hanandeh, E.S.: A new feature selection method to improve the document clustering using particle swarm optimization algorithm. J. Comput. Sci. **25**, 456–466 (2018)
3. Han, C., Zhou, G., Zhou, Y.: Binary symbiotic organism search algorithm for feature selection and analysis. IEEE Access. **7**, 166833–166859 (2019)
4. Ghosh, M., Guha, R., Sarkar, R., Abraham, A.: A wrapper-filter feature selection technique based on ant colony optimization. Neural Comput. & Applic. **32**, 7839–7857 (2020)
5. Allam, M., Nandhini, M.: Optimal feature selection using binary teaching learning based optimization algorithm. J. King Saud. Univ. Comput. Inf. Sci. **34**(2), 329–341 (2022)
6. Al-Tashi, Q., Kadir, S.J.A., Rais, H.M., Mirjalili, S., Alhussian, H.: Binary optimization using hybrid grey wolf optimization for feature selection. IEEE Access. **7**, 39496–39508 (2019)
7. Widiyanti, E., Endah, S.N.: Feature selection for music emotion recognition. In: 2nd International Conference on Informatics and Computational Sciences, pp. 1–5. IEEE (2018)
8. Byun, S.W., Lee, S.P., Han, H.S.: Feature selection and comparison for the emotion recognition according to music listening. In: International Conference on Robotics and Automation Sciences, pp. 172–176. IEEE (2017)
9. Er, M.B., Aydilek, I.B.: Music emotion recognition by using chroma spectrogram and deep visual features. Int. J. Comput. Intell. Syst. **12**(2), 1622–1634 (2019)
10. Ateş, O.K., Aydemir, O.: Classification of EEG signals recorded during imagery of hand grasp movement. In: Medical Technologies Congress, pp. 1–4. IEEE, Antalya, Turkey (2020)

Comparative Analysis of EEG Sub-band Powers for Emotion Recognition

Muharrem Çelebi, Sıtkı Öztürk, and Kaplan Kaplan

1 Introduction

Emotions are electrical activities that occur in the human brain and are revealed by the interaction between nerves against stimuli in the outside world. Emotions have an indispensable importance in human life, which gives the individual the ability to feel, understand, and like/dislike something [1]. Researchers have been working on human emotion recognition for a long time. The aim of emotion recognition studies is to develop a system that can distinguish human emotions by computer. The utility of this system has a wide range of applications, such as assisting the elderly person, helping people with disabilities, knowing about the human psychological state, and using it in PC games [2].

Emotion recognition is not only done using EEG signals but also with various signals such as ECG, audio signals, and facial pictures. However, EEG-based emotion recognition systems are more trustworthy than the others. Emotion recognition systems supply the correct way to the brain-computer interface (BCI). The BCI systems aim to classify human thoughts and feelings for controlling the other machine [2]. BCI is studied in various interdisciplinary areas, including computer science, signal processing, and machine learning.

Feature extraction is the first and most important step in studies related to EEG. The results obtained using only raw EEG data are not sufficient because the EEG signal has a structure that is constantly changing and contains noise. In EEG-based emotion recognition systems, there are two separate groups as time-based features and frequency-based features [3]. The time-based features are extracted without any

M. Çelebi (✉) · S. Öztürk · K. Kaplan
Kocaeli University, Kocaeli, Turkey
e-mail: muharrem.celebi@kocaeli.edu.tr; sozturk@kocaeli.edu.tr;
kaplan.kaplan@kocaeli.edu.tr

M. N. Seyman (ed.), *2nd International Congress of Electrical and Computer Engineering*, EAI/Springer Innovations in Communication and Computing,
https://doi.org/10.1007/978-3-031-52760-9_28

transformation on the time axis of the EEG signal and time-based features lack of the ability to generalize classification. If it is frequency-based features, the EEG signal needs to be converted to the frequency axis, and then, the features are generated in certain frequency ranges. A certain standard of these frequency ranges has been established in the literature [4]. The aim of the study is to compare the accuracy of each pair of different frequency sub-bands in order to examine the relationship between EEG signals and emotions, so as to eliminate unimportant pairs of frequency bands. The contribution of the study is advantageous to the creation of an emotion classification system that uses real-time, minimal EEG features.

This document is structured as tracks. In Sect. 2, brief backgrounds of connected studies are summarized. Section 3 introduces the materials and methods, databases used, attribute generation, and classification algorithms. In Sect. 4, testing results are delivered with discussions of the results. Finally, Sect. 5 shows the conclusion and recommendations for future work.

2 Related Works

Chatchinarat et al. [5] extracted features using Discrete Wavelet Transform (DWT) for the effect of frequency bands for emotion recognition. They used to the DEAP dataset, extracted features and applied them to support vector machines (SVM). Jatupaiboon et al. [6] collected EEG data using the EMOTIV device with 14 channels. They broke into sub-band EEG recordings by the Wavelet Transform. Finally, they classified using the SVM machine learning algorithm. Moretti et al. [7] performed the comparative analysis of band power in Alzheimer's disease using statistical analysis. Zheng and Lu [8] investigated the success of five frequency bands in emotion recognition. They calculated differential entropy as features and applied deep belief networks (DBNs) and tested on the SEED dataset. Zhang et al. [9] utilized the deep learning model as DBN and machine learning classifiers such as kNN, LR, and SVM for emotion sub-band analysis. Candra et al. [10] extracted features such as Wavelet energy and Wavelet entropy for emotion classification. They tested with the SVM classifier for the DEAP dataset. Chen et al. [11] submitted a model using the feature extraction strategy based on attributes such as Lempel–Ziv complexity, Wavelet coefficients, and approximate entropy. The model was evaluated at the DEAP recordings using the LIBSVM classifier and achieved 82.63% and 74.88% for valence and arousal axes. Li et al. [12] proposed ensemble learning using the base classifier. They used features such as the Hjorth indicators, spectral densities, various entropies, and the Lyapunov indicator. The mean accuracies for valence and arousal were 64.22% and 65.70%, respectively, for the DEAP recordings. Xu et al. [13] presented a mixed GRU and CNN deep learning (GRU-Conv) based on the DEAP dataset for person-independent experiments. The model extracted spatial and temporal behaviors of EEG signals using GRU from raw EEG signals and categorized them with CNN. The model provided a mean accuracy of 67.36% for valence and 70.07% for arousal.

3 Materials and Methods

The first step in emotion recognition is to obtain the appropriate dataset. In this study, the SEED and the DEAP datasets were preferred. EEG signals are separated by a 10-second and 50% small window. This process is called sliding window analysis. After this step, attributes are extracted on each window. Finally, these obtained features are applied to the classifiers and the success rate is obtained. Tests and research were carried out in the MATLAB 2022a packaged software environment. The block diagram of the test procedure is shown in Fig. 1.

3.1 *EEG Emotion Recognition Datasets*

The SJTU Emotion EEG Dataset (SEED) was recorded using the ESI NeuroScan device, which consists of 62 EEG channels. While recording the data, it was formed from the signals obtained from 15 people with an average age of 23.27 years. Ten different Chinese films were used as stimuli. Sampling frequency is 200 Hz. Label values are available for three states: positive, neutral, and negative [14].

The Database for Emotion Analysis using Physiological Signals (DEAP) was recorded with the Biosemi ActiveTwo EEG device containing 32 channels. The DEAP dataset includes 32 individuals, 16 females and 16 males, aged between 19 and 37. Each subject was evaluated in terms of arousal, valence, liking, familiarity, and dominance for the music video they watched. Sampling frequency is 128 Hz [15, 16].

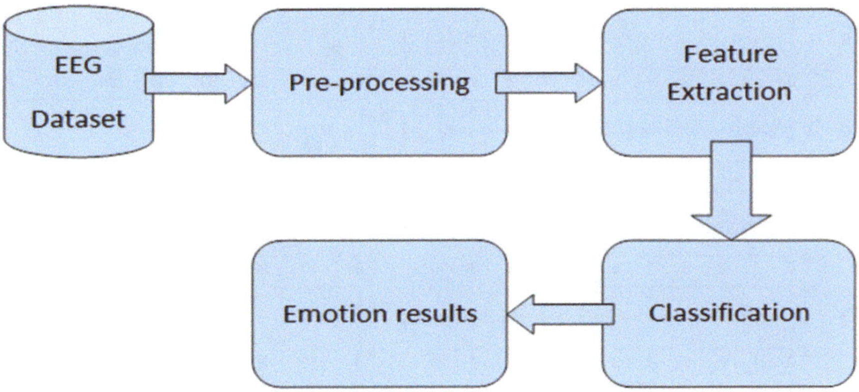

Fig. 1 Block diagram of the process flow

3.2 Feature Extraction

EEG signals can generally go up to the frequency value of 0.1 Hz to 128 Hz, and the amplitude level varies in the range of 5–400 μV. EEG signals are divided into sub-band frequency ranges. Delta (δ) sub-band in the range of 0.1-4 Hz, Theta (θ) sub-band in the range of 4–8 Hz, Alpha (α) sub-band in the range of 8–13 Hz, Beta (β) sub-band in the range of 13–30 Hz, Gamma (γ) sub-band in the range of 30–60 Hz are divided into five frequency bands [17, 18].

The characteristics of these sub-band frequencies are summarized shortly. The amplitude change in the delta sub-band is recorded in situations where the brain has very low activity, such as deep sleep, general anesthetic state. The amplitude changes in the theta band are encountered while dreaming in normal individuals, low activity states of the brain such as medium depth anesthetic conditions and also when the individual is under stress. Changes in the alpha band are observed when awake individuals are physically and mentally resting, in the absence of external stimuli, and when the eyes are closed. Changes in the beta band are observed in individuals in the phases when the individual's attention is high and focused on something. Changes in the gamma band are observed in high-level mental activities [17, 18].

As seen in the EEG signal spectrum in Fig. 2, it has high amplitude values at low frequencies, whereas it has low amplitudes at high frequencies. For this reason, the frequency ranges are separated by 4 Hz intervals at low frequencies and the range is gradually widened at high frequencies. In Fig. 2, the black represents the 10-second

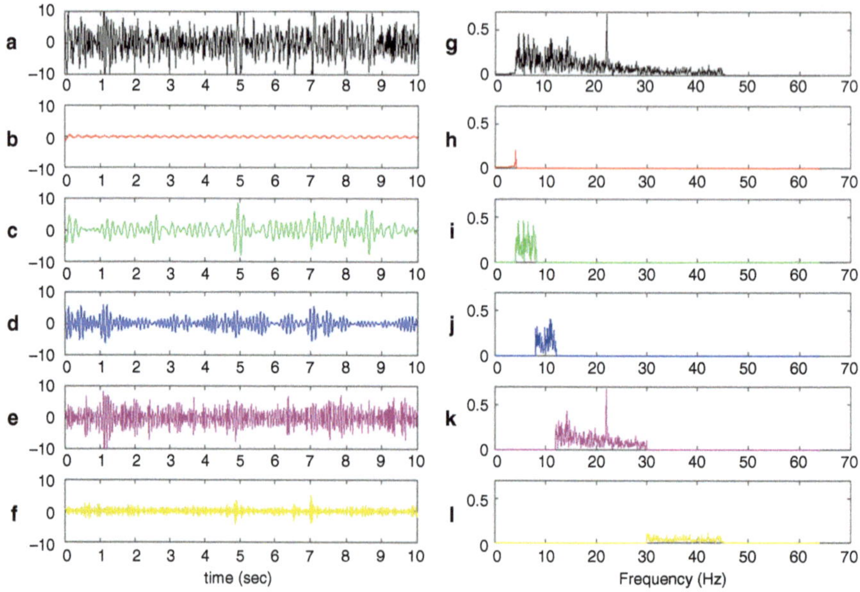

Fig. 2 The values of the 10-second EEG signal on the time and frequency axis

EEG signal for the time domain on (a) and the frequency domain on (g). The red represents the delta sub-band of the 10-second EEG signal for the time domain on (b) and the frequency domain on (h). The green color represents the theta sub-band of the 10-second EEG signal for the time domain on (c) and the frequency domain on (i). The blue represents the alpha sub-band of the 10-second EEG signal for the time domain on (d) and the frequency domain on (j). The magenta represents the beta sub-band of the 10-second EEG signal for the time domain on (e) and the frequency domain on (k). The yellow represents the gamma sub-band of the 10-second EEG signal for the time domain on (f) and the frequency domain on (l).

3.3 *Classification Algorithms*

In this study, popular machine learning methods, which are k-Nearest Neighbors (kNN), Random Forest (RF), and Support Vector Machines (SVMs), are used three classifiers. The kNN algorithm is a nonparametric approach, which means it does not perform any calculations or regulations on training data. During the training phase, it only stores the training data, and classifies all data points. In the testing phase, the distance between an input data and other previously classified data is calculated using various distance criteria, and then, a decision is made based on the calculated distances [19].

The RF algorithm was developed by Leo Breiman. The Random Forest is an ensemble algorithm that consists of many independent decision trees and chooses the appropriate one. The Random Forest algorithm uses multiple randomly generated decision trees to classify data [20, 21]. In this algorithm, the number of trees chosen is 50 for the RF classifier.

The aim of the SVM is to find out best segregate of class in decision boundary. The basis of the SVM algorithm is to create a hyperplane between the classes to separate the two classes from each other. It searches and uses support points in the data for this hyperplane [22]. If the structure of the dataset is nonlinear, SVM creates a nonlinear decision region by using kernel functions. In this study, the Gaussian kernel function is preferred.

3.4 *Performance Metric*

The success rate of each classifier was calculated according to the accuracy value. The mathematical expression of the accuracy rate is given in Eq. 1:

$$\text{Accuracy} = \frac{\text{TN} + \text{TP}}{\text{TN} + \text{FN} + \text{TP} + \text{FP}} \quad (1)$$

In this study, the 10-part cross-validation method was used and in order to test the success of the classifiers, the dataset is divided into training and test sets.

4 Results and Discussion

By dividing the problem into small parts, the approach was systematically advanced. In the first step, the energy value in the time axis was subtracted and tested for the window duration of 10 s and 50% overlap. In the second step, using the 10-second window time, power values of 2-Hz interval were extracted and the performance rates of the 2-Hz intervals were examined. In the third step, testing of the 5-piece sub-band frequency ranges was done such as delta, theta, alpha, beta, and gamma which are widely used and well known in the literature and examined singularly and combined with each other.

In Table 1, accuracy results for energy, 2-Hz sub-bands, and 5-Piece sub-bands are displayed according to three different methods. The highest success rate for the SEED dataset, the RF classifier achieved 91.44% success rate. The energy attribute alone offers low performance.

In Table 2, accuracy results for energy, 2-Hz sub-bands, and 5-Piece sub-bands are displayed according to three different methods. For the DEAP dataset, the RF classifier achieved a success rate of 78.80% for the valence axis and 79.10% for the arousal axis. The energy attribute alone offers low performance. On the other hand, 2-Hz interval values have a low success rate.

Table 1 Results of energy, 2-Hz, and 5-piece sub-bands for SEED dataset

	kNN	RF	SVM
Energy	78.96	70.76	68.63
2-Hz sub-bands	77.42	91.12	83.62
5-Piece sub-bands	71.72	**91.44**	82.37

Table 2 Results of energy, 2-Hz, and 5-piece sub-bands for DEAP dataset

	kNN	Valence RF	SVM	kNN	Arousal RF	SVM
Energy	64.09	69.47	63.69	65.80	72.15	65.54
2-Hz sub-bands	67.43	76.72	67.36	68.45	77.42	68.68
5-Piece sub-bands	62.93	**78.80**	66.87	65.70	**79.10**	67.29

Table 3 Results of single sub-band's powers for the SEED dataset

	kNN	RF	SVM
Delta	67.37	59.63	57.25
Theta	59.77	63.65	63.79
Alpha	68.53	69.86	67.60
Beta	84.65	87.16	79.86
Gamma	87.77	**92.56**	79.77

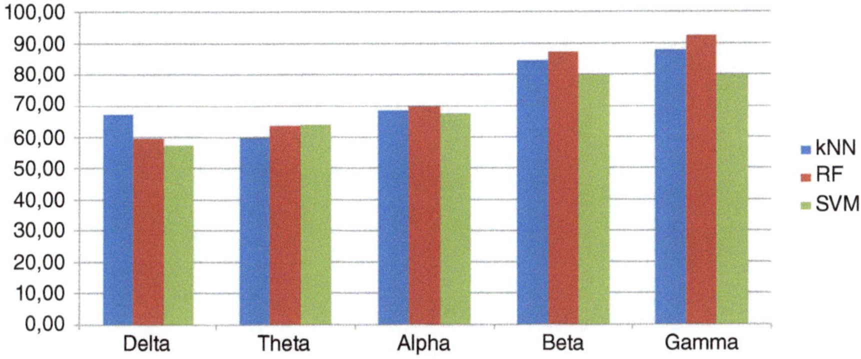

Fig. 3 Results of performance of sub-bands for the SEED dataset

Table 4 Results of single sub-band's powers for DEAP dataset

	kNN	Valence RF	SVM	kNN	Arousal RF	SVM
Delta	56.04	60.46	59.10	59.49	64.83	62.66
Theta	58.54	64.57	60.72	62.80	68.46	64.33
Alpha	62.02	66.11	62.13	64.93	70.07	66.08
Beta	71.59	75.23	68.59	73.08	76.64	70.45
Gamma	77.90	**80.81**	70.92	78.47	**81.57**	72.38

Table 3 lists the individual performance ratios of the power values of the 5-Piece sub-bands. The SEED dataset offers a 92.56% success rate for the RF classifier. Although the success rate in Table 1 is 91.44%, only the gamma band alone is more successful in Table 3.

In Fig. 3, the individual performance ratio of the frequency sub-bands for the SEED dataset is indicated. As can be seen, the most successful frequency region is the gamma band and the lowest success band is the delta band.

In Table 4, the individual performance ratios of the power values of the five sub-bands are listed. For the DEAP dataset, the success rate was obtained for the RF classifier, 80.81% for the valence axis, and 81.57% for the arousal axis. While the success rate in Table 2 is 78.80% for the valence axis and 79.10% for the arousal axis, only the gamma band alone is more successful in Table 4.

In Figs. 4 and 5, the individual performance ratio of the frequency sub-bands for the DEAP dataset is presented both on the valence and arousal axis. It has been shown that the most successful frequency region is the gamma band, and the second successful frequency band is the beta band.

In Table 5, double and triple combinations are combined and compared according to the individual success rates obtained in Table 3. For the RF classifier on the SEED dataset, the Gamma-Beta band together offers a 92.17% success rate, whereas the performance rate is slightly reduced. In double, triple, quadruple, and quintuple combinations, the success rate gradually decreased.

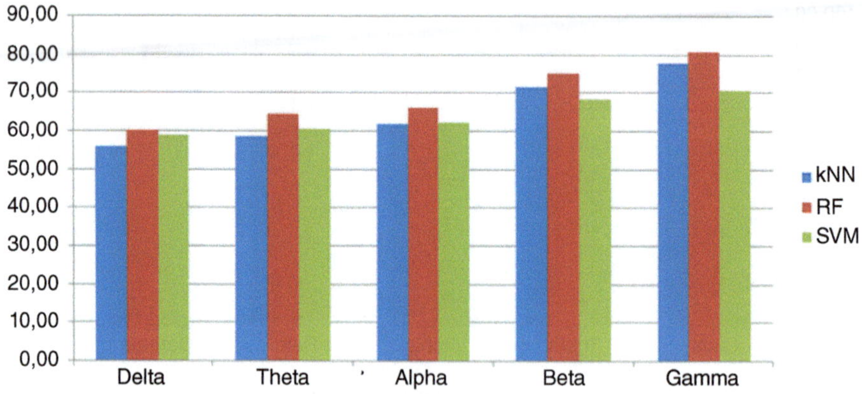

Fig. 4 Results of performance of sub-bands for DEAP dataset valence axis

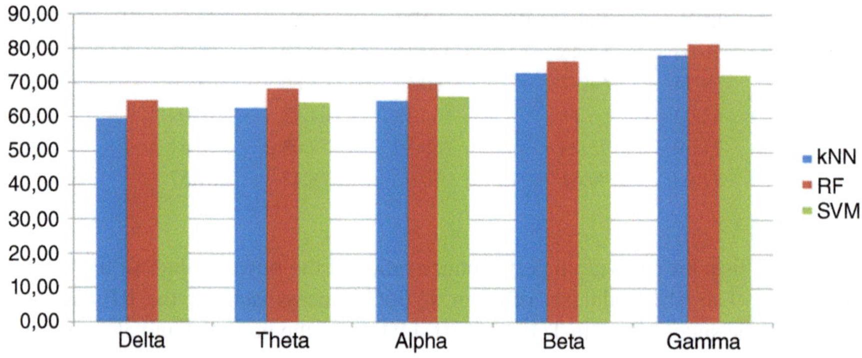

Fig. 5 Results of performance of sub-bands for DEAP dataset arousal axis

Table 5 Results of combination of sub-band's powers for SEED dataset

	kNN	RF	SVM
Gamma-Beta	87.43	**92.17**	81.91
Gamma-Beta-Alpha	85.27	91.81	81.63
Gamma-Beta-Alpha-Theta	80.30	92.05	82.23
Gamma-Beta-Alpha-Theta-Delta	71.72	91.44	82.37

Table 6 Results of combination of sub-band's powers for DEAP dataset

	kNN	Valence RF	SVM	kNN	Arousal RF	SVM
Gamma-Beta	75.38	**81.24**	71.29	75.77	**81.68**	71.36
Gamma-Beta-Alpha	67.23	80.32	68.52	69.07	81.04	69.55
Gamma-Beta-Alpha-Theta	62.81	79.85	67.53	65.50	80.98	69.00
Gamma-Beta-Alpha-Theta-Delta	62.93	78.80	66.87	65.70	79.10	67.29

In Table 6, double and triple combinations were combined and compared according to the individual success rates obtained in Table 4. For the RF classifier on the DEAP dataset, the Gamma-Beta band two-in-one offers a success rate of 81.24% for the valence axis and 81.68% for the arousal axis. The performance ratio is slightly increased compared with the individual gamma performance. In double, triple, quadruple, and quintuple combinations, the success rate is gradually decreasing.

5 Conclusion

Emotion classification is a popular field that increasingly attracts the attention of researchers. Because of the tests, only the energy value of the EEG signal is not sufficient, the power values of different frequency regions must be subtracted. In the second test phase, after subtracting the power values of the EEG signal only for the 2-Hz range, the experiments are carried out but this is not sufficient. Finally, it is separated according to five different frequency regions, and it has been seen that this result is more successful than the other two approaches.

In the second stage, individual sub-band analysis is carried out to understand which of the above 5-different frequency parts are successful. The most successful band is the gamma frequency band, and the second is the beta frequency region. The most unsuccessful is the delta frequency band. We also obtained that high-frequency sub-bands give higher accuracy compared with low-frequency sub-bands, especially Gamma and Beta sub-bands. In the third stage, in addition to the individual success rates, the combination of the gamma-beta band is more successful than the double and triple combinations of sub-bands.

Conflict of Interest The authors have no relevant financial or nonfinancial interests to disclose.

Data Availability Training and testing processes have been carried out using the SEED and DEAP datasets.

DEAP: (A Database for Emotion Analysis using Physiological Signals), can be reachable at https://www.eecs.qmul.ac.uk/mmv/datasets/deap.

SEED: (The SJTU Emotion EEG Dataset) can be reachable at https://bcmi.sjtu.edu.cn/home/seed/

References

1. Saganowski, S.: Bringing emotion recognition out of the lab into real life: recent advances in sensors and machine learning. Electronics, MDPI. **11**, 496 (2022)
2. Li, X., Zhang, Y., Tiwari, P., Song, D., Hu, B., Yang, M., Marttinen, P.: EEG based emotion recognition: a tutorial and review. ACM Comput. Surv. **55**, 1–57 (2022)

3. Torres, E.P., Torres, E.A., Hernández-Álvarez, M., Yoo, S.G.: EEG-based BCI emotion recognition: a survey. Sensors, MDPI. **20**(18), 5083 (2020)
4. Houssein, E.H., Hammad, A., Ali, A.A.: Human emotion recognition from EEG-based brain–computer interface using machine learning: a comprehensive review. Neural. Comput. Appl., Springer. **34**, 12527 (2022)
5. Chatchinarat, A., Wong, K.W., Fung, C.C.: A comparison study on the relationship between the selection of EEG electrode channels and frequency bands used in classification for emotion recognition. In: 2016 International Conference on Machine Learning and Cybernetics (ICMLC), pp. 251–256. Korea (South) (2016)
6. Jatupaiboon, N., Pan-Ngum, S., Israsena, P.: Emotion classification using minimal EEG channels and frequency bands. In: The 2013 10th International Joint Conference on Computer Science and Software Engineering (JCSSE), pp. 21–24. Thailand (2013)
7. Moretti, D.V., Babiloni, C., Binetti, G., Cassetta, E., Dal Forno, G., Ferreric, F., Rossini, P.M.: Individual analysis of EEG frequency and band power in mild Alzheimer's disease. Clin. Neurophysiol. **115**, 299–308 (2004)
8. Zheng, W.L., Lu, B.L.: Investigating critical frequency bands and channels for EEG-based emotion recognition with deep neural networks. IEEE Trans. Auton. Ment. Dev. **7**, 162–175 (2015)
9. Zheng, W.L., Guo, H.T., Lu, B.L.: Revealing critical channels and frequency bands for emotion recognition from EEG with deep belief network. In: 2015 7th International IEEE/EMBS Conference on Neural Engineering (NER), pp. 154–157. France (2015)
10. Candra, H., Yuwono, M., Handojoseno, A., Chai, R., Su, S., Nguyen, H.: T.: recognizing emotions from EEG subbands using wavelet analysis. In: 2015 37th Annual International Conference of the IEEE Engineering in Medicine and Biology Society (EMBC), pp. 6030–6033. Italy (2015)
11. Chen, T., Ju, S., Ren, F., Fan, M., Gu, Y.: EEG emotion recognition model based on the LIBSVM classifier. Measurement, Elsevier. **164**, 108047 (2020)
12. Li, R., Ren, C., Zhang, X., Hu, B.: A novel ensemble learning method using multiple objective particle swarm optimization for subject-independent EEG-based emotion recognition. Comput. Biol. Med. **140**, 105080 (2022)
13. Xu, G., Guo, W., Wang, Y.: Subject-independent EEG emotion recognition with hybrid spatio-temporal GRU-Conv architecture. Med. Biol. Eng. Comput. **61**, 61–73 (2023)
14. Duan, R.N., Zhu, J.Y., Lu, B.L.: Differential entropy feature for EEG-based emotion classification. In: 2013 6th International IEEE/EMBS Conference on Neural Engineering (NER), pp. 81–84. USA (2013)
15. Koelstra, S., Muhl, C., Soleymani, M., Lee, J.S., Yazdani, A., Ebrahimi, T., Patras, I.: Deap: a database for emotion analysis; using physiological signals. IEEE Trans. Affect. Comput. **3**, 18–31 (2011)
16. Scherer, K.R.: What are emotions? And how can they be measured? Soc. Sci. Inf. **44**, 695–729 (2005)
17. Wang, J., Wang, M.: Review of the emotional feature extraction and classification using EEG signals. Cognitive Robotics. **1**, 29–40 (2021)
18. Blinowska, K.J., Żygierewicz, J.: Practical Biomedical Signal Analysis Using MATLAB, 2nd edn. CRC Press (2021)
19. Bishop, C.M., Nasrabadi, N.M.: Pattern Recognition and Machine Learning. Springer, New York (2006)
20. Noroozi, F., Sapiński, T., Kamińska, D., Anbarjafari, G.: Vocal-based emotion recognition using random forests and decision tree. Int. J. Speech Technol. **20**, 239–246 (2017)
21. Gharsalli, S., Emile, B., Laurent, H., Desquesnes, X., Vivet, D.: Random forest-based feature selection for emotion recognition. In: 2015 International Conference on Image Processing Theory, Tools and Applications (IPTA), pp. 268–272. France (2015)
22. Alpaydin, E.: Introduction to Machine Learning, 4th edn. MIT Press (2020)

Index

M. N. Seyman (ed.), *2nd International Congress of Electrical and Computer Engineering*, EAI/Springer Innovations in Communication and Computing,
https://doi.org/10.1007/978-3-031-52760-9

MIX
Papier aus verantwortungsvollen Quellen
Paper from responsible sources
FSC® C105338

If you have any concerns about our products,
you can contact us on
ProductSafety@springernature.com

In case Publisher is established outside the EU,
the EU authorized representative is:
Springer Nature Customer Service Center GmbH
Europaplatz 3, 69115 Heidelberg, Germany

Printed by Libri Plureos GmbH
in Hamburg, Germany